# Policy and Strategic Behaviour in Water Resource Management

# Policy and Strategic Behaviour in Water Resource Management

Edited by Ariel Dinar and Jose Albiac

publishing for a sustainable future

London • Sterling, VA

First published by Earthscan in the UK and USA in 2009

Copyright © Ariel Dinar and José Albiac, 2009

This book is the outcome of the International Expo 'Water and Sustainable Development' held in Zaragoza, Spain, in 2008. Support from the Spanish Ministry of Environment, Caja Rioja, Government of Aragon and the World Bank is acknowledged.

**All rights reserved**

ISBN:    978-1-84407-669-7

Typeset by JS Typesetting
Cover design by Susanne Harris

For a full list of publications please contact:

**Earthscan**
Dunstan House
14a St Cross St
London, EC1N 8XA, UK
Tel: +44 (0)20 7841 1930
Fax: +44 (0)20 7242 1474
Email: earthinfo@earthscan.co.uk
Web: www.earthscan.co.uk

22883 Quicksilver Drive, Sterling, VA 20166-2012, USA

Earthscan publishes in association with the International Institute for Environment and Development

A catalogue record for this book is available from the British Library

Library of Congress Cataloging-in-Publication Data

Dinar, Ariel, 1947–
  Policy and strategic behaviour in water resource management / Ariel Dinar and José Albiac.
      p. cm.
  ISBN 978-1-84407-669-7 (hardback)
  1. Water-supply–Economic aspects. 2. Water resources development–Environmental aspects. 3. Water quality management–Economic aspects. I. Albiac, José. II. Title.
  HD1691.D563 2009
  333.91–dc22
                                      2008034729

At Earthscan we strive to minimize our environmental impacts and carbon footprint through reducing waste, recycling and offsetting our $CO_2$ emissions, including those created through publication of this book. For more details of our environmental policy, see www.earthscan.co.uk.

This book was printed in the UK by TJ International Ltd, Padstow, UK, an ISO 14001 accredited company. The paper used is FSC certified.

# Contents

| | |
|---|---|
| *List of Figures, Maps, Tables and Boxes* | *vii* |
| *List of Contributors* | *xi* |
| *Foreword* | *xv* |
| *List of Acronyms and Abbreviations* | *xxi* |

| | | |
|---|---|---|
| 1 | Policy and Strategy in Water Resource Management: Can We Do Better When Both Are Coordinated?<br>*Ariel Dinar and Jose Albiac* | 1 |

## Part I Issues in Water Resource Policy

| | | |
|---|---|---|
| 2 | Issues in Water Resources Policy in Jordan<br>*Munther J. Haddadin* | 13 |
| 3 | Water Scarcity, Quality and Environmental Protection Policies in Jordan<br>*Helena Naber* | 29 |
| 4 | Water Management in Urbanizing, Arid Regions: Innovative Voluntary Transactions as a Response to Competing Water Claims<br>*Bonnie G. Colby* | 47 |
| 5 | Groundwater Management Issues and Innovations in Arizona<br>*Katharine L. Jacobs* | 67 |
| 6 | Water Policy in Australia: The Impact of Change and Uncertainty<br>*Lin Crase* | 91 |
| 7 | The Policy Challenge of Matching Environmental Water to Ecological Need<br>*Terry Hillman* | 109 |

| 8 | Water Management in Spain: An Example of Changing Paradigms<br>*Alberto Garrido and M. Ramón Llamas* | 125 |
|---|---|---|
| 9 | Policy Issues Related to Climate Change in Spain<br>*Ana Iglesias* | 145 |

## Part II Issues in Water Resource Strategy

| 10 | Water Conflicts: Issues in International Water, Water Allocation and Water Pricing with Focus on Jordan<br>*Munther J. Haddadin* | 175 |
|---|---|---|
| 11 | Good and Bad Forms of Participation in Water Management: Some Lessons from Brazil<br>*Jerson Kelman* | 189 |
| 12 | Issues of Balancing International, Environmental and Equity Needs in a Situation of Water Scarcity<br>*Barbara Schreiner* | 207 |

## Part III Interaction between Policy and Strategy

| 13 | Modelling Negotiated Decision Making under Uncertainty: An Application to the Piave River Basin, Italy<br>*Carlo Carraro and Alessandra Sgobbi* | 233 |
|---|---|---|
| 14 | Strategic Behaviour in Water Policy Negotiations: Lessons from California<br>*Rachael E. Goodhue, Leo K. Simon and Susan E. Stratton* | 257 |
| 15 | Strategic Behaviour in Transboundary Water and Environmental Management<br>*George B. Frisvold* | 279 |
| 16 | Climate Change and International Water: The Role of Strategic Alliances in Resource Allocation<br>*Ariel Dinar* | 301 |

| Index | 325 |
|---|---|

# List of Figures, Maps, Tables and Boxes

## Figures

| | | |
|---|---|---|
| 3.1 | International comparison of water availability | 30 |
| 3.2 | Safe yield and total abstraction for groundwater basins in Jordan | 34 |
| 6.1 | Diversions for consumptive uses in the Murray-Darling Basin | 95 |
| 6.2a | Water consumption by sector | 101 |
| 6.2b | Industry gross value added | 101 |
| 7.1 | Hypothetical hydrograph | 116 |
| 8.1 | Farm employment trends | 131 |
| 8.2 | Total agricultural output and surface | 132 |
| 8.3 | Cost-effective programme with three independent water bodies | 140 |
| 8.4 | Least cost programme, integrating the standards of three connected water bodies | 141 |
| 9.1 | Changes in available water resources, reservoir inflow and irrigation water demand in the hydrological basins in Spain | 153 |
| 11.1 | Location of the Sao Francisco River Basin | 193 |
| 11.2 | Schematic description of the water supply problem at the recipient region | 193 |
| 11.3 | Developed hydropower as a percentage of potential hydropower | 199 |
| 13.1 | The Piave River Basin and the case study area | 239 |
| 13.2 | Varying the relative importance of the negotiated variables: comparing equilibrium utilities | 245 |
| 13.3 | Comparing individual and social welfare under different water allocation rules – quantities versus shares | 249 |
| 14.1 | California hydrologic regions and regional water flows | 261 |
| 14.2 | California rivers and water facilities | 262 |
| 15.1 | An equal cost-sharing rule can discourage cooperative solutions to transboundary pollution problems | 289 |
| 16.1 | Annual flow in the Gandes at Akkaraf | 311 |

| 16.2 | Annual flow in the Jordan at Lake Terenik | 312 |
| --- | --- | --- |
| 16.3 | The Lara River basin geography and hydrology | 314 |
| 16.4 | Annual flow in the Aral Sea at Lugotkot | 315 |
| 16.5 | Cooperation level as a function of climate | 320 |
| 16.6 | Switching coalitions as an adaptation to flow variability | 321 |

# Maps

| 2.1 | The Hashemite Kingdom of Jordan | 14 |
| --- | --- | --- |
| 2.2 | Surface water basins | 15 |
| 2.3 | Groundwater basins | 16 |
| 4.1 | Native American reservation lands in Arizona, US | 52 |
| 5.1 | Arizona groundwater basins and active management areas | 69 |
| 12.1 | Map of the Inkomati Water Management Area | 217 |

# Tables

| 4.1 | Arizona water transfers, 1987–2007 | 49 |
| --- | --- | --- |
| 4.2 | Arizona tribal water right quantification by courts and by settlements | 54 |
| 5.1 | A comparison of the five active management areas in Arizona | 70 |
| 9.1 | Effects of climate change on main water management determinants and expected social and ecological consequences | 149 |
| 9.2 | Proposed priority options for a flexible four-pronged EU approach to adaptation | 156 |
| 9.3 | Summary of the national adaptation strategies in the EU-27 and other European countries | 159 |
| 9.4 | EU Water Framework Directive | 164 |
| 10.1 | Water needed in various uses by income categories of Middle East countries | 176 |
| 11.1 | The Sao Francisco River Basin | 192 |
| 13.1 | Summary of players' utility function parameters | 243 |
| 13.2 | Summary of results – ex post assessment | 247 |
| 15.1 | Institutional arrangements for transboundary water management on the US–Mexico border water management | 282 |
| 15.2 | Border institutions and their potential roles in facilitating cooperative transboundary water management solutions | 290 |
| 16.1 | Regional arrangements | 316 |
| 16.2 | Coalitional payoffs for various climates without investments in water infrastructure | 318 |

| | | |
|---|---|---|
| 16.3 | Coalitional payoffs for energy swap and various investment arrangements | 319 |
| 16.4 | Grand coalition payoffs and Shapley allocation shares to the states without investments in water infrastructure | 319 |
| 16.5 | Grand coalition payoffs and Shapley allocation shares to the riparians with investments in water infrastructure | 319 |

# Boxes

| | | |
|---|---|---|
| 9.1 | Potential impacts of climate change in irrigation in the Mediterranean region | 151 |
| 11.1 | Description of the Sao Francisco River Inter-basin Diversion Project | 191 |
| 16.1 | Tajik energy crisis deepens as Uzbekistan cuts down natural gas supplies | 306 |

# List of Contributors

**Jose Albiac** is research fellow at the Agrifood Research and Technology Centre (CITA-DGA, Spain). He is an economist working on water resources management and forest economics.

**Carlo Carraro** holds a PhD from Princeton University and is currently chairman of the Department of Economics and professor of Econometrics and Environmental Economics at the University of Venice. He is director of research of the Fondazione ENI Enrico Mattei and research fellow of The Centre for European Policy Studies (CEPS), the Centre for Economic Policy Research (CEPR) and CESifo. He is also director of the Climate Impacts and Policy Division of the EuroMediterranean Center on Climate Change (CMCC) and belongs to the Scientific Advisory Board of the Potsdam Institute for Climate, of the Research Network on Sustainable Development, Paris, and to the High-Level Network of Environmental Economists set up by the European Environmental Agency.

**Bonnie Colby** is a professor at the University of Arizona, where she has been a faculty member since 1983 in the Departments of Resource Economics and Hydrology and Water Resources. Her expertise is in the economics of inter-jurisdictional water disputes, drought and climate change adaptation, water rights valuation, water transactions and water policy. She has authored over 100 journal articles and seven books.

**Lin Crase** is the executive director of La Trobe University's Albury-Wodonga Campus, Victoria, Australia. He is also an associate professor of economics with La Trobe and is a director of the Victorian Government's North East Regional Water Corporation. He holds a PhD in economics and has published widely on water policy formulation in Australia.

**Ariel Dinar** is a professor of environmental economics and policy, and director of the Water Science and Policy Center, University of California, Riverside, US. This book was prepared while he was lead economist of the Development Research Group at the World Bank, US. His research focuses on international water and cooperation, approaches to stable water allocation agreements, water and climate change, economics of water quantity/quality, and economic aspects of policy interventions and institutional reforms. His most recent undertaking is the Resources for the Future (RFF) book series 'Issues in Water Resource Policy' that aims to produce publications on contemporary water policy issues in various countries and states.

**George Frisvold** is a professor in the Department of Agricultural and Resource Economics at the University of Arizona. He is also an investigator with the Climate Assessment for the Southwest project (CLIMAS) of the University of Arizona's Institute for the Study of Planet Earth.

**Alberto Garrido** is an associate professor of agricultural and resource economics at the Technical University of Madrid. His work focuses on natural resource and water economics and policy. He has conducted consultant work for the Organisation for Economic Co-operation and Development (OECD), the Inter-American Development Bank (IADB), the European Parliament, the European Commission, the Food and Agriculture Organization of the United Nations (FAO), and various Spanish Ministries and Autonomous Communities. He is the author of 85 academic references.

**Rachael Goodhue** is associate professor in the Department of Agricultural and Resource Economics, University of California, Davis, and a member of the Giannini Foundation of Agricultural Economics. Her research addresses a number of areas, including contracting, agricultural organization and policy, pesticide use and regulation, industrial organization, and natural resource use policies and property rights.

**Munther Haddadin** is an author, consultant and former minister of Water and Irrigation of Jordan. He headed the Jordan Valley Authority as chairman and chief executive officer (CEO) and was Jordan's senior negotiator in the Middle East Peace Process on water, energy and the environment. Dr Haddadin is a courtesy professor at Oregon State University and an affiliate professor at the University of Oklahoma and at the University of Central Florida.

**Terry Hillman** was educated at Dookie Agricultural College and Australian National University. He has carried out scientific research at the Commonwealth Scientific and Industrial Research Organization (CSIRO) (Division Entomology), the Australian National University (ANU) and Murray Darling Freshwater Research Centre (MDFRC), and has supervised postgraduate students at ANU, Charles Sturt University and La Trobe University (where he currently holds an honorary position as adjunct professor). He retired in 2001 as director of MDFRC and deputy director of the Cooperative Research Centre for Freshwater Ecology.

His research interests centre on the ecology of floodplain rivers and wetlands and their management.

**Ana Iglesias** is a professor of agricultural economics at the Universidad Politécnica de Madrid. Her research focuses on understanding the relevant strategies of society for adaptation to drought and climate change. She has contributed to the research programmes of the United Nations Environment Programme (UNEP), the United Nations Educational, Scientific and Cultural Organization (UNESCO), the United States Agency for International Development (USAID), the United States Environmental Protection Agency (USEPA) and the European Union (EU). She leads the development of Drought Management Guidelines within the MEDA programme of the EU. Her collaborative work has been published in over 100 research papers.

**Kathy Jacobs** is the executive director of the Arizona Water Institute and a professor in the University of Arizona Soil, Water and Environmental Science Department. She has more than 20 years of experience as a water manager for the state of Arizona, including 14 years as the director of the Tucson Active Management Area. Her research interests include groundwater management, water policy, connecting science and decision making, stakeholder engagement, and use of climate change and climate variability information for water management applications. She has served on six National Academy panels, including several related to the federal climate change science program and climate change issues.

**Jerson Kelman** is the general-director of the Brazilian electricity regulatory agency – Agência Nacional de Energia Elétrica (ANEEL) and board member of the Brazilian Foundation for Sustainable Development (FBDS). Previously he was president of the Brazilian water regulatory agency – ANA. Since 1976 he has been professor of water resources at the Federal University of Rio de Janeiro. In 2003, he received the King Hassan Great World Water Prize.

**Helena Naber** is a young professional with the World Bank. Her work experience includes water resources economics and management, environmental valuation and environmental governance.

**Fioravante Patrone** is professor of game theory at the Faculty of Engineering, University of Genoa (Italy), and is author of more than 50 publications in mathematical analysis and game theory. His recent research interests have focused on applications of game theory in different fields: natural resources, medicine, biochemistry and industrial organization. He has been the promoter of the 'Game Practice' meetings that started in 1998 and has served as director of the Interuniversity Centre for Game Theory and Applications for many years.

**Manuel Ramón Llamas** is emeritus professor of hydrogeology at the Complutense University of Madrid and fellow of Spain's Royal Academy of Sciences, where he chairs the Section of Natural Sciences and the International Relations Committee. He is the author of 100 books or monographs and almost 200 scientific papers. He has been president of the International Association of Hydrogeologists (1984–1989), vice-president of the International Water Resources Association (2001–2003) and fellow of the European Academy of Sciences and Arts (2004).

**Barbara Schreiner** is a director at Pegasys Strategy and Development in Pretoria, South Africa. She is the former Deputy Director General: Policy and Regulation in the Department of Water Affairs and Forestry, and former advisor to the Minister of Water Affairs and Forestry. She was a member of the Global Water Partnership Steering Committee and is on the Board of the Challenge Programme for Water and Food.

**Alessandra Sgobbi** holds a PhD in analysis and governance of sustainable development from the School of Advanced Studies of the University of Venice (Italy). She is collaborating with the Fondazione Eni Enrico Mattei in the Natural Resources Management and Climate Change Modelling and Policy Programmes. Her main research interests are the integration of 'hard' and 'soft' sciences in modelling the economy–environment system, and the analysis of negotiation processes. Previously, she worked as a senior economist in the Ministry of Finance, Planning and Economic Development of the Republic of Uganda.

**Leo Simon**'s research interests range from theoretical game theory to simulation modelling to applied political economy. His most recent theoretical work introduces a new class of incomplete information games, called aggregation games. His current work in political economy compares the agri-environmental policy formation processes in Europe and the US, and the role played by the World Trade Organization (WTO) in these two regions.

**Susan Stratton**, Smith College and the University of California, Berkeley, conducts research on the intersection of environmental and resource economics with political economics. Her current work uses game theoretic bargaining models to understand the development of water policy, with a particular focus on groundwater policy.

# Foreword

The collection of chapters in this book aims to emphasize the relevance of the game theoretical tool as a support for decision making in the context of water management and, more broadly, in water-related policies.

The widespread influence of game theory in the social sciences can also be felt in this field. However, since game theory usually does not offer 'ready-made' solutions for problems, often its contribution, or the relevance of its contribution, may look somehow obscure, or can be objected to.

My viewpoint, as a game theoretical practitioner, could be viewed with some suspicion, or could be considered as biased, to say the least. I would like to provide some arguments, anyway, in favour of the (appropriate) use of game theory.

As a starting point, let me stress the relevance of the use of 'principles'. There are a whole bunch of game theoretical contributions, which make an essential use of principles, which are translated (hopefully, in a faithful way) into specific properties in a specific mathematical model. This model serves as a basis from which one may infer consequences, aiming to characterize the 'solution' for a given situation in which decision makers interact. I have in mind the widespread use of the 'axiomatic' approach to solutions in game theory, whose most well known representatives are the Nash bargaining solution[1] and the Shapley value;[2] see also Parrachino et al.[3]

Interestingly, some 'principles' end up with quite different translations according to the specific mathematical model that is used. The best known case is, perhaps, the principle of 'Independence of Irrelevant Alternatives', which appears in the context of the celebrated impossibility theorem by Arrow[4] and in the approach used by Nash to the bargaining problem. But this fact is also true for less critical principles: for example, both the Shapley value and the Nash bargaining solution are subject to a symmetry condition, but the precise formal statement is different in the two cases (a way to rejoin the two approaches is via non transferable utility (NTU) games, on which one could quite usefully see the axiomatic characterizations given by Aumann[5] and Hart[6]).

But the presence of 'principles' is much more pervasive, going beyond their use in the 'axiomatic' approach. There are many results that appear to be formally distinct, being possible, at the same time, to be imputed for general underlying ideas, or themes.

The effect of this is that one can extract some kind of 'general guidelines', some kind of 'experienced common sense', which is one of the most relevant contributions that arises from the (appropriate) use of game theory in the field of water to which this book is devoted.

I will try to highlight a few occurrences of this point of view, discussing briefly the following themes:

- long-term interaction;
- stochasticity, uncertainty, predictability;
- the relevance of details, including consequences and preferences.

The effect of a long-term interaction between the same decision makers (being individuals, institutions or groups, such as states) is a typical theme of game theory. The celebrated 'folk theorem' (again, more than a single theorem, a general principle, which is translated into many different specific theorems, with many variants) guarantees that a long enough interaction among the agents allows for efficiency gains with respect to the one-shot interaction, thus breaking the inefficiency traps that are usually present due to externalities (the prisoner's dilemma, the tragedy of the commons and the like) or to difficulties that decision makers find in coordinating their choices (the battle of the sexes, or congestion games).

It is also well understood in this case that the role of the discount factor needs to be appropriately low to enhance the long-term effects, viz. the discussion about the value of the discount factor stemming from the Stern[7] report on climate change.

Added to this, repeated games allow for the creation of reputation, which is quite relevant in a context of incomplete information (e.g. on the preferences of the players, or on their discount factor, or on players' risk aversion). The role of this long-standing reciprocal interaction is well emphasized in the contribution by Frisvold, who describes, in this book, the importance of the agreement between Mexico and the US, which worked as a frame for more than 300 specific agreements ('Minutes') over transboundary water-related issues.

Of course, incomplete information means uncertainty with relation to some relevant parameters: this brings us to the general role of uncertainty. The stochastic component is quite relevant when the focus is on water availability: mainly its quantity, but also quality can be affected. It seems fair to say that this issue is not yet well addressed by the game theoretical tools, in particular cooperative games, despite the existence of formal approaches (Suijs[8]) that already have a few applications (Dinar et al[9]).

The problems that arise from uncertainty are not limited to finding an appropriate formal tool to deal with risky events. The issue is much deeper, and is connected with the presence of high levels of uncertainty, the possibility that very extreme events occur (severe drought or rain) and the possibility that some events of which we are not aware can also happen ('states of the world' that we ignore).

At this level, approaches like the precautionary one (even if it is full of difficulties), or the adoption of a code of conduct, could be best suited rather than the usual allocation-like approaches that are used by game theory in the context of uncertainty and risk (let's say: ordinary risk). What is quite interesting is that in corporate social responsibility, and more precisely on ethical codes for firms and institutions, the need to deal in an appropriate way with unforeseen contingencies has been evoked, using game theoretical ideas, as an explanation of the role of such codes (Sacconi[10]). I would say that this idea shares a common ground with that of a 'strategic alliance' – this is developed in the contribution from Dinar (Chapter 16) who looks for useful institutional approaches to cope with increasing uncertainty on water availability (and quality).

Another typical, recurrent, theme is the sensitivity to details that non-cooperative game theoretical models exhibit. It is not by chance that game theory has had striking success in dealing with auctions and their design: a relevant contribution for this success is the availability of sophisticated formal tools, but it is not secondary the fact that the rules of interaction in an auction are (should be) quite well specified.

A comparison with bargaining and negotiation makes a clear difference. Game theory has been always in trouble, when requested to analyse a negotiation process. Despite the relevant and deep insights that were provided (among which Rubinstein[11] and Myerson[12] deserve to be mentioned, together with the previously quoted Nash (see note 1)), experienced negotiators know very well how relevant can be details, not to say small tricks, in influencing the final outcome. On the topic of details, I suspect that anyone old enough remembers the preliminary discussions about the shape of the table to be used for the peace negotiations in Paris on the Vietnam War, and the clever solution that was found. More to the point, to quote Goodhue et al (this book): '… uncertainty among game theorists about how best to model the complicated real world interactions that lead to a bargaining solution'.

Of course the difficulties to deal with an issue are not a valid reason to give up. So, Goodhue et al and Carraro and Sgobbi (this book) offer their contribution to see how some details of a bargaining situation can be incorporated into the model: see, for example, the approach by Carraro and Sgobbi to model the different bargaining power of the 'players'.

The need for identifying which is the game that is really being played seems to be fairly obvious, but it is difficult to underestimate it. A couple of quick references show how relevant are the true preferences of the players.

One example is the influence that Dom Luis has had in the issue of the Sao Francisco river inter-basin transfer, described in Kelman's chapter. A project that is evaluated, taking into account payoffs for the players that (being connected with money or not) could be somehow amenable to a comparison, has found a non-trivial difficulty that was (also) due to the irreducibility of the goals of one 'player'. The influence of cultural and religious factors is also emphasized in the contribution by Haddadin, in connection with the water pricing issue.

We can see another instance of this issue in the difficulties (largely unexpected) that are coming from the gender issue in the process of implementation of water reform in South Africa in Schreiner's chapter. At a level that is somehow even deeper, we can see the difficulties connected with the identification of the relevant consequences for a situation of strategic interaction. It is not always easy to fix the borders of a model, and this is especially true when, at the level of the so-called 'game form', we try to envisage which consequences should be considered in the model. An instance of this problem is provided, once again, from the experience coming from South Africa: a side effect of the restoration of legal rights that is being pursued could be a reduction in employment levels. Such a drop could affect the social strata that are the target of this policy of right restoration.

I will stop here, leaving the floor to the contributions to be found in this book, hoping that these short notes offer some support to the usefulness of game theory in the context of the implementation of water policies and strategic behaviour, without trying to brush over the fact that difficulties do exist when one tries to apply game theory properly.

Fioravante Patrone
University of Genoa, Italy

## Notes

1  J. F. Nash, Jr. (1950) 'The bargaining problem', *Econometrica*, vol 18, 155–162.
2  L. S. Shapley (1953) 'A value for n-person games', in H. W. Kuhn and A. W. Tucker (eds), *Contributions to the Theory of Games*, vol. II, Annals of Math. Studies, 28, Princeton University Press, Princeton (NJ), 307–317.
3  I. Parrachino, S. Zara and F. Patrone (2006a) 'Cooperative Game Theory and its Application to Natural, Environmental, and Water Resource Issues: Basic Theory'. World Bank, Policy Research Working Paper, no. WPS 4072; I. Parrachino, A. Dinar and F. Patrone (2006b) 'Cooperative Game Theory and its Application to Natural, Environmental, and Water Resource Issues: Application to Water Resources'. World Bank, Policy Research Working Paper, no. WPS 4074.
4  K. J. Arrow (1951) *Social Choice and Individual Values*, Wiley, New York; (2nd edition, 1963).
5  R. J. Aumann (1985) 'An axiomatization of the non-transferable utility value', *Econometrica*, vol 53, 599–612.
6  S. Hart (1985) 'An axiomatization of Harsanyi's nontransferable utility solution', *Econometrica*, vol 53, 1295–1313.
7  N. Stern (2007) *The Economics of Climate Change: The Stern Review*, Cambridge University Press, Cambridge.
8  J. Suijs (1999) *Cooperative Decision-Making Under Risk*, Kluwer, Dordrecht.

9   A. Dinar, S. Moretti and S. Zara (2006) 'Application of stochastic cooperative games in water resources', in R. Goetz and D. Berga (eds), *Frontiers in Water Resource Economics*, Kluwer, Dordrecht.
10  L. Sacconi, (2006) 'A social contract account for CSR as an extended model of corporate governance (I): Rational bargaining and justification', *Journal of Business Ethics*, vol 68, 259–281; L. Sacconi (2007) 'A social contract account for CSR as an extended model of corporate governance (II): Compliance, reputation and reciprocity', *Journal of Business Ethics*, vol 75, 77–96.
11  A. Rubinstein (1982) 'Perfect equilibrium in a bargaining model', *Econometrica*, vol 50, 97–109.
12  R. B. Myerson (1979) 'Incentive compatibility and the bargaining problem', *Econometrica*, vol 47, 61–73.

# List of Acronyms and Abbreviations

| | |
|---|---|
| AC | Autonomous Community |
| ADWR | Arizona Department of Water Resources |
| AGUA | Initiative for Water Management and Utilization (Spanish acronym) |
| AMA | Active Management Areas (Arizona, US) |
| ANA | Brazilian water regulatory agency |
| ANEEL | Agência Nacional de Energia Elétrica (Brazil) |
| ANU | Australian National University |
| ASEZ | Aqaba Special Economic Zone (Jordan) |
| ASGISA | Accelerated and Shared Growth Initiative of South Africa |
| AWBA | Arizona Water Banking Authority |
| AWS | assured water supply (Arizona, US) |
| AWSA | Arizona Water Settlements Act |
| BBBEE | broad based black economic empowerment |
| BECC | Border Environmental Cooperation Commission |
| BEIF | Border Environmental Infrastructure Fund |
| BOD | biological oxygen demand |
| CAGRD | Central Arizona Groundwater Replenishment District |
| CAP | Central Arizona Project |
| CAP | Common Agricultural Policy (EU) |
| CAWCD | Central Arizona Water Conservation District |
| CBA | cost–benefit analysis |
| CBDX | LRB of Destra Piave |
| CBPB | LRB of Pedemontanto Brentella (Italy) |
| CEO | chief executive officer |
| CEPR | Centre for Economic Policy Research |
| CEPS | Centre for European Policy Studies |
| CES | Center for Economic Studies |
| CESifo | Center for Economic Studies (CES), the Ifo Institute for Economic Research and the CESifo GmbH (Munich Society for the Promotion of Economic Research) |
| CFSP | Common Foreign and Security Policy |
| CIRCLE | Climate Impact Research Coordination for a Larger Europe |

| | |
|---|---|
| CLIMAS | Climate Assessment for the Southwest (project, University of Arizona) |
| CMA | Catchment Management Agency |
| CMCC | EuroMediterranean Center on Climate Change |
| CoAG | Council of Australian Governments |
| COMRIV | riverside communities |
| CSD | Commission for Sustainable Development (UN) |
| CSIRO | Commonwealth Scientific and Industrial Research Organization |
| CSIS | Center for Strategic and International Studies |
| DAHNGO | dam-hating NGO |
| DWAF | Department of Water Affairs and Forestry (South Africa) |
| DWR | Department of Water Resources (California) |
| EC | electrical conductivity |
| EC | European Commission |
| EEA | European Environment Agency |
| EIA | Environmental Impact Assessment |
| EPA | Environmental Protection Agency |
| EPCC | European Climate Change Programme |
| ESA | Endangered Species Act (US) |
| EU | European Union |
| FAO | Food and Agriculture Organization (United Nations) |
| FBDS | Brazilian Foundation for Sustainable Development |
| FP7 | Seventh Framework Programme (EC) |
| GAEC | Good Agricultural and Environmental Condition (EU) |
| GCEP | General Corporation for Environmental Protection (Jordan) |
| GDP | gross domestic product |
| GEF | Global Environment Facility |
| GES | good ecological status |
| GHGs | greenhouse gases |
| GMA | Groundwater Management Act (Arizona, US) |
| GMES | Global Monitoring for Environment and Security |
| GNEB | Good Neighbor Environmental Board |
| GNP | gross national product |
| GPP | gross primary productivity |
| GRIC | Gila River Indian Community, |
| GTZ | (Deutsche) Gesellschaft für Technische Zusammenarbeit |
| IADB | Inter-American Development Bank |
| IBC | International Boundary Commission |
| IBWC | International Boundary and Water Commission |
| ICZM | Integrated Coastal Zone Management |
| IMF | International Monetary Fund |
| IPCC | Intergovernmental Panel on Climate Change |
| IRN | International Rivers Network |

| | |
|---|---|
| IRP | integrated resource planning |
| ISIIMM | Institutional and Social Innovations in Irrigation Mediterranean Management |
| IUCN | International Union for Conservation of Nature |
| JAGS | Jordanian agricultural strategy |
| JSSD | Jordan Society for Sustainable Development |
| JVA | Jordan Valley Authority |
| KAC | King Abdullah Canal (Jordan) |
| KAFY | kilo-acre-feet per year |
| KTD | King Talal Dam (Jordan) |
| LRB | Land Reclamation and Irrigation Boards (Italy) |
| M&I | municipal and industrial |
| MAF | million acre-feet |
| MCM | million cubic metres |
| MDFRC | Murray Darling Freshwater Research Centre (Australia) |
| MEDA | (programme that is the principal financial instrument of the European Union for the implementation of the Euro-Mediterranean Partnership) |
| MGD | million gallons per day |
| MOE | Ministry of Environment |
| MRSP | Metropolitan Region of Sao Paulo |
| MWD | Metropolitan Water District (California) |
| MWF | minimum water flow |
| MWI | Ministry of Water and Irrigation (Jordan) |
| MWWG | Murray Wetland Working Group (Australia) |
| NADBank | North American Development Bank |
| NAPA | National Adaptation Programmes of Action |
| NCART | National Center for Agricultural Research and Technology Transfer |
| NetSyMoD | Network Analysis, Creative System Modelling and Decision Support |
| NGO | non-governmental organizaton |
| NHP | National Hydrological Plans (Spain) |
| NSDS | national strategies for sustainable development |
| NSF | National Science Foundation (Tucson, AZ) |
| NSW | New South Wales (Australia) |
| NTU | non transferable utility |
| NWI | National Water Initiative (Australia) |
| NWRS | Nation Water Resource Strategy |
| O&M | operation and maintenance |
| OECD | Organisation for Economic Co-operation and Development |
| OM&R | operation, maintenance and replacement |
| OPA | Oferta pública de adquisición de derechos (Offer of Public |

|   |   |
|---|---|
|  | Purchase, Spain) |
| PAH | polycyclic aromatic hydrocarbons |
| PDP | Project Development Program |
| PEAG | Especial Plan of the Upper Guadiana (Spanish acronym) |
| PRB | Piave River Basin (Italy) |
| PRBL | Province of Belluno (Italy) |
| PRODES | River Basin Pollution Abatement Program (Brazil) |
| RBAs | River Basin Authorities |
| RFF | Resources for the Future |
| SA | South Australia |
| SADC | Southern African Development Community |
| SAHRA | Science and Technology Center for Sustainability of Semi-Arid Hydrology and Riparian Areas (NSF) |
| SAWRSA | Southern Arizona Water Rights Settlement Act |
| SCERP | Southwest Consortium for Environmental Research and Policy |
| SEA | Strategic Environmental Assessment |
| SET-PLAN | Strategic Energy Technology Plan (Europe) |
| SFMPs | Stream-flow Management Plans |
| TLM | The Living Murray (Murray-Darling Basin, Australia) |
| TSS | total soluble solids |
| UN | United Nations |
| UNDP | United Nations Development Programme |
| UNEP | United Nations Environment Programme |
| UNESCO | United Nations Educational, Scientific and Cultural Organization |
| UNFCCC | United Nations Framework Convention on Climate Change |
| USAID | United States Agency for International Development |
| USEPA | United States Environmental Protection Agency |
| WAF | water allocation framework (South Africa) |
| WAJ | Water Authority (for Jordan) |
| WAP | Water Allocation Plan |
| WAR | Water Allocation Reform (South Africa) |
| WFD | Water Framework Directive (EU, 2000) |
| WL | Water Law |
| WMA | water management area |
| WSSD | World Summit for Sustainable Development |
| WTO | World Trade Organization |
| YDP | Yuma Desalting Plant |

# 1

# Policy and Strategy in Water Resource Management: Can We Do Better When Both Are Coordinated?

*Ariel Dinar and Jose Albiac*

In the summer of 2006, a group of economists and game theorists met in Zaragoza, Spain, to present their works as part of the 'sixth meeting of game theory and practice dedicated to development, natural resources and the environment'.[1] The purpose of that meeting was to demonstrate the usefulness and policy relevance of game theory applications in natural resources and the environment. Indeed, as extraction rates and utilization of natural resources such as land, water and other resources are exceeding sustainable levels, the likelihood of disagreements among stakeholders arises. In such situations policies are urgently needed, but policy makers are faced with the difficult task of accommodating opposed interests. Thus, a need for trade-offs and political decisions is unavoidable. This is a set of conditions where game theory and strategic behaviour are very valuable.

At the conclusion of the Zaragoza meeting, the participants challenged the organizers to bring 'real' policy makers and economists and applied game theorists together to talk about their experiences in policy and strategy. While this is an innovative idea, many still raise their eyebrows when told about the likely usefulness of applying game theory to real life situations.

Indeed, not many applications exist that demonstrate to non-technical people and policy makers how an analytical tool such as game theory can be of use. The literature includes an increasing body of work demonstrating the policy relevance of game theory in natural resources and the environment (e.g. Ostrom et al, 1997; Carraro and Filar, 1999; Hanley and Folmer, 1999; Patrone et al, 1999; Finus, 2001; Carraro and Fragnelli, 2004; Dinar et al, 2008). In addition there are the papers presented at the 2006 Zaragoza conference (Dinar et al, 2008; Patrone et al, 2008; Sumaila et al, 2009).

But all the above-mentioned literature is biased towards economists' and game theorists' point of view – the policy makers' perspective is missing. To rise to the challenge of bringing policy makers, economists and game theorists to the podium, a conference took place in Zaragoza in 2008 ('Agua: Economía, Política y Agricultura').[2] The conference was part of the International EXPO in Zaragoza on 'Water and Sustainable Development'. This book assembles works from that conference that together demonstrate the importance of the policy–strategy nexus in managing water resources. The various chapters cover a variety of issues that high-level policy makers, as well as policy analysts have faced during various stages of the policy making process. They include experiences from Jordan, Arizona, Australia, Spain, Brazil, South Africa, Italy and California, and discuss transboundary issues between the US and Mexico, and transboundary issues in the Jordan, Ganges and Aral Sea basins. Among the specific issues addressed are pricing, environmental quality, groundwater quantity and quality, climate change and inter-basin water transfers.

The book's chapters are assembled into three parts. Part I focuses on water policy issues in Jordan, Arizona, Australia and Spain. Through a very detailed discussion on the thread of water policy issues addressed in each of these countries, the fine line of the policy–strategy connection is identified. Part II highlights extremely important strategic issues from Jordan, Brazil and South Africa, as experienced by high-level policy makers in the design and implementation of very specific issues such as pricing, allocation, international water and equity. Part III provides the perspective of the strategic framework as it is reflected by the application of negotiation and game theory approaches to actual policy issues such as water allocation, investment in water infrastructure, transboundary environmental issues and transboundary water treaties stability in the era of climate change.

A detailed description of each chapter of the book is provided in the next three sections. This is followed by a summary of the lessons learned from the interaction between the policy makers' experience and the economists' and game theorists' approaches and assessments.

## Part I: Issues in water resource policy

Policy makers are concerned with decisions related to the management, allocation and use of water by different sectors and strata of society. As such, many trade-offs exist that necessitate evaluation and comparison.

We begin with the issues in water resource policy in Jordan – one of the most water scarce countries in the world. In Chapter 2, Haddadin addresses the critical role of the human resources problem that the water sector has to cope with. In terms of quantity, water has traditionally been allocated such that the majority is given over to the irrigation sector. With increased economic growth and increased population both from internal growth and immigration, the demand for water by the urban sector has increased exponentially. This situation may lead to conflicts among the irrigation and the urban users unless 'external' water resources are introduced. In addition, a good and comprehensive policy could also benefit from the allocation of water resources from a friendly neighbouring country. A third component that could alleviate water scarcity is the possible integration of groundwater resources into the water system. Water policy in Jordan encompasses all three pillars: wastewater, groundwater and water imported from good neighbours. The chapter reviews the development of the policy over time, and the problems and difficulties it has encountered. One dimension not fully covered – quite deliberately – is the environmental dimension of severe water scarcity. This is addressed in Chapter 3.

A major question that is always at the centre of the policy debate is whether or not environmental amenities suffer in the struggle between conflicting water needs. The case of Jordan is a classic one (see Chapter 2), as the level of scarcity may lead to the environment being assigned the lowest priority. In Chapter 3, Naber describes the drivers behind the increased pressure on Jordan's ecosystems. Chapter 3 complements Chapter 2 by considering the environmental dimension of Jordan's water and other related policies. Clearly, social welfare is affected by poor water quality and environmental degradation. However, as described in Chapter 3, the implementation of the various policies and regulations face difficulties mainly because of private interests that do not internalize the value of the ecosystem and its services. Undoubtedly, there is a possible internal conflict between existing water and agricultural policies (wastewater reuse, use of groundwater) and environmental and water quality policies. The chapter concludes that bringing environmental mainstreaming into the planning and policy making process, through the use of impact assessment studies and public awareness policies, may lead to better implementation of environmental policies and harmonization among users.

A similar pressure on the water resource base faces Arizona. In Chapter 4, Colby focuses on water management challenges and policy responses in Arizona where irrigated agriculture consumes large quantities of water, which are now also needed for growing urban areas, indigenous peoples and the environment. The chapter describes an innovative policy intervention to ease potential conflicts whereby voluntary water transactions are introduced as a response to competing water claims. One important aspect of the policy framework is the inclusion of all sources of water – groundwater, treated wastewater, imported surface water and others – in the pool, subject to claims and voluntary transactions. An additional sector entering the equation is the indigenous (aboriginal, native) peoples. In extrapolating from Arizona's experience to other countries, caution must be exercised. Some necessary conditions are needed, including strong institutions, a legal framework and conveyance infrastructure.

As in many other arid regions of the world, groundwater is one of Arizona's main sources of water. It is subject to many demands and many conflicting use needs. To address such conflicting demands, the State of Arizona has instituted several water management innovations, which are analysed by Jacobs in Chapter 5. They include state-mandated water management plans within watersheds, investments in artificial recharge and water banking, a permit system for groundwater withdrawals, and a mandated 100-year renewable supply guarantee for all new housing developments. While each of these management aspects looks autonomous, they all act together as an integrative policy that aims to both sustain the groundwater resource and suppress present and possible future conflicts by existing and potential users.

And yet, in another corner of the world, Australia provides many lessons that could help other countries handle competing demands for water. In Chapter 6, Crase focuses on policy making in an era of hydrological uncertainty. Against competing demands for water, and along with growth, historical allocation and future climatic uncertainty, this chapter (as do Chapters 4 and 5) reviews the mechanisms of market instruments to address the competing demands of agricultural, urban and environmental needs. One application of such mechanisms includes options contracts between various users. For the environment, option contracts can support environmental flow regimes during low flow years; for urban communities, which are already physically connected to irrigation systems, option contracts can support the movement of water resources from agriculture to urban users. Options contracts are politically neutral because they bypass the untenable issue of agricultural interests having to forgo water access rights to satisfy urban users' needs. The Australia experience is quite encouraging. Not only do such market innovations potentially defray the political costs of change, they provide a mechanism for more accurately addressing economic, hydrological and climatic uncertainties.

The Murray-Darling Basin is one of the richest and most complex ecosystems in Australia. Over the years, as in the case of many other countries, the basin waters have been over-allocated and agricultural uses have left their mark on the ecosystem, leading to an increased risk of its collapse. In Chapter 7, Hillman reports on attempts to supply 'environmental' water allocations to sustain the river ecosystem in the Murray-Darling Basin. In addition to still existing scientific unknowns, policies to recover the ecosystem also face regulatory and political difficulties. But as water entitlements in Australia are handled and exchanged by market mechanisms, the strategic goal of saving the environment in the Murray-Darling can be easily achieved. The chapter provides an example of one means of approaching the need to make tactical decisions about deploying environmental water at the regional scale. By identifying and establishing local stakeholder groups, they may provide well-researched, technically sound and community-endorsed environmental allocations. Allocations based on such mechanisms are more likely to be sustainable and politically accepted.

Moving to Spain, the issues are remarkably similar to those of Jordan, Arizona and Australia. The policies are also quite similar. Two ways in which Spain differs when compared to these other countries are the European Water Framework Directive, and the 2001 national hydrological plan and its subsequent reform in 2004, leading to several inter-basin water transfer plans. In Chapter 8, Garrido and Llamas review the history of water policy in Spain. While they describe issues facing the water sector in Spain that are very similar to those in Jordan, Arizona and Australia, the policy–strategy nexus in the case of Spain is much more complicated due to the very close links between sectors and regions. For example, a super inter-basin water transfer between the Ebro Basin to the south-eastern part of the country created a political debate not only among the said regions (contributing and receiving) but also among environmentalists in other regions of the country. Similar inter-basin water transfer plans were also shelved due to politically opposed views and values.

In Chapter 9, Iglesias provides an explanation for the potential implications of climate change to policy development in Spain. Part of the EU, Spain is obliged to operate under the umbrella policy of the EU. As in the case of the Water Framework Directive, complications may arise. It is apparent that there is a potential conflict between the EU and Spanish national climate change policies. For example, adaptation to climate change is unlikely to be facilitated through the introduction of new and separate policies at the national level, but rather by the revision of existing local policies that undermine adaptation and the strengthening of policies that enhance it. Iglesias suggests that if adaptation is to become 'mainstreamed', it will be preferred from a transaction cost point of view for relevant EU-wide polices, such as the Common Agricultural Policy and the Water Framework Directive, to address the issue more directly. Iglesias concludes that adaptations often involve a concerted effort across several sectors due to the links between sectors. For example, water resources are sensitive to the responses in other sectors, particularly agriculture, tourism and biodiversity conservation. Therefore, adaptation policies in water resources should have to consider policies in other sectors.

The chapters in Part I suggest that the various aspects they consider, including pricing, allocation, equity, environment and investments, may all be addressed by policies that are planned and implemented with a top-down approach. However, by introducing strategic considerations, a balanced policy–strategy approach achieves much more and in a quicker and less costly way. Part II looks at very specific cases that focus mainly on the strategy used to apply the policy.

## Part II: Issues in water resource strategy

In central decision making, strategic decisions are required on a daily basis, especially when designing and/or implementing policies. Strategic issues that face policy makers include the prices of water and water services, water allocation, investments, etc. The chapters in this part of the book include the experiences of policy makers from Jordan, Brazil and South Africa in dealing with regulation policies, aspects of equity, environmental amenities and transboundary water issues.

In Chapter 10, Haddadin discusses conflicting issues emanating from water allocation within Jordan. Issues such as equity, reallocation and the risks thereof, and the augmentation of water resources are discussed. Issues in water pricing and the cultural, economic, awareness, water quality, income disparities and demographic factors affecting water pricing are presented. In addition, Haddadin uses examples from international water disputes Jordan has with its neighbours to reviews various conflicts, for example over water sharing, over compliance, over territory and over water quality. The chapter analyses the repercussions there could be from the conflict, perhaps affecting diplomatic relations among states of the region and their international alliances. Haddadin is a great believer in the power of negotiation (see also Chapters 13 and 14). The main conclusion from his experience is that a centre for conflict prevention and management in the Middle East may be a good starting point toward debating and resolving water conflicts.

The need for the strategic behaviour of a regulator in the water and hydropower sector of Brazil is described in Chapter 11 by Kelman. A set of very pragmatic questions are raised in this chapter, referring to investments, the operation of infrastructure and the impact on efficiency and equity, for example who are the stakeholders and to what extent should their interests be taken into account? By looking at several cases, Kelman identifies the bare truth that the search for (nonexistent) unanimous decisions often paralyses the democratic decision making process. This is an issue common to many countries that lack strong democratic traditions and institutions. The author proposes a number of operational conclusions such as that in new democracies the elected government is the only institution capable of reconciling the full range of interests in complex decision making processes. Therefore, for a regulator, it is essential to have teams that are allowed to strategically approach various stakeholders and work with them directly.

A unique opportunity to share the experience faced by a South African water resource policy maker is provided by Schreiner in Chapter 12. South Africa faces several challenges that affect the strategic implementation of water policies. It is a water scarce country with a large proportion of its water coming from international sources; a majority of its population is poor and with no adequate water services; and it has special ecosystem needs. Against these constraints, South African national policy requires the government to allocate water such that it meets international requirements, the requirements of the aquatic ecosystems and addresses the special needs of race and gender. The mechanism by which policy is implemented, as described in the chapter, is a consultation with key stakeholders, subject to a set of principles. This set of principles includes, among other things, meeting ecological reserves and international agreement; allocating to the disadvantaged population;

and using water efficiently. Schreiner concludes that the desired consultation process was administratively costly (in terms of time, effort and resources), which can negatively affect the ability to implement a good policy.

How many of the issues addressed in chapters from Part I and II of the book can actually be translated into analytical frameworks and be quantified? Surprisingly, many of the issues have been part of analyses that allow more generalization and quantification. Several examples are provided in Part III.

## Part III: Interaction between policy and strategy

The chapters in Part III focus on applications to actual situations that faced policy makers. This part includes two negotiation applications, one non-cooperative game theory application and one cooperative game theory application.

In Chapter 13, Carraro and Sgobbi apply a non-cooperative negotiation model to the Piave River Basin, northeastern Italy, where players negotiate in an alternating-offer manner over the sharing of water resources. The suggested framework builds upon existing non-cooperative, multilateral, multiple issues bargaining models and captures players and policy space, various characteristics of the region. In addition, the framework introduces stochastic variations in water supply on to players' strategies. And lastly, it is inclusive of all players in determining management policies.

In Chapter 14, Goodhue et al apply bargaining theory to aid policy makers in the current policy debate regarding investment in California's water system through the issuance of a state bond. With population growth, an increased frequency of droughts, increased environmental needs that are anchored in Federal laws, and economic growth, the deteriorating water infrastructure reduces the ability of California's existing infrastructure to capture the Sierra Nevada snowmelt, hence reducing available supplies. The policy debate is about the size of the bond issue, the allocation of the resulting funds across various prospective uses and the sharing of the financial burden between Californian taxpayers (via the bond) and water users (via fees). The analysis suggests that the negotiation issues, the definition of a successful outcome and the default outcome are critical for the process. Broadening the set of alternatives provides space for negotiation and compromise. Setting the bar of expected success low enough reduces the rigidity in the negotiation and allows an agreement to be reached more quickly and with a lower transaction cost. Defining the default such that it will be the least desired may lead to faster and cheaper transaction cost related processes.

Chapter 15 by Frisvold views the history of environmental management on the US–Mexico border through the lens of game theory. Frisvold claims and shows that game theory can help to improve funding mechanisms for border environmental infrastructure; facilitate issue links as way of resolving multiple environmental disputes; and reduce pollution cost. Frisvold concludes that game theory can help policy makers improve the design of transboundary water and environmental policies in various ways. It encourages institutions that operate under repeated game rather than one-shot rules, has formal side-payment mechanisms among the negotiating parties, and has superior interconnected game outcomes.

Finally, in Chapter 16, Dinar addresses vulnerability that might be placed on international water treaties by climate change, for example the increased variability of water supply. The likelihood of not meeting treaty specifications may lead to performance failure on the part of one or more riparians. Using cooperative game theory concepts, Dinar demonstrates how pooling resources, not subject to a treaty, can provide the needed expansion cushion in situations where climate change is expected. Because of the additional transaction costs associated with water supply variability, a departure from the existing treaty, forming strategic alliances between a subset of riparians is also demonstrated.

## Lessons learned and policy implications

While the 'successful' outcome of a policy intervention is not defined in comparative terms, the various chapters of this book suggest that policy interventions work better if the implementation process is designed to address strategic issues that stakeholders may use against it. Several factors that individually and collectively have to be in place in order to ensure a plausible outcome can be observed from the various case studies.

First, policy implementation will be more successful if there is economic rationality in its design and political sensitivity during its implementation. Second, the timing of the implementation may be such that it creates conditions under which it is politically possible to undertake the policy intervention. Finally, what is common to many of the chapters in the book is that policy that is narrow or has been implemented on a sub-sectoral basis may be more difficult to implement and less likely to succeed. Policies that are designed and implemented in a comprehensive manner have a greater likelihood of succeeding. Addressing such aspects in the policy design and implementation will be likely to lead to lower transition costs, safety nets for the poor and the unaccounted for (environment), and compensation packages to potential 'losers'.

It also can be quite safely said that additional aspects mentioned in the various chapters make strategic behaviour more apparent in policy making. Such aspects include: institutions, fairness and equity, power, asymmetry of information, transaction costs, comprehensive considerations, adequate distribution of benefits, participation/education, coalitions, financial crisis, external shocks and the stochastic nature of natural phenomenon.

## Notes

1  The papers presented at the meeting can be found in www.iamz.ciheam.org/GTP2006/home1.htm. A collection of selected papers from the meeting was also published (Dinar et al, 2008; Patrone et al, 2008, Sumaila et al, 2009).
2  www.cita-aragon.es/index.php/mod.eventos/mem.detalle/idevento.22/chk.0da56afd444e2af07c95aa229a40c190.html.

## References

Carraro, C. and J. A. Filar (eds) (1999) *Control and Game-Theoretic Models of the Environment*, Birkhauser, Boston
Carraro, C. and V. Fragnelli (eds) (2004) *Game Practice and the Environment*, Edward Elgar, Cheltenham, UK
Dinar, A., J. Albiac and J. Sanchez-Soriano (eds) (2008) *Game Theory and Policy Making in Natural Resources and the Environment*, Routledge, London
Finus, M. (2001) *Game Theory and International Environmental Cooperation*, Edward Elgar, Cheltenham, UK
Hanley, N. and H. Folmer (eds) (1999) *Game Theory and the Environment*, Edward Elgar, Cheltenham, UK
Ostrom, E., R. Gardner and J. Walker (1997) *Rules, Games and Common Pool Resources*, University of Michigan Press, Ann Arbor
Patrone, F., S. Tijs and I. Garcia-Jurado (eds) (1999) *Game Practice: Contributions from Applied Game Theory*, Kluwer Academic Publishers, Boston
Patrone, F., J. Sanchez-Soriano and A. Dinar (guest eds). (2008) *International Game Theory Review*, vol 10, no 3, Special Issue, September
Sumaila, R. Y., A. Dinar and J. Albiac (guest eds). (2009) *Environment and Development Economics*, vol 14, no 1, Special Issue, February

*Part I*

# Issues in Water Resource Policy

# 2
# Issues in Water Resources Policy in Jordan

*Munther J. Haddadin*

Jordan, a kingdom in the heart of the arid and semi-arid Middle East (Map 2.1, Plate 1), is so water strained that its blue water resources (1016mcm/yr) make a per capita share of 180 cubic metres per year in the year 2007. The green water resources add 866mcm per year on average (155m$^3$/cap/year) and the grey water of 90mcm per year add another 16m$^3$/cap/year.[1] The total is about 351m$^3$/cap/year. Compared to the need per capita, computed at 1700m$^3$/cap/year, in 2007 Jordan possessed only 21.8 per cent of the water resources it needed. The balance, or about 1349m$^3$/cap/year, was closed by food and industrial imports in the form of what the author calls 'shadow water' (Haddadin, 2006, 2007). Maps 2.2 and 2.3 show the surface and groundwater basins of the country respectively.

The pressure of population on the water resources at 2850 persons per unit flow (defined as a million cubic metres per year), compared to the optimum average of 588 persons, calculated on the basis of 1700m$^3$/cap/year as reported by Haddadin (2006, 2007) indicates that Jordan's water is stressed at 4.85 times the level it can safely endure, a high water stress indeed. The second most water stressed country in the region of Jordan's economic category, Egypt at 1050 persons per unit flow, is stressed at only 1.8 times the safe level. Syria, Jordan's neighbour to the north and the third ranking of the Lower Middle Income Economy for water stress, is actually water relaxed as its stress is about 0.93 times the safe level. This is, therefore, the highest priority water resources policy issue in Jordan. The population pressure on natural resources was not just the result of biological population growth, as this was equalled by man-made factors manifested by three waves of involuntary displacement of Palestinians: two from their own homeland in 1947–1948 and 1967 and one from the Gulf States in the wake of Iraq's invasion of Kuwait in 1990. Voluntary movement of Palestinians from their own homeland

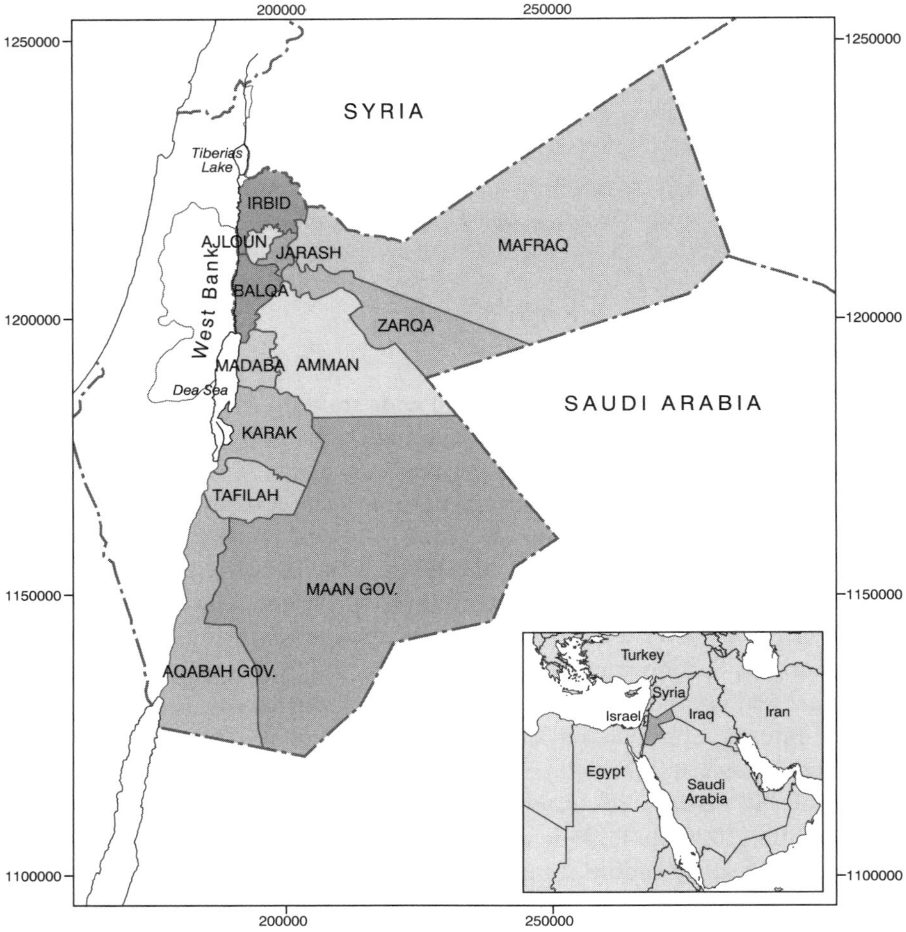

Note: See Plate 1 for a colour version.

**Map 2.1** *The Hashemite Kingdom of Jordan*

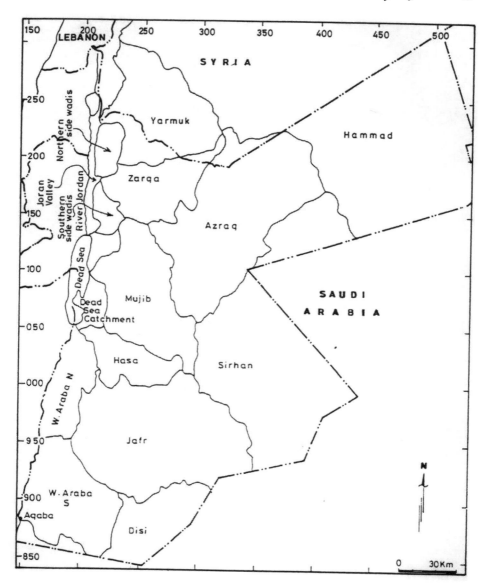

**Map 2.2** *Surface water basins*

to Jordan had also taken place between 1950 when the West Bank united with the Hashemite Kingdom of Jordan and 1988 when Jordan severed all legal and administrative ties with the West Bank.

The Hashemite Kingdom of Jordon shares the waters of the Jordan River Basin with other riparian parties: Lebanon, Israel, Syria and the Palestinian Authority. Attempts to work out a plan to use these waters for the development of the

16   *Issues in Water Resource Policy*

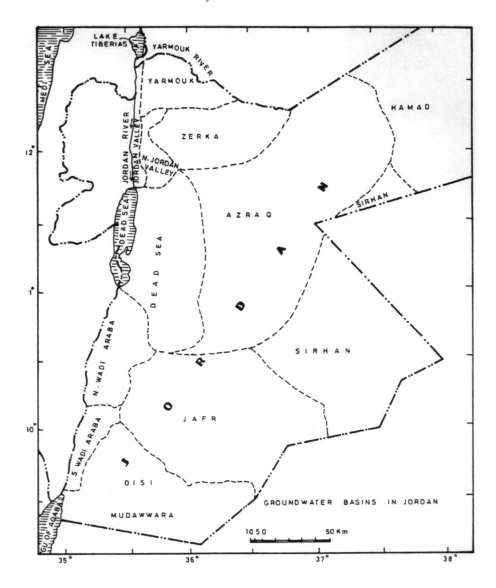

**Map 2.3** *Groundwater basins*

Jordan Valley succeeded at the technical level in 1955 after two years of shuttle diplomacy conducted by a presidential US envoy, Ambassador Eric Johnson. The shares of each riparian were determined. Jordan got the indigenous side wadis that discharge into the Lower Jordan (Wadis Arab, Ziglab, Abu Zayyad, Yabis, Kufrinja, Rajb, Zarqa, Shueib, Kafrein and Hisban) totalling 175mcm per year and the residual share in the Yarmouk River after Syria's and Israel's shares are deducted at 90mcm and 25mcm per year respectively. The West Bank shares with Jordan this residual flow, which averages 377mcm per year after the entire flow of the Yarmouk is controlled. The split is 296mcm and 81mcm for Jordan and the West Bank respectively. The West Bank also shares with Israel water from the Upper Jordan to be released to the West Bank from Lake Tiberias at the rate of 100mcm/yr with a ceiling of 15mcm of brackish water drawn from saline water springs around the lake.

Associated with securing the Jordan's water shares has been the regulation of the Yarmouk River so that a total flow of 506mcm (including 26mcm return flow from Syrian use and 13mcm of historic use) can be used (Haddadin, 2006, 2007; plus working sheets given in 1984 to the author by Dr Wayne Criddle, the expert engineer on the Johnston's mission). The construction of regulating structures on the Yarmouk entails spanning Jordanian territories in the southern bank area and Syrian territories in the northern bank area. This means that the two countries have to cooperate to have such a structure built and, indeed, they entered into a bilateral treaty in 1953 for that purpose. However, diplomatic relations between Jordan and Syria since that time have alternated between strained relations and cooperative action. In 1967 Syria violated the terms of the treaty by impounding flows of the Yarmouk tributaries in its territories, thereby increasing its share over and above its entitlement according to the treaty. The US brokered plan worked out by Ambassador Johnston allocated to Syria the same flow from the Yarmouk that had been stipulated in the Jordan–Syria treaty of 1953, which was the discharge of all springs in Syrian territories above an elevation of 250m above sea level (Haddadin, 2001).[2] Another complication has been the need to build a diversion structure across the Yarmouk to divert Jordan's share of its waters to a canal Jordan had built with assistance from the US for the irrigation of the East Jordan Valley. The state of war that had prevailed between the Arab States – including Jordan – and Israel since 1948 upon its proclamation stood in the way of building such a structure. Syria is the upper riparian, Jordan is the middle and Israel has been the lower riparian party on the Yarmouk.

An important characteristic of the East Mediterranean climate is the frequency of droughts. This entails the need for multi-year storage of floods, thus raising the cost of dam yields above what would be the case in a stable climatic situation. This becomes a problem for Jordan, a Low Income economy country promoted in the mid-1970s to the status of a Lower Middle Income economy country.

Finally, human resources development, although achieved at very impressive rates in Jordan, have not been in synchrony with market demand and the water sector is no exception. On-the-job training is crucial and overseas training will further enhance the capacities of the human resources working in the water sector.

As such, Jordan's water policy is that of an arid and semi-arid country that witnesses pronounced inter-annual variability in its water resources that are primarily supplied by rainfall, and that it is the middle riparian on an international river, the Yarmouk, which is stochastic both physically and politically. Water supply and Jordan's water policy have to be adjusted to a multi-parameter water situation.

## Emergence of the need for water policy

The water (and wastewater) sector in Jordan has passed through various stages of administration, legislation and institutional arrangements. Prior to 1950 it was administered by municipalities for municipal water and by the Department of Lands and Surveys for irrigation water. In the 1950s the Ministry of Public Works became involved in water irrigation projects. The distinct organizations that managed water in one way or another started in 1957–1958 when the Central Water Authority and the East Canal Authority were established to care, respectively, for the management of water resources in the country and to administer the irrigation development in the Jordan Valley. The Central Water Authority did the investigations for water resources and supplied municipal water in bulk to municipalities. The two organizations were then merged in a newly established Natural Resources Authority in 1965, only to be split again in 1973 when the Jordan Valley Commission was established to manage the integrated development of the Jordan Valley. A Domestic Water Corporation was established in 1974 to manage municipal water supplies. In 1977 the Jordan Valley Authority succeeded the Jordan Valley Commission and was empowered to care for the operation and maintenance of the irrigation and municipal water systems in the Jordan Valley. In 1983 the Water Authority of Jordan was established to succeed the Domestic Water Corporation, the Amman Water and Sewage Authority and the Jordan Valley Authority in so far as the latter is involved in the management of municipal water systems. Finally, the Ministry of Water and Irrigation (MWI) was established, with both the Water Authority and the Jordan Valley Authority under its wings.

The frequency of institutional changes and the dichotomy in the allocation of responsibilities did not allow an overall planning vision for water needs in the country to emerge, especially municipal and industrial water needs. Municipal water shortages prompted the National Planning Council, the predecessor of the Ministry of Planning and International Cooperation, to have a study performed for the municipal and industrial water needs in northwest Jordan where 91 per cent of the population lived. The result was surprising. The needs for municipal and industrial water in northwest Jordan surpassed all expectations and were beyond the capacity of indigenous water resources without infringing on the planned allocation for irrigation in the Jordan Valley.

It was at that time, 1978, and as a result of the aforementioned study and the municipal water shortages, especially in Amman, that four primary pillars of water policy emerged and were accepted by the Cabinet and the King:

1. Allocating surface water from the Jordan Valley, otherwise meant for irrigation development there, to supply the capital city of Amman and the city of Sult with municipal water in a total amount of 90mcm per year;
2. Formalizing the reuse of treated wastewater to make up for the deficit in irrigation water budget after the freshwater resources are diverted to municipal use, and;
3. Approaching the Republic of Iraq for the possible supply to Jordan from the Euphrates River of 160mcm per year, forecast as the deficit of municipal and industrial supplies by the year 2000.
4. In the meantime, Amman's water shortage was to be met by pumping water from a valuable aquifer feeding an environmentally important desert oasis at the rate of 12mcm/year.

Action started on all four fronts: (i) a major municipal water supply project was adopted to convey Jordan Valley water to Amman in two stages; (ii) a wastewater treatment plant using stabilization ponds was decided for Amman and its effluent would flow back to the Jordan Valley to be reused for irrigation; (iii) a study was commissioned to look into the feasibility of diverting 160mcm of water from the Euphrates inside Iraq at Qaim to Amman (Haddadin, 2006); and (iv) a project was initiated to augment Amman's water with water from the Azraq basin.

The Iraqi government agreed to supply Jordan with that annual amount from its own share in the Euphrates River. The study, commissioned by the National Planning Council and conducted by Howard Humphrey Consultants at the time, indicated that a pipeline of about 2m in diameter and 605km in length would be needed to transfer 5m$^3$ per second of Euphrates water to Amman. The static head would be 830m and the total dynamic head would be 1380m. The cost of supplying water to the outskirts of Amman from the Euphrates was roughly US$2.00, too expensive for Jordan's economy to afford. The project was shelved in 1985.

The United Nations (UN) declared the 1980s the decade of drinking water and sanitation, and Jordan abided by the call and expanded the coverage of municipal water networks. The demand from the networks therefore increased and supplies fell below demand levels. Wastewater treatment plants were built to serve cities and towns, and the reuse of treated wastewater came into vogue.

One serious handicap accompanied the emergence of the water policy pillars. Each pillar was cared for by a different institution: diversion of irrigation water to municipal use was entrusted to the Jordan Valley Authority; the wastewater treatment plant was assigned to the Amman Water and Sewerage Authority; the study of water imports from Iraq was entrusted to the National Planning Council, and the transfer of Azraq water was entrusted to the Amman Water and Sewerage Authority. This diversity in responsibility created operational problems that affected the quality of water supplies.

Late in 1983 a Water Authority was established that merged the Domestic Water Supply Corporation, the Amman Water and Sewerage Authority, the water department in the Natural Resources Authority, and the water responsibilities of all municipalities and village councils in the country. The projects supplying Amman and Irbid from the Jordan Valley were transferred, on completion, from the Jordan Valley Authority to the Water Authority for operation and maintenance.

While the collection of wastewater and its treatment was entrusted to the Water Authority, the reuse of the majority of the treated effluent in the Jordan River catchment was entrusted to the Jordan Valley Authority. Wastewater of urban centres such as Amman, Zarqa, Sult, Ajloun, Kufrinja, Karak, all lying in the Jordon River catchment, was entrusted for reuse to the Jordon Valley Authority. Again, this duality created quality problems that impacted the potential for reuse. Surface water in the Jordan Valley, under the jurisdiction of the Jordan Valley Authority, became a source of municipal water supply to Amman. A failure in the operation of the raw water from that source, performed by the Water Authority, was blamed on the Jordan Valley Authority, and that resulted in the resignation of high ranking officials in the administration of water.

Concurrently, the management of groundwater resources was shifted from the Natural Resources Authority in 1984 to the Water Authority. The control over abstraction and even over drilling was relaxed between 1967 and 1984. When it was time for the Water Authority to exercise strict control in the late 1990s, there had been a score of illegally drilled wells and a culture of non-compliance with the regulations had emerged making control very difficult. Finally, the government legislated for better control of groundwater abstraction. It issued a by-law that rationed a specified annual quantity to be abstracted from each well and assigned a tariff for whatever annual quantity was pumped in excess of the permitted quota. Control has improved since this time through the creation of a special surveillance unit for monitoring abstraction and the illegal drilling of wells. In 2007, the government took a bold step in the Jordan Valley where illegal well-drilling was in progress and trespassing on government land was practised. It demolished whatever illegal wells were in existence and obliterated unauthorized crops or buildings that were on government land.

## Centralization of institutions

Prior to 1984, the following organizations shared the management of the Jordanian water sector:

- From 1965, the Natural Resource Authority: in charge of groundwater investigation and data collection and management, and for the environmental affairs of water resources.

- From 1973, the Jordan Valley Authority (JVA): in charge of the integrated development of the Jordan Valley with irrigated agriculture as its backbone. In terms of water the JVA was in charge of the waters of the Jordan River basin including surface and groundwater resources, and was the recipient of the treated effluent from wastewater treatment plants in the catchment of the Jordan River.
- From 1974, the Domestic Water Supply Corporation: in charge of supplying municipalities and towns with municipal water in bulk.
- From 1908, municipalities and village councils: in charge of distributing municipal water inside city limits.
- From 1954, the National Planning Council: in charge of seeking foreign assistance funds to develop water resources; it also supervised the preparation of the Euphrates to Amman project feasibility.

In late 1983, the government legislated for the creation of the Water Authority (WAJ) and transferred to it responsibilty for all of the above, except for irrigation in the Jordan Valley, which was retained by the JVA. During the operation of the Water Authority, certain clashes occurred between the responsibilities of the JVA and the newly created WAJ, especially over the management of dams and diversion structures and in the management of international shares of water. After the disputes were settled by a new government in 1985, it was considered essential to bring the two organizations in charge of the water sector under a ministry, and that two boards of directors chaired by the Minister of Water and Irrigation be made responsible for their operation. The ministry was created in April 1988.

## Integration of water policy

Jordan faced an economic and financial crisis in 1989 that prompted the government to seek the assistance of the International Monetary Fund (IMF) to help manage their debt servicing and design and apply an economic restructuring programme. Several adjustment loans were extended to Jordan by the World Bank to help restructure a number of economic sectors, the water sector included.

The economic and financial shock, coupled with the inflow of people from the Gulf States in the wake of the Iraqi invasion of Kuwait in August 1990, burdened the government in an unprecedented way. The idea of an integrated water policy for the country was entertained, with encouragement from the World Bank who worked in tandem with the IMF on the structural adjustment loans. However, this dragged on because the MWI were relying on consultants to draw up a proposal for such a policy, financed by grants from an interested donor (the United States Agency for International Development (USAID) in this instance).

The Cabinet that was formed in March 1997 included a water Minister who focused on the need to formulate a centralized water policy and to follow up on the implementation of the water annex agreed with Israel in 1994. The Minister formulated a 'Water Strategy' for Jordan under which he wrote documents for four 'Water Policies' that the minister worked out single-handedly:

- irrigation water policy;
- water utilities policy;
- groundwater management policy;
- wastewater management policy.

The drafts were forwarded to officials of the World Bank for comments, and the final drafts were forwarded to the Council of Ministers (Cabinet) for discussion and approval. The strategy and the four policies were approved and issued by the Council of Ministers as official government documents (Haddadin, 2006).

The same minister directed and supervised an 'Investment Program' (MWI, 1997) for the water sector covering the period (1997–2011) that formed the basis of future projects and gave the donors an official document to work with. The Investment Program was presented at a donors' conference convened at Petra, Jordan, in 1997 and formed the basis of future work.

## Topics addressed in the water policies

The strategy, supplemented by the policy papers, has set objectives for each of the following: resource development; resource management; legislation and institutional set-up; shared water resources; public awareness; performance; health standards; private sector participation; financing; and research and development.

On resource development, the strategy maintains that (i) the full potential of surface water and groundwater shall be tapped wherever economics permit; (ii) marginal water (wastewater and brackish water) is to be considered as a valuable resource that cannot be treated as waste; and (iii) the first priority in allocating new water resources shall be given to satisfying basic human needs. In this respect, the MWI initiated (and completed in 2000) the construction of a diversion weir on the Yarmouk River, and completed the Wehda Dam on the Yarmouk in 2005. The Ministry further conducted feasibility studies and built a small number of dams on streams to increase supplies for municipal and industrial water, recharge aquifers, and for irrigation. The Wala, Mujib and Tannur dams were built in 2005 as was Wadi Abu Barqa Dam in Wadi Araba. Two other feasibility studies are being attempted at the time of writing: a dam on Wadi Kufrinja and another on Wadi Ibn Hammad.

The development of brackish water resources was also pursued. The waters of the Mujib, Zara and Wadi Zarqa Maien were developed and desalination of the brackish blend was made. The desalinated water is being pumped to serve tourism development on the East Shore of the Dead Sea and also to Amman.

On resource management, the strategy gives priority to the sustainability of existing and future water resources. It calls for the highest possible efficiency in water resource management on both the supply side and the demand side of water management. The strategy specifically articulates the need for wastewater management in its use as a resource. In this regard, penalties have been enforced on the violators of groundwater abstraction after the status of illegal wells was clarified pursuant to a by-law that the government passed in 2002. Prior to that date, reducing the number of illegal wells and controlling abstraction from groundwater were among the objectives of water policy and agricultural policy. Old municipal water networks have been renewed in Amman and elsewhere. Water tariffs for municipal, industrial and tourism uses were formulated and enforced, enhancing demand management measures. The reuse of treated wastewater has been organized and specifications for its implementation adopted.

On legislation and institutional set-up, the strategy calls for institutional arrangements to be updated, and maintains the need for assuring cooperation and coordination among public and private entities involved in the development and management of water resources. In this regard, MWI adopted a Management Contract approach to manage the water and wastewater services of Greater Amman. This contract was initially for three years, later extended to six years, after which a water company was founded to succeed the operators. Miyahuna is a new water company, founded in 2006, that is to take charge of the provision of the water and wastewater services of Greater Amman. In Aqaba, a water company was set up directly, without passing through the phase of a Management Contract. Both are doing well. Attempts to shift toward water companies in the northern governorates are underway.

On shared water resources, the strategy appeals for cooperation on bilateral and multi-lateral levels with neighbouring states within the provisions of a Regional Water Charter. It also calls for the protection of Jordan's share in international water courses. In this regard, cooperation with Israel continued and the transfer of water from Israel to Jordan proceeded without much hardship. The same cannot be said about Syria's use of the Yarmouk. Syria has been overusing the Yarmouk surface flow and also the groundwater feeding springs that form an integral part of Jordan's share. The impact of that has been negative on Jordanian investment in the Wehda Dam and in irrigated agriculture. However, political factors have hampered cooperation and understanding. More recently (December 2007) contacts on the matter have been renewed and attempts to adhere to the terms of the bilateral treaty between the two countries were seriously undertaken, but no encouraging conclusions were reached.

On the objective of public awareness, this something that is given an important role in the management of water resources, and economic measures are to be adopted to reinforce public awareness. Activities by the Ministry, non-governmental organizations (NGOs) and some donors boosted public awareness efforts and the impact of that has been positive.

On performance, the strategy stipulates the monitoring of the efficiency of water and wastewater utilities, data collection and entry, and of human resources development. In this regard, the Ministry and its daughter organizations have achieved success with assistance rendered by the German GTZ and by USAID.

Health standards are to be set and reinforced, especially where the reuse of treated wastewater is concerned. In this regard, the seriousness by which government views health impacts led to the resignation of two ministers in late 2007, the Minister of Water and the Minister of Health, because of reported water pollution in a town northwest of Amman.

On private sector participation, the objective is that the role of the private sector shall be expanded through management contracts, concessions and other forms of private sector participation. Participation by the private sector is also to be encouraged in irrigated agriculture. In this respect, farmers' participation in irrigation water distribution has started and the results are encouraging (Haddadin et al, 2008). In the case of municipal water and wastewater services, the forms of management reviewed above, the Management Contracts and the establishment of companies, have also yielded good results. The amount of water that is unaccounted for has been reduced from 55 per cent to 45 per cent and the collection of dues from subscribers covered the operation and maintenance cost and more (PMU, 2007).

On financing, the cost recovery of utilities in their provision of services is set as a target in the strategy. In this regard, the water tariff enables the recovery of the operation and maintenance costs of municipal and industrial water, and collection showed that revenues covered, and at times exceeded, the cost of operation and maintenance. It also covers a good percentage of the operation and maintenance costs of irrigation projects.

On research and development, the strategy encourages research and development in issues relevant to water resource management and in the applied aspects of the research. In this regard, the National Center for Agricultural Research and Technology Transfer (NCART) has been established and also faculties of agriculture in the government universities. Serious research has and is being done on the possible irrigation use of treated wastewater with assistance from the Arab Fund for Economic and Social Development, and meaningful soil research for Jordan Valley soils is also under investigation.

## Jordanian agricultural strategy 2001 (JAGS)

In 2001, Jordan issued its agricultural strategy (MoAg, 2001). The basic content on irrigated agriculture is compatible with the Water Strategy and with the Irrigation Water Policy. It is also compatible with the Wastewater Management Policy, and highlights the fact that treated wastewater is going to be a primary irrigation water source. The estimated flow of treated wastewater is 170mcm, 202mcm and

231mcm for the years 2010, 2015 and 2020 respectively. Wastewater treatment in Jordan is based on secondary treatment methods and about 20 cities and towns are served with secondary wastewater treatment plants. The JAGS stresses the imperative of using treated wastewater in conformity with the health, environment and technical requirements and guidelines.

The JAGS lists the continuous decrease in the quantities of potable surface water available for irrigation, and the groundwater depletion and quality degradation that result from over-pumping along with the main problems facing the development of agriculture in the country. The weakness of the involvement of the Ministry of Agriculture in planning for irrigation projects is also cited.

The cited objectives of the JAGS in relation to irrigated agriculture in the Jordan Valley include:

1 Ensuring the sustainability of irrigated agriculture in the Jordan Valley through the development of available water resources, including treated wastewater, long-term planning, desalination of brackish water resources, and enabling Jordan to obtain its full water rights.
2 Protecting agricultural lands and irrigation water quality by increasing the efficiency of wastewater treatment plants; decrease, as far as possible, pollution of surface and groundwater resources.
3 Maximizing irrigation water use benefits through private sector participation in the management of irrigation projects, expanding the establishment of water user associations, an undertaking that has shown some success (Haddadin et al, 2008); the application of modern irrigation and agriculture technologies, a practice that has been successfully undertaken by farmers; and implementation of proper irrigation water pricing mechanisms. While water pricing is limited in its ability to maximize efficiency, it does play a role in maximizing the dollar gains per unit flow of irrigation water.

The cited objectives of the JAGS in relation to irrigated agriculture in the highlands include:

1 Protection of groundwater and regulating its extraction through a reduction in the numbers of unlicensed wells. The legislation of the groundwater control by-law in 2002 facilitated the achievement of this objective.
2 Protection of land and water resources from pollution. (This could be partly achieved by controlling pollution from point sources and by prohibiting irrigation of lands lying in the recharge areas of aquifers.)
3 Development of agricultural production in terms of quantity and quality.

Thus a long awaited set of policies for water and for agriculture came into being and became official government documents to which the Ministry of Water and the Ministry of Agriculture have adhered as closely as has been possible.

## Linkage imperative

While JAGS links up with the Water Strategy in its consideration of irrigated agriculture and irrigation water, both fail to establish a similar link in rain-fed agriculture.

The link between rain-fed agriculture and water is simple. It stems from the fact that flowing water of all colours (blue, green and grey water) is used in irrigation to produce a similar effect in the soil that rainwater creates. In both instances, the root zone is wetted and water in excess of surface tension and capillary action remains free and travels down to the groundwater table. In both forms of agriculture (irrigated or rain-fed), the objective is the same, that is the production of food for humans, livestock and birds.

Attention has been focused on irrigated agriculture more than rain-fed agriculture in such fields as research and extension, agricultural credit and human resources deployment. Rain-fed agriculture receives less attention, probably because of the centuries of practice among Jordanian farmers and the mode of operations.

The availability of soil water (soil moisture) in Jordan exceeds in its equivalence to irrigation water (866mcm/yr) the flows of surface water resources of the country (800mcm/yr). Soil water, therefore, warrants a focus of attention that at least approaches the attention given to irrigated agriculture.

To elaborate, users of irrigation water are charged a water tariff whereby they are penalized for excessive or wasteful use, but the owner of a rain-fed tract of land that is left fallow for no good reason is not even questioned. That tract of land stores rainwater that is capable of supporting winter or summer crops; if left fallow, that soil water is lost to evaporation. If the property is, say, 200 dunums in area, the annual irrigation water equivalent is 40,000m$^3$, a quantity that is worth, in terms of Jordan Valley water costs, JD144 at the rate of 35 fils real cost per cubic metre. The value added is, of course, a lot more.

A policy statement on irrigation water policy and rain-fed agriculture could stipulate that, unless there is an acceptable technical excuse, an owner of rain-fed land will be fined JD1.00 per dunum$^3$ of his land that is left fallow. Another approach is to treat soil water in rain-fed land as an integral part of the water resources and, as such, consider that that water is public property while the land in which it exists is private property. Hence, the utilization of both resources, land and water, is an ideal area for private–public partnership, at least in as much as preserving and utilizing the water resource is concerned. The mode of such a partnership can be defined to maximize the utilization of rain-fed lands, and to develop supplementary irrigation in lands where rainfall is too low to meet cropping requirements. In the process, a host of social and environmental gains can accrue, and the lot of many families may be substantially improved.

## Human resources development

The Water Strategy stresses the importance of human resources development and the MWI has paid attention to that aspect by founding a department for training in each of the JVA and the WAJ. However, despite their achievements in training water technicians and engineers from neighbouring countries, greater focus has to be placed on training Jordanian professionals to cope with the challenges facing the water and wastewater sector. In particular, planners, directors of departments and section chiefs should be accorded on-the-job training and sent out to advanced countries to receive advanced training.

## Conclusion

Water stress and escalating demands encouraged the Jordanian authorities to map out a water strategy and formulate water policies . This chapter has presented the contents of the water resources policies and of the agricultural policy as it relates to water. The evolving policies for development, the environment and management of shared water resources were outlined, as was the role of the development of human resources in managing water resources. Attempts to import water from the Euphrates to Amman were considered and the role of trade, which provides an invisible component of Jordan's water resources, was discussed. Some of the measures required to deal with Jordan's high level of water stress have been listed.

## Notes

1 Blue water is the freshwater that flows, the surface and groundwater resources; grey water is treated municipal and industrial wastewater; and green water is rainwater that gets stored in the topsoil and supports rain-fed agriculture, forests, natural green areas like pastures, swamps and other wild vegetation.
2 The author had managed, on Jordan's behalf, the shared international water resources. By 1998 when he left office, the Syrian uses of the Yarmouk were in the order of 230mcm (95mcm of springs and 125mcm of surface water, compared to 90mcm as their share in the US brokered plan. The Syrian abstraction from these sources by 2006 totalled between 260mcm and 300mcm.
3 1 dunum = 0.1 hectare.

## References

Haddadin, M. J. (2001) *Diplomacy on the Jordan: International Conflict and Negotiated Resolution*, Kluwer Academic Publishers, Norwell, MA

Haddadin, M. J. (2006) *Water Resources in Jordan: Emerging Policies for Development, the Environment and Conflict Resolution*, Resources for the Future Publishers, Washington, DC

Haddadin, M. J. (2007) 'Shadow water: Quantification and significance for water strained countries', *Water Policy*, vol 9, pp439–456

Haddadin, M. J. et al (2008) 'Participatory irrigation water management in the Jordan valley', *Water Policy*, vol 10, pp305–322

MWI (Ministry of Water and Irrigation) (1997) *Water Sector Investment Program 1997–2011*, Ministry of Water and Irrigation, Amman, Jordan

MoAg (Ministry of Agriculture) (2001) *Agricultural Sector Strategy, 2001*, Ministry of Agriculture, Amman, Jordon

PMU (Project Management Unit) (2007), Project Management Unit of the Ministry of Water and Irrigation, Jordan, Ministry of Water and Irrigation, Amman, Jordon

# 3

# Water Scarcity, Quality and Environmental Protection Policies in Jordan

*Helena Naber**

Jordan is one of the most arid countries in the world at 169m$^3$ per capita per year of annual renewable fresh water resources from internal and external sources (see Figure 3.1 and WRI, 2003). Population increases coupled with economic growth are further straining water resource availability, as well as increasing competition among sectors for water resources. In this race for valuable water resources, ecosystems have been losing out to agriculture, industry and drinking water requirements, and as impacts of decreasing flows and quality degradation became apparent, so has the need to do something about it, so as not to lose these valuable ecosystems. An example of ecosystem degradation is the Azraq oasis, a Ramsar site, which due to abstractions for domestic and irrigation purposes now receives only 10 per cent of its historical flow of water – which has reflected negatively on the size and health of this historical oasis. Similarly, the flow of water into the Dead Sea has substantially decreased from both surface and groundwater flows, which resulted in a decrease of sea level by an average of a metre annually. The degradation of the Zarqa and Jordan rivers is obvious as well, where these rivers are only a shadow of what they used to be at the beginning of the 20th century.

This chapter presents Jordan's legislative background in environmental and water protection, the main quality issues facing Jordan's water resources, illustrates some of the negative impacts on ecosystems that have resulted from quality degradation and decreased quantity of water, tries to provide policy recommendations for mitigating the situation, and presents an overview of recent efforts by the government to enhance the water situation in the country.

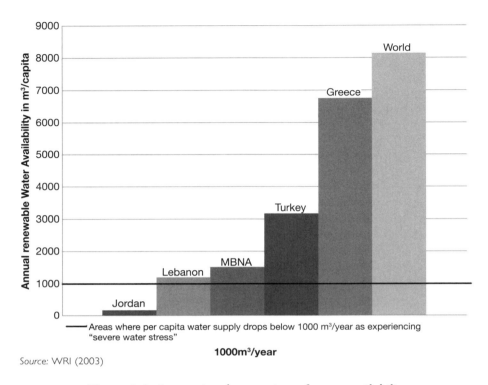

Source: WRI (2003)

**Figure 3.1** *International comparison of water availability*

## Legislative and institutional background

Jordan is signatory to several global environmental conventions, and as a consequence, the government developed strategy documents to reflect Jordan's strides towards achieving the conventions' objectives. In 1997, Jordan developed its water strategy and issued four related water policies pertaining to the management and use of municipal water, wastewater, groundwater and irrigation water resources.

In 2005, Jordan developed its National Agenda document that sets the policy roadmap for Jordan's development for the years 2006–2015. The National Agenda tackles the sectors of water and environment among others. For water, it identifies the problems of scarcity of water and depletion of groundwater resources and sets Jordan's policy towards increasing supply by relying on unconventional sources, improving the efficiency of water distribution networks, reducing subsidies in the water sector and reusing treated wastewater in agriculture and industry. Within the environmental challenges that the National Agenda identifies are: legislative and regulatory frameworks, waste management, air pollution, desertification, natural reserves and land use and the protection of the Dead Sea and the Red Sea ecosystems. It may be noted that the National Agenda regards water as an input for agricultural, domestic and industrial uses, and the link is not being made between water availability and ecosystems' well-being apart from the case of Dead Sea ecosystem. In a recent development, a Royal Commission is to be established in 2008 to develop a comprehensive strategy for the water sector in Jordan.

Jordan's first law dedicated to environmental protection was enacted in 1995, by which Jordan's first environmental protection agency was established. The General Corporation for Environmental Protection, GCEP, under the umbrella of the Ministry of Municipal and Rural Affairs and the Environment and reporting to the Minister was thus set up. Subsequently, in 2003, a ministry of environment was established in lieu of GCEP, in accordance with a new environmental protection law. In 2006, the interim law was ratified by Jordan's parliament to become Law no. 52 for year 2006 for the environment.

According to Jordan's Environmental Law, the Ministry of Environment (MOE) has the overall responsibility for environmental protection in the Kingdom, as well as setting Jordan's environmental policy; planning and executing the necessary actions for the realization of sustainable development; setting standards for environmental quality; monitoring environmental quality; and, most importantly, coordinating national efforts to preserve the environment. It also mandates the Ministry with protecting water resources, and to take action to prohibit any activities that cause pollution to, or degradation of, water resources in the Kingdom.

In line with the Law, the Council of Ministers issued several environmental protection regulations pertaining to the protection of nature; protection of the environment against hazardous interventions; air protection; sea environment and shores protection; natural reserves and national parks; management of harmful and hazardous substances; management of solid wastes; environmental impact assessment; and soil protection.

The role of the MOE is distinct from the role of the Ministry of Water and Irrigation and its affiliate authorities, the Water Authority of Jordan (WAJ) and the Jordan Valley Authority (JVA). While the MOE is responsible for protecting water resources, MWI retains the responsibility for setting and implementing water policies in Jordan, developing water resources and undertaking the monitoring and enforcement of water quality standards. Any responsibility assigned to any agency under a previous law in Jordan is amended by a subsequent law. In other words, the MOE authority over environmental protection supersedes any previous assignment of such a role to any other agency. This does not negate the responsibility of water agencies to take measures to prevent any threat to the water quality falling under their jurisdiction.

Nevertheless, a few grey areas remained in terms of delineating these responsibilities in the work on the ground, and since its inception, the MOE has had several negotiations with the MWI and its related authorities in order to enhance their cooperation and outline the grey areas of on-the-ground work vis-à-vis protection of water resources in the Kingdom. A formal memorandum of understanding is yet to be signed between the agencies in order to formalize their cooperation.

In addition, the Ministry of Health is empowered under its law to take any measure the Minister sees necessary to protect public health. As such, the Ministry of Health monitors and controls the quality of water pumped for domestic use, and checks on sewage water networks, internal sanitary installations, and on wastewater treatment plants, to ensure conformity with health standards.

In 2006, the environmental rangers department was established within the Public Security Directorate with the aim of enforcing environmental laws and regulations. The environmental rangers' directorate has been functioning for over a year; however, the vision for and details of their work are being ironed out at present with the participation of relevant institutions.

In terms of quality standards, four main sets of water quality standards have been developed based on local needs and international experience for the reuse and disposal of reclaimed domestic wastewater; the reuse and disposal of industrial wastewater; the use of treated sludge in agriculture; and standards for drinking water.

A number of institutions carry out monitoring to ensure standards: the Water Authority of Jordan, through its laboratories and quality control department, is at present the main institution responsible for monitoring various water related standards, and is implementing several diversified programmes for quality monitoring of surface water, groundwater and treated wastewater. The Ministry of Health has its own laboratories that are mainly used to ascertain the fitness of municipal water for drinking purposes. The Ministry of Environment, through a subcontract with the Royal Scientific Society, monitors the quality of industrial effluents, wastewater treatment plant effluents, water quality in dams and groundwater quality in hotspot areas (such as those located in the vicinity of municipal landfills). The Ministry's contract with the Royal Scientific Society is renewed on annual basis, and therefore the specific monitoring locations can be modified as per new developments. Finally, quality monitoring is sometimes performed by the many research centres within the Jordanian universities as part of their educational and research programmes.

## Water resources and quality issues

Jordan faces acute water shortages, which makes water quality of vital importance as it directly affects the availability of water. Threats to water resources are mainly pollution from point and non-point sources and quality degradation that results from excessive abstraction. Climate change is posing a longer-term threat to the availability of water resources in Jordan.

## Sources of water quality degradation in Jordan

The main sources of water pollution in Jordan are pollution through municipal and industrial wastewater. Municipal wastewater reaches water sources through seepage from municipal wastewater treatment plants, as well as through direct discharge from these plants when they overflow. Approximately 61 per cent of Jordan's population are connected to the sanitary sewer system (WAJ, 2007), which means that there is potential seepage of wastewater into groundwater aquifers in areas that are not covered by municipal wastewater collection networks.

Industry is another source of pollutants. One example is the Amman–Zarqa basin, which is home to approximately 50 per cent of Jordan's industries and lying within the catchment area of the Zarqa River. Primary pollutants from this point source are fat, oil and grease and heavy metals such as mercury, manganese, cadmium, chromium and copper. Leaks from municipal landfills are also implicated in pollution: the Russeifa within the Amman–Zarqa conurbation and Ekader, in northern Jordan, are major solid waste landfills, and both are associated with groundwater quality degradation (Dorsch Consult and ECO Consult, 2001).

Non-point pollution sources include irrigation water runoff that carries pollutants such as pesticides, herbicides, fertilizers, organic chemicals, heavy metals and salts. Rain-fed agriculture makes similar contributions to pollution. Animal manure, fertilizers, pesticides and other chemical pollutants find their way to groundwater aquifers during the rainy season.

Among the negative impacts of urbanization in Jordan are the impacts on aquifer recharge rates and changes in the natural water balance of precipitation between surface runoff, soil water and groundwater recharge to the disadvantage of the latter. Over-abstraction from virtually all the groundwater aquifers, notably in the Jafr, Dhuleil, Zarqa, Azraq and Mafraq areas has been causing increased salinity and a decline of the groundwater water table. As a result, many wells in the above areas have dried up; however, the legally drilled ones are replaced by deeper wells (Figure 3.2).

The Amman–Zarqa aquifer, for example, has lost a good part of its recharge area in the west and south to the urban expansion of Amman and satellite towns around it. The result has been a decline in the recharge area, and the diversion of a good part of the precipitation water that historically recharged the aquifer to surface runoff. No quantification of this effect has been made, nor has it altered the practice by urban planners to preserve such areas to sustain their function. At the same time, pumping rates from groundwater in the Amman–Zarqa aquifer have progressively increased unabated over the past three decades. In attempting to quantify the impact of urban expansion on aquifer recharge, analysts are faced with several factors that are difficult to account for. One such factor is the identification of the effect of inter-annual variability of rainfall; another is the contribution of irrigation return flow including flows from domestic gardens and green areas in the recharge area and others.

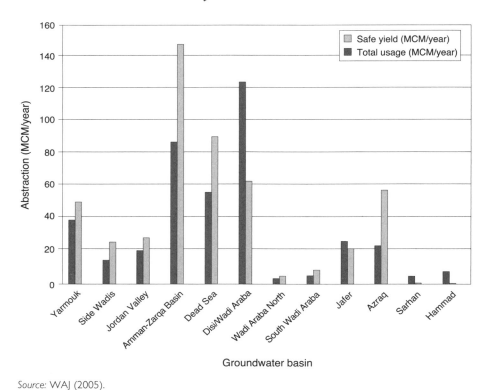

Source: WAJ (2005).

**Figure 3.2** *Safe yield and total abstraction for groundwater basins in Jordan*

Jordan's rapid population and economic growth has naturally created a corresponding increased demand for water. The conflicts in the region, mainly the Arab–Israeli wars of 1948 and 1967, as well as the Gulf Wars of 1991 and 2003 resulted in Jordan's population rising. For example, in 1991 over 300,000 Jordanians residing in the Gulf States returned to settle in Jordan as a result of the Gulf War or what constituted at the time a 10 per cent increase of population within one year (Fariz and Hatough-Bouran, 1998). As sudden as the influx of population has been, the additional demand was fulfilled through increased abstraction from wells already in use, or from wells drilled in aquifers utilized already. Agricultural abstractions from groundwater have increased in particular in the highlands with a short time lag behind the expansion of irrigated agriculture in the Jordan Valley. In 2005, there were about 2779 operational groundwater wells abstracting water for all purposes beyond the sustainable yields of aquifers (WAJ, 2005).

## Wastewater treatment and reuse

Wastewater treatment plants in Jordan receive relatively high strength domestic wastewater, which has high levels of biological oxygen demand (BOD) and total soluble solids (TSS) compared to rates that would normally be expected. This is because the per capita municipal water served, averaging 50m³/year, is below the desired average on the one hand, and its salinity level is on the higher side of the allowable levels on the other. Sixty per cent of the water served to subscribers returns to the treatment plants; the rest is lost to leakage and to uses that do not send wastewater to sewers, such as gardening.

Jordan has 22 wastewater treatment plants, the largest of which is the As-Samra plant that serves the Amman and Zarqa areas and treats almost 75 per cent of all wastewater. The treatment plants provide secondary treatment either through extended aeration, stabilization ponds or trickling filters. In 2006, wastewater treatment plants in Jordan received 110mcm of wastewater from 61 per cent of Jordan's population connected to sanitary sewer system. About 87mcm of treated wastewater was produced from Jordan's wastewater treatment plants, of which 80mcm were reused for irrigation and industrial purposes. According to the WAJ Annual Report for 2006 (2007), WAJ had 115 agreements with various entities for the use of reclaimed water, and its revenue from selling reclaimed water amounted in 2006 to approximately US$2 million.

## Curbing over-abstraction

Jordan's water strategy and its associated water policies emphasize the need to protect groundwater resources from over-abstraction and related quality degradation and pollution and to give priority in the use of these resources to municipal and industrial uses. The groundwater management policy considers the monitoring of groundwater resources and the protection of recharge areas of aquifers as important factors in safeguarding the quality of groundwater resources. In terms of minimizing over-abstraction from aquifers, the policy stresses the need to stop illegal drilling, meter all water wells and use financial instruments for deterrence.

In line with these policies, several measures have been implemented to protect aquifers from degradation and over-abstraction. These measures include the delineation of groundwater protection zones, the preparation of groundwater vulnerability maps, the establishment of a groundwater monitoring directorate within the Ministry of Water and Irrigation and issuance of the groundwater by-law in 2002. The Groundwater Control By-law regulates groundwater well licensing, drilling and water abstraction. A tariff was set in the regulation for water abstracted over and above the permitted annual abstraction rate. The tariff differentiates between licensed agricultural wells, unlicensed agricultural wells, agricultural wells in Azraq basin and brackish water wells. There is no charge for water abstraction below 150,000m³ from licensed wells, and for water that is abstracted within the limits of licences from Azraq basin's wells. The unlicensed wells are charged higher tariffs with no free of charge water allocation, in order to encourage the owners of these wells to rectify their situation and license their wells.

The implementation of the provisions of the above by-law has been relatively effective. By the end of 2006 the rate of metering reached 97 per cent of all licensed wells, and the MWI received over 943 requests for rectification of well legal status. The implementation of the groundwater by-law was challenged by the farmers, especially in the Disi-Mudawwara area where agricultural companies are utilizing fossil water under contract with the government. Imposition of water tariffs, they claimed, violated the provisions of their contracts. Facing insistence from the MWI, the companies resorted to litigation in courts. The new government formed in the spring of 2005 decided to honour the contracts signed with these companies and stopped charging them for water.

Moreover, initial results indicate very modest success: the total groundwater abstraction from licensed irrigation wells decreased by 25mcm from 202mcm in 2002 to 177mcm in 2004, and total groundwater abstraction for irrigation purposes from unlicensed wells decreased from 40mcm in 2002 to 36mcm in 2004. Nevertheless, even though the groundwater by-law was an important step towards managing the use of groundwater resources, data are not yet sufficient to evaluate its long-term impact on abstraction, especially for irrigation purposes.

The MWI is following up on the implementation of the groundwater by-law through its groundwater protections studies directorate, and it publishes the outcomes of its work in an annual report. The Ministry has also embarked on a project, with support from USAID, to enhance groundwater data collection and analysis, and integrate existing groundwater databases. It is also studying the impacts of the implementation of the groundwater by-law on the Zarqa–Azraq basins.

## Climate change: An emerging threat

Climate change poses a new challenge to water resources management. As predictions are made for changes in precipitation and an increase in extreme weather events, it is a matter of time before the availability of Jordan's water resources is impacted by climate change. Freimuth et al (2007) provide a summary of expected climate change impacts in the region, which include among other things a decline in precipitation coupled with an increase in evapotranspiration, reducing the stream flow and groundwater recharge, as well as a possible shift of biomes leading to desertification of areas that enjoy a Mediterranean climate.

Jordan developed its initial national communication report to the United Nations Framework Convention on Climate Change (UNFCCC), and is in the process of developing its second national communication that, in addition to an enhanced methodology, focuses more on adaptation to the impact of climate change, and only on an inventory of greenhouse gases and mitigation measures. Nevertheless, the scarcity of Jordan's existing resources poses a challenge to researchers in terms of calculating the additional impact that climate change will have on water resources versus the current use of water under the water budget deficit conditions.

Given that the Government's priorities will be for drinking, industrial and irrigation water, any additional stress on water resources will impact ecosystems, unless a concerted effort is made towards securing the necessary water resources for nature's needs.

## Water resources and ecosystems

Jordan is home to four distinct bio-geographic regions: (i) the Mediterranean region comprising the most fertile part of Jordan and home to most of Jordan's main cities and towns; (ii) the Irano–Turanian region, which is phyto-geographically a narrow strip of variable width that surrounds the Mediterranean eco-zone from the east, south and west; (iii) the Eastern Desert; and (iv) an Afro-tropical region located in the Jordan Rift Valley.

These bio-geographic regions give rise to several unique ecosystems in Jordan (MOE, 2003): the Desert Ecosystem, which covers approximately 75 per cent of Jordan's area and forms Jordan's main rangeland; the Scarp and Highlands Ecosystem that is characterized by the mountainous strip that adjoins the Jordan Valley to the east; a subtropical ecosystem in the Jordan Valley; the Dead Sea Basin Ecosystem; and the Jordan River Basin Ecosystem. In addition, Jordan's southernmost tip is home to the Gulf of Aqaba ecosystem, a marine ecosystem that is unique due to its coral reefs and its position on a major seasonal bird migratory route. Jordan is also home to the Azraq Oasis, a freshwater ecosystem and a Ramsar site.

These different ecosystems have been influenced to various degrees by the quality degradation and over-abstraction of water. The following subsections provide an illustration of these ecosystems and the impacts that they have endured, as well as the efforts that have been made in some cases to alleviate these impacts.

## Azraq Oasis

Azraq Oasis is located northeast of Amman, and was historically fed by Soda and Qaysiya springs in Azraq South, and Aora and Mustadhema springs in Azraq North. The Oasis is considered an important wetland protected by the Ramsar Convention, to which Jordan has been a signatory since 1977. The oasis has also been designated by Bird Life International as an important area for birds because it falls on a major migratory route.

The Azraq aquifer started to be utilized for municipal water supplies in the mid-1960s to supply the northern city of Irbid and its environs, and in 1978 it also started supplying Amman with drinking water. Drilling of illegal wells in the area began and increased in the aftermath of the 1967 war when law and order were hard to uphold and maintain. Over the years, abstraction increased to reach 25mcm per year for municipal supply while an estimated annual quantity of 25mcm is abstracted for irrigation purposes. When a further estimate of 8mcm abstracted for other uses is added, the total abstraction rate amounts to about 58mcm per year, which is almost double the recharge rate – estimated at 32mcm per year.

This over-abstraction has had an impact: the water levels in the surroundings of the oasis have dropped by a few metres, resulting in the drying out of the discharges of Soda and Qaysiya springs feeding the oasis, and in increased salinity of the groundwater. In 1958, water salinity in observation wells in the Azraq basin ranged between 340 and 970mg/l. By the mid-1990s salinity had increased to around 1500mg/l in deep wells, and to approximately 10,000mg/l in shallow wells. The groundwater in the central part of the Azraq Basin is mineralized and sulphurous and is generally of poor quality. Total dissolved solids concentrations range from 800mg/l to 2500mg/l. In the western and northwestern rims of the basin, the quality is good with total dissolved solids concentrations between 200mg/l and 500mg/l.

Drying out of the springs feeding the oasis led to drying out of vast surface areas of the oasis. A project undertaken by the Royal Society for the Conservation of Nature and supported by the Global Environment Facility – UNDP – attempted to rescue and restore what could be saved, and a nature reserve has been established with an area of 12km$^2$. The rescue effort resulted in partial improvement, and the return of around 160 migratory bird species. However, and in spite of all the efforts, the water flow into the Azraq wetland is barely enough to support 10 per cent of the wetlands that once existed (Azraq Oasis Conservation Project, 1999; Gouede, 2002). It is unfortunate that Jordan's environmental awareness and the resulting legislation and administrative arrangements came too late to preserve the magic of the Azraq Oasis.

The recent groundwater by-law focuses on agricultural wells located in the Azraq basin, and the WAJ decreased its abstraction from the Azraq aquifer for municipal purposes as a lead action. The availability of incremental supplies from the Jordan Valley makes this possible, and it is hoped that WAJ will decrease its abstraction from the Azraq basin as more supplies become available from additional sources. Along with strict law enforcement over the Azraq wells, the reduction in abstraction will give the aquifer a chance to recuperate.

## Amman–Zarqa basin

The Amman–Zarqa basin is the preferred area for Jordan's population and the centre of its modest industry. According to the latest statistics, Amman and Zarqa Governorates are home to over 50 per cent of Jordan's population (over 3 million) as well as home to more than 50 per cent of its industries, including some of the largest ones such as the Hussein Thermal Power Plant, Jordan's Petroleum Refinery, Russeifa phosphate mines and consequently a landfill (closed in 2003) and the As-Samra wastewater treatment plant (Dorsch Consult and ECO Consult, 2001).

The speed with which these industries sprang up overtook in many cases the zoning for industries in national plans, with consequent shortages in public utilities, which encouraged many industries to dispose of their wastes by simply dumping them into neighbouring wadi beds. Subsequently, the government prohibited the disposal of untreated industrial wastes into the environment, and exercised control on the polluting inputs of industries. A study (Al-Jundi, 2000) targeting pollutants in the Zarqa River sediments showed elevated concentrations of heavy and trace elements, particularly zinc (Zn), chromium (Cr), arsenic (As), vanadium (V), cobalt (Co) and zirconium (Zr). The locations of these elevated concentrations corresponded with the locations of industries along the river and were attributed to discharges of industrial waste, sewage and industrial pre-treatment facilities.

Due to the concentration of population and industry, the groundwater quality of this basin is most vulnerable to quality degradation in comparison to other groundwater basins in Jordan. The salinity of many wells and springs has increased substantially to levels far above the standard value for domestic and even agricultural uses. Comparing water quality monitoring records for the period 1985–1989 with the period 1995–1999 shows that among 53 wells in the Amman–Wadi Sir aquifer B2/A7, 40 wells show a substantial increase in electrical conductivity (EC), a measure of salinity. The average increase is 324.4μs/cm or 23.1 per cent above the EC level of the 1985–1989 period.

The upper aquifer in the Amman–Zarqa basin (Amman–Wadi Sir aquifer) is also mostly affected by industrial activity, particularly in Russeifa and Awajan districts. The main contaminants in the Amman–Wadi Sir aquifer, which is the nearest to the surface, include nitrate and selenium. Elevated concentrations of polycyclic aromatic hydrocarbons (PAH) were measured in the vicinity of the now closed Russeifa landfill (Jiries and Rimawi, 2005).

At present, a national effort is being led by the Ministry of Environment, and with the cooperation of other line ministries and the International Union for Conservation of Nature (IUCN), to restore the Zarqa River basin through the application of principles of integrated water resources management and water governance and infrastructure improvements.

## Jordan River Basin

The Jordan River system is an important wetland area in the Middle East because it maintains many globally valuable species such as the brown fish owl, the common otter, Arabian leopard, rock hyrax, freshwater turtle, several endemic fresh water fish, fresh water snake and many other endangered species. It is also an important migratory bird route, with an estimated one million birds annually passing through this narrow corridor, for example the black and white stork, Dalmatian and common pelican, kingfisher, herons, sandpipers, shanks, francolin and other globally threatened waterfowl.

The main problem for the Jordan River basin is the diversion of its entire upstream freshwater base flow for irrigation and other uses, and the diminished flood inflow below Lake Tiberias.

According to JVA monitoring data at nine stations along the Jordan River, the water shows very high salinity in the summer months (average EC = 4480–6663µs/cm). The concentration of boron, which can cause crop damage and a reduction in the yields of sensitive crops, is occasionally over 10mg/l. The sensitivity of various crops to boron varies, for example, the maximum concentration tolerated without yield or vegetative growth reductions for lemons (a very sensitive crop) was estimated at 0.5mg/l, for tomato (a tolerant crop) 4–6mg/l, while the range for very tolerant crops such as cotton and asparagus is estimated at 6–15mg/l (Ayers and Westcot, 1985).

The quality of Yarmouk water diverted to the King Abdullah Canal (KAC) in Jordan is showing signs of recession, while water input from the King Talal Dam (KTD) has a significant impact on water quality downstream in the canal especially in terms of increases in EC and $N-NO_3$ values.[1] This is particularly evident during the summer months when Yarmouk water does not cross the Zarqa River siphon and the entire flow of the canal downstream consists of KTD water (ECO Consult and PA Consulting Group, 2003). Upstream from the Zarqa River inflow into the canal, the water quality is much better and is pumped for municipal uses in Amman via the Deir Alla–Amman project.

A World Bank/Global Environment Facility (GEF) funded project has been implemented by the Royal Society for the Conservation of Nature to develop a network of well managed protected areas in the Jordan Valley to meet the environmental and social needs of the country. The project aims to establish a network of four protected areas and seven special conservation areas, one of which would be in the Yarmouk area where it aims to protect its deciduous forest. Implementation of the project began in September 2007.

## Dead Sea and Side Wadis

The Dead Sea is an internal lake, naturally fed by the Jordan River and its tributaries, and by the direct inflow of the runoff of side wadis, the most important of which, on the Jordanian side, are Wadi Mujib and Wadi Hasa. The Dead Sea's shore forms the lowest dry contour on earth, and is important for the uniqueness of its natural environment and for its historical and biblical significance. The Dead Sea has been exposed to a dual jeopardy with a cumulative effect on its surface level. On the one hand, water inflow has been diminishing as cited above and, on the other, evaporation from its surface has intensified. Water budget studies indicate that the average water inflow, surface and subsurface, into the Dead Sea has decreased significantly over the past 30 years from an average total surface inflow of 1670mcm/year to 407mcm/year. Similarly, current groundwater annual inflow from the Jordanian side to the Dead Sea went down from an average 220mcm to 140mcm within the same time period (Salameh and El-Naser, 2000).

The result has been a drop in the level of the Dead Sea from 392m below sea level in 1920 to 407m in 1990, and down to 418m below sea level in 2006. Subsequently, the surface area of the Dead Sea has shrunk from 1050km$^2$ in 1920, to 515km$^2$ in 2000, and the geometry of its coastline has changed as well.

The government's policy on this issue has been to import water to the Dead Sea from the Red Sea. Its negotiators in the bilateral peace talks with Israel succeeded in promoting this option, and a separate article on this matter was included in the Peace Treaty between Jordan and Israel.[2]

## Ecosystems occupy a place in water allocation

Although water allocation policies often ignored the requirements of ecosystems, water for the environment has scored some successes in recent years. The diversion dam on Wadi Mujib, meant to divert its entire regulated flow from a point upstream of its discharge in the Dead Sea, is a case in point. Such a diversion would have abolished the sustainability of unique aquatic life in the wadi water downstream of its location, and upstream of the discharge point in the Dead Sea. Additionally, the dam would have been built in the heart of a natural reserve managed since 1986 by the Royal Society for the Conservation of Nature. The wadi is an important bird area and is home to at least 10 globally threatened flora and fauna species. It is, also, one of cleanest and least disturbed river systems that have perennial water flow.

Protests by the Royal Society for the Conservation of Nature, and subsequent campaigns that it conducted, convinced the JVA to modify the project and move the diversion dam a considerable distance downstream. The Mujib Diversion Dam was completed, in 2004, in the agreed location, downsteam from its originally planned location to allow for water flow within the wadi (*Jordan Times*, 1999; Khatib, 1999; Shehadeh, 2005).

Another example is the voluntary measure taken by the MWI to reduce the pumping of water from the Azraq basin to Amman after incremental supplies were secured from other sources. The reduction is expected to increase as more incremental water become available.

Another measure was the construction of the Wala Dam as a recharge dam to resurrect the base flow of Wadi Heedan after it had dried up due to pumping from the groundwater reservoir to Amman for municipal uses. The water table has bounced back and the springs are discharging freshwater into the wadi.

The attention paid to migratory birds that were attracted by the wetland created by the Aqaba wastewater treatment plant is another testimony to the place ecosystems have occupied in water resources allocation.

Finally, the giant Red Sea–Dead Sea Conduit is clear testimony to the priority given to ecosystems in Jordan and in Israel and Palestine.

## Aqaba

Aqaba Governorate has been transformed into the Aqaba Special Economic Zone (ASEZ), with plenty of investment incentives. A master plan that aims to promote industrial, business and tourist activity in the area has been formulated for ASEZ. Even though environmental impacts and considerations have been taken into account in the plans for development of the master plan, it is a challenge for ASEZ Authority to maintain environmental quality, and protect Aqaba's fragile environmental resources in relation to the planned economic activity.

Aqaba's environmental resources include both land species and marine habitats. Aqaba's marine coastal ecosystem is unique with its coral reefs and several endemic species (Mir and Abbadi, 1995). Aqaba also lies on the major routes for seasonal bird migration, and its wastewater treatment plant's stabilization pond system formed a wetland where migrating birds take rest.

The Jordan Society for Sustainable Development (JSSD) and the Royal Marine Conservation Society have been active in advocating the maintenance of environmental quality in Aqaba. The JSSD together with the Friends of the Earth Middle East worked to establish a bird observatory in Aqaba in an effort to highlight the issue of migratory soaring birds in the region, and to help in maintaining the important bird habitat in Aqaba.

## Conclusion and policy implications

Even though there are several strategies, laws and regulations that address the issue of water resources management and conservation, the role of water in the maintenance of ecosystems and the allocation of water for the needs of nature, is not specifically addressed or emphasized, except in the role of irrigation in combating desertification and land degradation. The role of water in the maintenance of ecosystems should be emphasized and valuation techniques applied to allow the true value of the services these ecosystems render to be taken into consideration in any cost–benefit analysis of water allocation and investment projects.

In terms of tackling pollution from industrial sources, advantage should be taken of the relative geographic concentration of industries in the Amman–Zarqa basin; the geographical proximity of industries and the high proportion of food industries allow beneficial economies of scale in building and operating an industrial wastewater treatment plant. In addition, waste management directly impacts water quality. The management of solid and liquid wastes, medical waste from hospitals and clinics, in particular, should receive special attention.

The reliance on command and control systems followed thus far for the protection of water resources in the Kingdom should be reconsidered with a view to integrating other technical and financial instruments into the system to improve the efficiency of controls and the implementation of legislation. The establishment by the MOE of an environment fund could contribute towards the implementation of such financial instruments. Moreover, monitoring and enforcement practices can be improved by establishing baseline data on environmental quality especially within sensitive areas. A database of operating industries linked to a geographical information system will also be helpful in identifying the priority industries. Initial activities for establishing such a database are being undertaken by the MOE; however, it is not yet established or functioning.

The reuse of treated wastewater in irrigation is increasing. In 2005, on average, the cost of secondary treatment in Jordan's wastewater treatment plants was 6.5 US cents/m$^3$, which is significantly higher than the average revenue per cubic metre (in 2005 a cubic metre of reused water was sold for 1.4 US cents), as well as the tariff charged according to the groundwater by-law. The cost of reusing treated wastewater should be made substantially cheaper than groundwater abstraction costs. However, it should be kept in mind that in addition to revenue from selling the treated wastewater, the producers of wastewater (mostly the domestic sector) are charged for wastewater collection and treatment.

There should be more effort made towards environmental mainstreaming, and environmental and ecosystem experts should participate in the planning and discussion of water projects, and in environmental impact studies in order to decrease the negative impact of reallocating water from nature to other consumptive uses. Environmental impact assessment should be used as a tool to evaluate the possible consequences of development projects on environmental and water resources. The use of strategic environmental assessment and other tools to evaluate policy impacts and mitigate any negative impacts should be exercised.

Public awareness campaigns should be continued and reinforced to protect and conserve Jordan's valuable and limited environmental resources. These include an awareness of water conservation, protection of water resources from pollution and protection of biodiversity.

Regional cooperation is essential in order to address the environmental issues of regional resources such as the Dead Sea, Jordan River and the environmental resources of the Gulf of Aqaba.

## Notes

\*Note: This paper is an updated and extended version of R. Daoud, H. Naber, M. Abu Tarboush, R. Quossous, A. Salman, and E. Karablieh (2006) 'Environmental issues of water', in M. J. Haddadin (ed), *Water Resources in Jordan: Evolving Policies for Development, the Environment, and Conflict Resolution*, Resources for the Future, Washington, DC.

Disclaimer: The findings, interpretations and conclusions expressed in this paper are entirely those of the author and should not be attributed in any manner to the World Bank, to its affiliated organizations, or to members of its Boards of Executive Directors or the countries they represent.

44  *Issues in Water Resource Policy*

1 Ayers and Westcot (1985) estimated that an electrical conductivity value of 3000μs/cm poses severe restrictions to the growth and yields of crops. The electrical conductivity values during the dry months exceed this limit. The tolerance of specific crops towards salinity varies from sensitive crops such as strawberries (700μs/cm) to squash, which can tolerate a salinity of 3100μs/cm. Some crops such as barley and cotton tolerate salinity levels up to 5000μs/cm. Nitrogen is a plant nutrient and stimulates crop growth, and is often added to plants as fertilizer. However, excess nitrogen may be harmful to the plan hindering its growth and leading to poor quality yields. The nitrogen in irrigation water thus poses additional challenges for farmers in calculating the necessary fertilizer amounts, especially if they are not aware of the concentrations of nitrogen in their irrigation water (Ayers and Westcot, 1985).
2 Article 6.3 states 'The parties recognize that their water resources are not sufficient to meet their needs. More water should be supplied for their use through various methods, including projects of regional and international co-operation', while article 18 on environment states: 'The Parties will co-operate in matters relating to the environment, a sphere to which they attach great importance, including conservation of nature and prevention of pollution, as set forth in Annex IV. They will negotiate an agreement on the above, to be concluded not later than 6 months from the exchange of the instruments of ratification of this Treaty.'

## References

Al-Jundi, J. (2000) 'Determination of trace elements and heavy metals in the Zarka River sediments by instrumental neutron activation', *Nuclear Instruments and Methods in Physics Research Section B: Beam Interactions with Materials and Atoms*, vol 170, nos 1–2, pp180–186

Ayers, R. and D. Westcot (1985) 'Water quality for agriculture', *Irrigation and Drainage Paper* 29 Rev 1, FAO, Rome

Azraq Oasis Conservation Project. (1999) Environmental Impact Assessment and Implementation of Ramsar Convention. *Comprehensive Report on the Monitoring Plan for the Most Important Wetland Sites in Jordan*, Jordan, report prepared for the General Corporation for Environmental Protection and UNDP, Amman

Daoud, R., H. Naber, M. Abu Tarboush, R. Quossous, A. Salman and E. Karablieh (2006) 'Environmental issues of water', in M. J. Haddadin (ed.), *Water Resources in Jordan: Evolving Policies for Development, the Environment, and Conflict Resolution*, Resources for the Future, Washington, DC

Dorsch Consult and ECO Consult. (2001) Feasibility Study for the Treatment of Industrial Wastewater in the Zarqa Governorate: Technical Proposal, Jordan, prepared under the Mediterranean Environmental Assistance Program, Amman

ECO Consult and PA Consulting Group. (2003) *Water Resources, Supply and Demand & Water Balance: Water Quality Issues Supplementary Report*, Report prepared for the Jordan Valley Authority, Amman

Fariz, G. and A. Hatough-Bouran (1998) 'Case study: Jordan', in A. De Sherbinin and V. Dompka (eds), *Population Dynamics in Arid Regions: The Experience of the Azraq*

Oasis Conservation Project, in *Water and Population Dynamics: Case Studies and Policy Implications*, AAAS, Washington, DC

Freimuth, L., G. Bromberg, M. Mehyar and N. Al-Khateeb (2007) *Climate Change: A New Threat to Middle East Security*, Friends of the Earth Middle East for the United Nations Climate Change Conference in Bali, Indonesia

Gouede, N. (2002) 'Restoring an oasis strengthens communities in Jordan', *Choices*, UNDP, New York

Jiries, A. and O. Rimawi (2005) 'Polycyclic aromatic hydrocarbons (PAH) in top soil, leachate and groundwater from Russeifa solid waste landfill, Jordan', *International Journal of Environment and Pollution*, vol 25, no 2, pp179–188

*Jordan Times.* (1999) 'Conservationists to bring awareness to schools with release of new teachers guide to reserves', 12 January

Khatib, A. (1999) 'Jordan signs JD66 million agreement for dam construction projects in the Southern Ghor region', *Jordan Times*, 16 January

Ministry of Environment (MOE). (2003) *National Biodiversity Strategy and Action Plan*, Amman

Mir, S. and M. Abbadi (1995) *Genetic Resources of Fish and Fisheries in Jordan*, UNEP, Amman

Salameh E. and H. El-Naser (2000) 'Changes in the Dead Sea level and their impacts on the surrounding groundwater bodies', *Acta Hydrochim Hydrobio*, vol 28, pp24–33

Shehadeh, Y. (2005) Personal communication between Yehya Shahadeh, Acting Director General, Royal Society for the Conservation of Nature, and the authors. 15 May

WAJ (Water Authority of Jordan). (2005) Unpublished data

WAJ (Water Authority of Jordan). (2007) *Annual Report for Year 2006*, Amman

WRI (World Resources Institute). (2003) *Earthtrends Country Profiles: Water Resources and Freshwater Ecosystems – Jordan*, WRI, Washington, DC

# 4

# Water Management in Urbanizing, Arid Regions: Innovative Voluntary Transactions as a Response to Competing Water Claims

*Bonnie G. Colby*

This chapter focuses on the water management challenges and policy responses faced by arid regions that contain an irrigated agricultural base, which consumes large quantities of water alongside a growing urban population and claims for water by native peoples and for environmental restoration. Detailed background on these issues in the State of Arizona, US, can be found in Colby and Jacobs (2006). This chapter draws on the experience of Arizona (and, occasionally, other arid urbanizing locations), focusing on innovative voluntary water transactions as a response to competing water claims. The chapter begins by summarizing reasons why policy makers need to consider encouraging voluntary transactions and key considerations in setting up a policy framework to govern such activities. The chapter reviews a series of experiences with such transactions and concludes with suggestions for other regions that are considering more widespread use of these tools.

In the western US, transactions historically motivated by a growing urban demand for water have now also become a valuable tool for environmental protection and for accommodating aboriginal (Native American) water claims. Some tribal and environmental advocates embrace voluntary transfers as a means to encourage water conservation, to stretch scarce regional water resources and to replace the hostility induced by litigation and forced administrative reallocations with more collaborative interactions (Colby et al, 2005).

Municipalities, Native American tribes, environmental interests and recreation advocates (seeking to assure water flow for boating and angling) have engineered many different types of water transfers to supply their year-to-year needs and to drought-proof their supplies. (See the *Water Strategist* (2003) for examples of specific transactions.)

Irrigated agriculture remains the primary consumptive use of water in Arizona and the western US, and is the sector others look to for acquiring water. Due to diversity in the economic value of water across different types of crops, particularly during dry years, water transfers also occur between irrigated farms throughout the west.

One force prompting changes in water use is the growing recognition that the ability to transfer water generates regional economic benefits by making water available for higher value uses. The various stakeholders have recognized that significant government funding for new water development is not forthcoming and that water transfers generally are the most cost-effective and environmentally acceptable approach to accommodating new growth and other needs. Another source of pressure for more adaptive water use and for responsive types of water transactions is prompted by concern that climate change is increasing the variability of regional water supplies and decreasing supply reliability for junior entitlement holders.

Water transactions involve individuals and organizations that buy, sell and lease water rights, the use of water supplies and access to water-related infrastructure (aqueducts, wells and reservoirs). Only a few regions in the west have what could be described as water 'markets', with regular transactions and multiple buyers and sellers. In most areas water trades largely involve neighbouring farmers, transactions occur sporadically and price information is difficult to obtain.

However, while market acquisitions involving water are seen as inevitable, they are also the subject of much debate and controversy. Complex regulatory systems have evolved to address concerns and impacts. Transferring the location and type of water use can affect supplies for water right-holders, diminish economic activity in areas from which water is taken, degrade water quality, fish and wildlife habitat, and recreation opportunities. The key challenge in developing policies to govern water markets is to utilize the flexibility that markets offer, while protecting third parties and public interests that can be impaired by water transfers. While urban growth is still the driving force behind water markets, water transfers to support wildlife, fisheries and recreation have become more common. Transfers have become more complex and innovative in order to respond to drought, and to environmental and community concerns. Water transfers can also exacerbate environmental degradation when they alter water use patterns in a manner that deprives water-dependent habitat of return flows and seepage from irrigation.

## Innovative water transactions

This section reviews different types of water transactions to illustrate the diversity of what is possible. Examples are drawn primarily from experiences in and near the State of Arizona, US. Table 4.1 summarizes the various types of transactions occurring in Arizona over the period 1987–2007.

**Table 4.1** *Arizona water transfers, 1987–2007*

| Purpose of water acquisition | Number of transactions | Total volume (acre-feet) | Average transaction size (acre-feet) |
|---|---|---|---|
| Municipal | 124 | 1,036,647.61 | 8360 |
| Agricultural | 24 | 2,346,728.50 | 97,780 |
| Environmental | 16 | 140,142.70 | 8759 |
| Industrial | 8 | 22,269.50 | 2784 |
| Water banking | 8 | 1,261,836.00 | 157,730 |
| Multiple uses* | 6 | 3,152,280.25 | 525,380 |
| Tribal settlement | 2 | 55,250.00 | 27,625 |
| Total | 188 | 8,015,154.56 | 42,634 |

Note: *Multiple uses includes agricultural and municipal; agricultural, municipal and industrial; and agricultural, municipal and environmental.

## Groundwater transfers in Arizona

Arizona water policy provides several different types of groundwater permits, each with differing considerations regarding transfers. Groundwater use is regulated and requires a pumping permit within Active Management Areas (AMAs), but is subject to little regulation or monitoring outside of AMAs. One notable exception is a prohibition on inter-basin groundwater transfers enacted in 1991 to protect rural groundwater from being moved to thirsty urban areas located within AMAs (Groundwater Transportation Act of 1991). Now, rural communities face rapid growth and drought, and the prohibition on inter-basin groundwater transfers may foreclose options to accommodate rural needs. There are exceptions to the prohibition, including one that allows transferring groundwater from the Big Chino groundwater sub-basin into the rapidly growing and water-short Prescott AMA.

The 1980 Groundwater Management Act (GMA) governs groundwater use within AMAs. Permits to use groundwater fall into multiple categories, classified as either grandfathered (established pre-1980, prior to GMA) or permitted (rights developed after 1980). Grandfathered rights are subdivided into various categories, of which Type II grandfathered non-irrigation rights (Type II) can be separated from the land and transferred within their AMA, making Type II rights more marketable than other groundwater within AMAs. Type II rights are routinely bought and sold in the Phoenix and Tucson AMAs and sales and leases of Type II rights have comprised the most regular form of water market activity in Arizona over the past 20 years. While these trades involve only small quantities of water, they are an important source of groundwater use flexibility within the AMAs.

## Transfers of public project water entitlements

When delivery of Central Arizona Project (CAP) water began in 1985, a new type of water became available for Arizona water users. CAP water is delivered to public municipal water providers, private water companies and irrigation districts by the Central Arizona Water Conservation District (CAWCD). CAP water users are divided into various classes that have different delivery priorities (and thus risks of shortfalls), and pay differing charges. In the event of insufficient CAP water to make customary deliveries, non-Indian agricultural users are cut off first, followed by Indian agricultural users, with the most secure deliveries made to municipal and industrial (M&I) users and non-agricultural Indian users. A host of complex exchanges involving CAP entitlements and other types of water has occurred, some of which is described below.

CAP entitlements can be transferred between users, subject to specific rules. The Arizona Department of Water Resources (ADWR) and the CAWCD have issued policies to guide transfers of CAP allocations. The director of ADWR reviews proposed transfers of CAP water and makes recommendations for approval to the Secretary of the Interior (who has ultimate jurisdiction, as the CAP is a federal water project). The ADWR facilitates a public review of proposed transfers and reviews them for consistency with water policy goals. To get their transfer authorized, the relinquishing contractor must demonstrate that their water supplies are adequate without the CAP allocation, and must demonstrate through modelling and other evidence that the transfer will not increase groundwater overdraft in their area. The CAWCD requires the relinquishing subcontractor to not make a profit from the transfer, although they can be reimbursed for accumulated past costs of holding a CAP subcontract. ADWR requires the party giving up a CAP allocation to use the proceeds from the sale to secure allowable alternative water supplies.

Many CAP allocations were relinquished by non-Indian agricultural districts. These agricultural users had signed CAP contracts but found, after delivery began, that CAP water was too expensive to be profitable in their farming operations. Their relinquishment of these agricultural contracts made CAP water available for settling Arizona Indian water rights claims. Leasing of Indian CAP allocations is another type of CAP transaction. These leases are usually incorporated into Indian water rights settlement agreements that are approved by Congress. Other CAP entitlement transfers include the city of Scottsdale acquiring CAP allocations from the cities of Prescott and Nogales (whose locations make it costly for them to take delivery of CAP water) and other subcontractors for a total of 17,823 acre-feet of CAP water. Scottsdale uses these additional CAP entitlements to meet its growth needs.

## Effluent transfers

Treated municipal wastewater (effluent) is another important type of water transaction in Arizona. In a 1989 court case, the Arizona Supreme Court established that until the legislature declared otherwise, effluent would not be regulated as either surface water or groundwater. Instead, effluent is 'owned' by the municipality that generates it and is available for transfer (Arizona Public Service Company v. John F. Long, 1989). For instance, Pima County, in the Tucson AMA, sells effluent to farmers for crop irrigation, thus reducing their need to pump groundwater. The City of Tucson requires new golf courses to be irrigated with effluent, and encourages commercial water users and older golf courses to purchase and use effluent. Treated effluent (about 140,000 acre-feet annually) from Phoenix-area municipalities is sold to the Palo Verde nuclear power station for use in cooling the plant's reactors. Effluent transfers are included in a number of water agreements with Arizona's tribes, described in the subsequent section.

## Water transactions involving native peoples

Tribal or aboriginal water rights are important not only to native peoples but to all other regional water users. All interest groups in regions where such claims exist have a stake in addressing them constructively. Experience in Arizona and the western US suggests that comprehensive water claim settlements can positively shape water policy and provide for innovations that improve the water future for both tribal and non-Indian communities. However, these issues are very contentious and can be quite expensive to resolve. For more detail on tribal water issues in the US see Colby et al (2005) and Thorson et al (2007). For more on Arizona tribal water issues, see Colby and Jacobs (2006).

Twenty Native American reservations account for about 28 per cent of Arizona's land base, with potentially vast associated water claims (see Map 4.1). Arizona tribes control large senior rights to Colorado River surface water, arguably the most drought-proof water rights in the region. Pressure to address tribal water claims comes from rapid urban population growth, scarcity of dependable surface water supplies and declining groundwater levels. There is also pressure to provide water for habitat restoration. Uncertainties regarding the extent and scope of Indian water rights in Arizona interfere with regional water management planning (Colby and Smith, 2006).

Native peoples' water rights are a factor in many parts of the US, and throughout the world. Native Hawaiians' traditional and customary water uses are recognized in Hawaii's Water Code. Aboriginal water rights are protected under Australia's Native Title Act of 1993. In New Zealand, the Treaty of Waitangi provides a basis for addressing the water rights of the Maori people. In Canada the water claims of First Nations are addressed by the doctrine of aboriginal title and treaty rights.

Historically, resolving US tribal water claims involved long and costly court battles. Tribal water rights are often addressed in the context of general stream

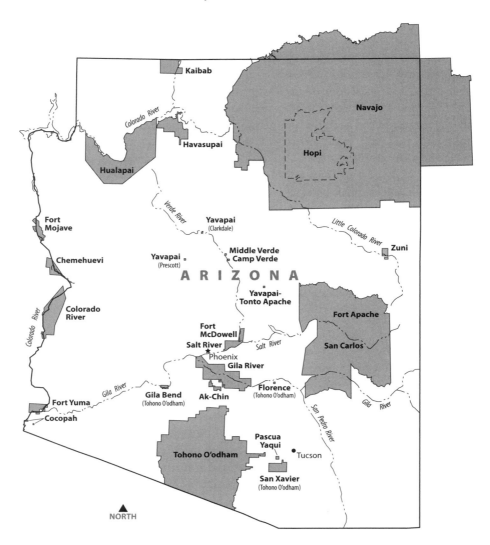

**Map 4.1** *Native American reservation lands in Arizona, US*

adjudications, which attempt to resolve the water rights of every user in a particular watershed through a complex court proceeding. These take decades to complete and involve thousands of water users. Parties in Arizona have taken another route, partially due to frustration with the costs and slow pace of adjudications. In Arizona, there have been eight Congressionally approved tribal water rights settlements, more than any other state in the US (see Table 4.2). Selected Arizona water rights settlements that highlight the use of innovative transactions are summarized below.

Settling aboriginal water claims can involve disrupting established non-Indian water uses that have relied for decades on water supplies that legally now need to be shared with tribes. Yet the consequences to tribal people of not having their water claims converted into usable 'wet' water are profound. On the Navajo Reservation (located in Arizona, New Mexico and Utah) approximately 40 per cent of the population lacks a potable domestic water supply.

Native tribes seeking to develop and use their long-ago reserved water rights can meet opposition from environmental advocates. Water development has advanced slowly on tribal reservations, and so the water sources on tribal land may provide the last remaining aquatic habitat for endangered species. The US Endangered Species Act prohibits the federal government from engaging in activities that jeopardize endangered species or their habitat. Federal agencies are involved in tribal water settlements and thus tribes may be constrained in using their water supplies by the Endangered Species Act or other environmental laws.

There is now over a quarter of a century of experience in negotiating Indian water rights settlements in Arizona. Water settlements with tribes have been shaped by the state's policy goal to eliminate groundwater overdraft in AMAs. One important component of settlements is limits on tribal groundwater pumping, and also on previously unregulated groundwater pumping by non-Indians near the reservation boundary. Most Arizona settlements provide water to tribes through the Central Arizona Project and authorize the tribe to lease their Central Arizona Project water in specific areas within the state. Settlements typically require the tribes to create a tribal water code to govern water use on the reservation.

Phoenix metropolitan area cities lease thousands of acre-feet of CAP water each year from central Arizona tribes, providing a long-term water supply for growing cities. Under these leases, the CAP water retains its Indian priority date, so that in times of shortage this water is a relatively reliable supply. When a tribe leases its CAP water to a city, the city does not have to pay the capital repayment charges it would have to pay if it were ordering the water in its usual capacity as a city.

Arizona tribes have been both acquirers of water through water settlement processes and lessors of water to others. The recent Zuni Settlement, for example, required surface water rights to be purchased on behalf of Zuni Pueblo to provide flows for wetland and stream restoration (Zuni Indian Tribe Water Rights Settlement Act of 2003). The Zuni Reservation is located in northwestern New Mexico, and members of the tribe make ceremonial pilgrimages to land known as Zuni Heaven, a riparian area located in eastern Arizona. The tribe has been working to reacquire and restore the Zuni Heaven area. Congress established the Zuni Indian Resource Development Trust Fund to provide money to restore the wetland area and acquire water rights to restore Zuni Heaven's wetland habitat.

In 2003, Congress passed the Zuni Indian Tribe Water Rights Settlement Act. This settlement extensively focuses on environmental restoration through a voluntary water rights acquisition programme. The tribe is purchasing up to 3600

Table 4.2 Arizona tribal water right quantification by courts and by settlements (chronological)

| Settlement or court case | Indian tribes | Entitlement (acre-feet annually) | Date of settlement/ case | References | Comments |
|---|---|---|---|---|---|
| Arizona vs. California Litigation | Chemehuevi, Cocopah, Colorado River, Fort Mohave, and Quechan (Ft. Yuma) | 905,496 | 1963 | 373 US 546 (1963) | Upheld the Winters Doctrine Established 'Practicable Irrigable Acreage' as a basis for quantifying Winters rights |
| Ak-Chin Water Rights Settlement | Ak-Chin Indian Community | 85,000 | 1978 (1984) (1992) | P.L. 95-328, 92 Stat. 409 (1978) P.L. 95-530, 98 Stat. 2698 (1984) P.L. 102-497, 106 Stat. 3255 (1992) | Original legislation modified due to impractical water supply plans No local cost share; fully federally funded |
| Southern Arizona Water Rights Settlement (SAWRSA) | San Xavier & Schuk Toak Districts, Tohono O'odham Nation | 66,000 | 1982 (1992) | P.L. 97-293, 96 Stat. 1274 (1982) P.L. 102-497, 106 Stat. 3255 (1992) | Allows limited off-reservation leasing Provides federal project water for tribe Title III of proposed Arizona Water Settlements Act of 2002 (Kyl/McCain, R-AZ) would settle Nation's litigation concerning implementation of 1982 settlement |
| Salt River Pima-Maricopa Indian Community Water Rights Settlement | Salt River Pima-Maricopa Indian Community | 122,400 | 1988 | P.L. 100-512, 102 Stat. 2549 (1988) | Complex multiparty water exchanges Significant local cost sharing at insistence of federal government |

| | | | | |
|---|---|---|---|---|
| Fort McDowell | Fort McDowell Indian Community | 36,350 | 1990 | P.L. 101-602, 104 Stat. 4480 (1990) | Water sources not finalized. Allows limited off-reservation leasing |
| San Carlos Apache Tribe Water Rights Settlement | San Carlos Apache Tribe | 77,435 | 1992 | P.L. 102-575, 106 Stat. 4600 (1992) | Entitlement comprising primarily CAP water. Allows limited off-reservation leasing. Portion of water source strongly opposed by Arizona's non-Indian agricultural community |
| Yavapai Prescott | Yavapai Prescott Tribe | Up to 16,000 | 1994 | P.L. 103-434 (1994) | Tribe has the right to pump groundwater within the boundaries of the reservation. Water contract with the City of Prescott. May divert water from nearby creek currently diverted by local irrigation district |
| Zuni Indian Tribe Water Rights Settlement | Zuni Pueblo | 1500afa groundwater; up to 3500afa surface water may be purchased | 2003 | Zuni Indian Tribe Water Rights Settlement Act of 2003, P.L. 108-34 (2003) | Settlement addresses Zuni Pueblo's land in Arizona, known as Zuni Heaven. $26.5 million will be used to acquire water and settle claims, implement the agreement, and restore Zuni Reservation land. Of that sum, $19.25 million will come from federal government |
| Arizona Water Settlements Act | Multiple tribal governments | Multiple provisions affecting entitlements | 2004 | Arizona Water Settlements Act of 2004, US Statutes at large 118: 3478 (2004) | Act is lengthy with multiple complex water management, water exchange and financial provisions. Refer to chapter text and references cited in text for details. |

acre-feet of water rights that will retain their early priority date and help maintain stream flows for the riparian area.

The settlement also deals with groundwater pumping and its effect on the water table and habitat in the Zuni Heaven area. Two large utilities signed 'pumping protection agreements' that limit the utilities' pumping in this area. Other parties also agreed to limit their pumping near the area, effectively creating buffer zones to protect the tribe's ceremonial area. The Zuni tribe is allowed to pump up to 1500 acre-feet per year on the reservation to supplement the surface water irrigation available to restore the wetland area. The negotiated settlement provides the tribe with the necessary resources to acquire land and water from willing sellers to restore an important religious site and also settles their water claim, removing a source of uncertainty for non-Indian water users.

# Arizona Water Settlements Act of 2004

The Arizona Water Settlements Act (hereafter referenced as AWSA) addresses separate settlements with three distinct tribes, as well as providing further finality to other water issues in Arizona. The AWSA is lengthy and complex. Only a brief overview of several of its features pertaining to water transfers are provided here (for a more detailed account of the Act, see McGinnis and Alberts, 2005 and Bark, 2006).

Water leases and exchanges are a key component of the AWSA (Title II, §205, 2004). Lease payments from cities to tribes provide a key means for funding economic development on the reservations. Tribal CAP allocations may be leased off-reservation (AWSA, §205a2) with several restrictions specified in the AWSA (§205f2): CAP water can only be leased within Arizona (AWSA, §205a2A; §205a8f1; Title I, §104e2), and water can be leased for a maximum of 100 years (AWSA, §205a2B). Tribal CAP water must be delivered through the CAP system (AWSA, §205a4A) and is subject to the CAP system's shortage-sharing arrangements during drought (AWSA, §205a4B). At the time of the AWSA, one of the tribes (Gila River Indian Community, GRIC) already had negotiated lease agreements with four central Arizona cities (AWSA, Exhibit 17.1 A-D). These leases provide 41 kilo-acre-feet per year (KAFY) of tribal CAP water to the cities for a period of not less than 100 years, thereby satisfying the state's AWS requirements so the lease water can be used to support new development. The price was determined by a water valuation study and includes consumer price inflation adjustments over the lease period. Cities that paid for the entire lease upfront paid $1743 for 1 acre-foot per year (AFY) of water provided each year for 100 years. This arrangement with the GRIC allows the cities to receive this CAP water without the otherwise required payment of water service capital charges. Operation, maintenance and replacement (OM&R) charges associated with CAP water still must be paid. This water is also subject to shortage sharing reductions that apply to M&I priority CAP water. The cities may re-lease this water to others within the CAP three county service area.

The leases provide the cities with affordable and secure water that meets AWSA standards and the tribe receives around $70 million capital for investment on the reservation. The federal budget receives payments from lease holders for CAP OM&R costs that would not have been paid if the water was used by the tribe on the reservation (AWSA, Title II, §205a6, 2004). In 2006, payments to the federal government for these costs would be around $2 million.

One shortcoming of the 100-year arrangement is that the lease pricing formula has little connection with forces affecting water demand and supply. Lease prices could readily become out of balance if the value of water in the region rises substantially, possibly leading to pressure to alter or terminate the agreement.

GRIC, in the AWSA, also has an agreement with a large mining corporation resolving all outstanding water rights litigation between the parties and incorporating provisions for lease and exchange (AWSA, Exhibit 10, 1). The initial lease involves 12 KAFY of high priority CAP water for 50 years, and 48 years into this lease the parties can negotiate renewal for 50 more years, with the new price determined by 'fair market value' (AWSA, Exhibit 8.2, §57). Like the city leases, the mining company will not pay CAP capital charges, but will pay OM&R charges (AWSA, Exhibit 8.2, §57). The corporation can use the water for direct use, recharge or exchange within the Central Arizona Water Conservation District (CAWCD) Service Area and has an option to lease an additional 10 KAFY within a 20-year option period (AWSA, Exhibit 8.2, §57). Further, GRIC reached an exchange agreement with the mining corporation allowing the corporation to divert water upstream (convenient to their mining operations), in lieu of CAP water. Another source of funds for GRIC is an $18M compensation fund (Exhibit 10.1, §4.1) paid by the mining company in return for tribal waivers and confirmation of the company's water rights.

Other agreements incorporated into the AWSA (Exhibit 18.1) authorize exchanges that give GRIC reclaimed water supplies from nearby cities for use in reservation agriculture (up to 45.1 KAFY). These exchanges give the cities access to a commensurate amount of CAP water beyond their contractual allocation (Arizona Water Settlements Act, Title I, §104d2Ei). Several of the cities receive 20 per cent less CAP water than the reclaimed water they provide in the exchange, an arrangement still attractive to them because the exchange of reclaimed water for CAP water postpones major investments in upgrading and expanding secondary and tertiary wastewater treatment facilities. Many central Arizona cities recharge treated wastewater, which mixes in the aquifer with groundwater and is later recovered. This expensive process requires large tracts of land and is bypassed in these agreements. The cities save money at their wastewater treatment facilities and GRIC secures reclaimed water, a relatively drought-proof supply for its farming operations.

AWSA also authorizes leases and exchanges, including reclaimed water, for the Tohono O'odham Nation located in the Tucson area (McGinnis and Alberts, 2005; Bark, 2006). One portion of the 2004 AWSA is amendments to the Southern Arizona Water Rights Settlement Act (SAWRSA). With respect to leases and exchanges, these amendments provide for the use of effluent in settling the Tohono O'odham Nation's claims. When originally enacted, SAWRSA envisioned providing about 22,000 acre-feet of municipal effluent per year to assist in satisfying the tribe's water claims, for use in reservation agriculture. However, the tribe has declined to accept effluent for its agricultural needs and so the effluent will be recharged in exchange for credits to use groundwater or CAP water. The Nation will have access to those credits under the rules of the Arizona Groundwater Management Act.

While most of Arizona's water transactions involving tribes rely on groundwater and CAP water, Colorado River water has been involved in a specialized transfer involving tribes. Congress ratified the Ak-Chin Settlement in 1978, with a large portion of the tribe's settlement water provided by transferring senior Colorado River water from a large western Arizona irrigation district. The Tribe received a senior water entitlement and the irrigation district received several types of financial benefits.

## Arizona Water Bank

The Arizona Water Banking Authority (AWBA), created in 1996 as a way to fully utilize Arizona's Colorado River allocation by storing water underground, plays an important role in innovative transfers and exchanges. The bank has a key role in interstate water banking through agreements with Nevada and California that allow these states to take more water from the Colorado River in times of need. During such times, Arizona will use the stored water in lieu of taking water from the Colorado River. The Bank's activities have a minor effect on flow levels in the river, but a large regional plan is being implemented to protect habitat and endangered species along the river (Reclamation, 2007). The Arizona Water Bank differs from California's drought banks in several respects, particularly in its objective of bringing Arizona's Colorado River entitlement off the river and into use and storage within the state. A primary reason for its creation was to protect Arizona's underutilized Colorado River entitlement from encroachment by neighbouring states. The water bank does not actually buy and sell water, but instead facilitates the substitution of one type of water for another in times of drought or shortage.

An Interstate Water Banking Agreement was reached in 2001 with the Southern Nevada Water Authority and the Colorado River Commission of Nevada, under which Arizona agrees to use its 'best efforts' to store 1.25 million acre-feet for Nevada's use after 2016. However, after years of drought, Nevada renegotiated the agreement to provide more security. The new 2005 agreement requires Arizona to store 1.25maf on behalf of Nevada. In exchange, Nevada pays Arizona $100 million above the actual cost of water delivery, storage and recovery. The agreement does not place Arizona water users at greater risk of shortage because the AWBA only stores water for Nevada that is not being utilized by Arizona water users.

## Transfers based on temporary land fallowing

Population growth, water-based recreation and environmental restoration programmes in the Lower Colorado River region all increasingly demand water. Meanwhile, extended drought and climate change make existing supplies less reliable. Concerns about the effects of climate change on junior entitlement holders, primarily Arizona's large urban centres, is prompting interest in new types of arrangements with agriculture.

Accommodating new demand and requirements for improved reliability necessarily will be accomplished primarily through transferring water out of agriculture. Irrigated agriculture is the largest consumptive water use in Arizona and much of this water is used to irrigate low profitability crops. Transfers of only a small fraction of water used in agriculture can fulfil changing demands and reliability needs. Irrigation forbearance programmes have been utilized throughout the western US and are being further explored in Arizona and the southwest. Several proposals to address differing regional water management challenges all draw upon irrigation forbearance as a voluntary reallocation mechanism.

One climate change challenge for water managers potentially addressed through forbearance is CAP water users' vulnerability as the first to experience reduced deliveries if a shortage were declared in the Lower Colorado River Basin. A proposal, 'Conservation before Shortage', was developed by several NGOs in 2005, to use part-year fallowing programmes, dry-year options and other similar voluntary arrangements to improve the reliability of CAP deliveries during extended drought and avoid extreme and uncompensated water shortages. The proposed strategy is triggered by the elevation of the major regional reservoir, Lake Mead, such that when the lake is drawn down to specific elevations, pre-negotiated small-scale reductions in use by irrigators occur. An economic study undertaken by the NGO Environmental Defense suggest that over 2.3 million acre-feet of irrigation water is currently being applied to crops in Arizona and California that yield profits under $100 per acre-foot. Of this, about 1 million acre-feet are being applied to crops that generate profits under $20 per acre-foot.

Another regional challenge involves the Yuma Desalting Plant (YDP) whose operation would probably have negative impacts on an important wetland habitat in Mexico. Operating the plant would reduce the bypass of saline drainage water to Mexico from a large irrigation district in southwestern Arizona and this in turn eliminates the need to release stored water from Lake Mead to make up for the bypass water and thus reduces the risk of shortage to Lower Basin water users. Such shortages are expected to become more frequent and severe under climate change. Operation of the plant will be costly and could have severe environmental consequences in the Cienega de Santa Clara, a large wetland in Mexico sustained by agricultural drainage water. One recommendation is a basin-wide programme to pay farmers to voluntarily reduce their use of Colorado River water for irrigation, and then credit the unused water to offset the obligation of the bypass flow. The irrigation forbearance could occur in the long term, on an annual basis, or temporarily through mechanism such as dry-year options. Potentially, irrigators in the US and in Mexico could participate in the programme, though major legal and political obstacles would need to be addressed.

The Lower Colorado Region of the Bureau of Reclamation (Bureau) has explored agricultural forbearance programmes, including a pilot programme in 2004 (not implemented due to political opposition) and a small scale programme in 2006 (partnering with the Metropolitan Water District, MWD) and 2007. Water saved through land fallowing makes water available to replace YDP bypass water, and to store against dry periods. Water not delivered due to forbearance remains in the Colorado River storage system. Participating irrigators are eligible to forbear water use on up to 33 per cent of their acreage.

Forbearance programmes tend to fall into several categories: irrigation suspension on a one-time 'as needed' basis, multi-year programmes in which irrigation will be suspended under specific conditions (dry-year triggers) over the lifetime of the agreement and multi-year programmes under which a specific amount of forbearance occurs each year, rotated among differing acreage so farm land is not permanently fallowed. A land fallowing rotation requirement can ensure that all willing farmers have a chance to participate in the programme. Programmes can ensure broader distribution of the benefits of participation through random selection among eligible farmers to provide the water needed each year, with each field limited to participate in forbearance only twice in every four years (IID, 2007b). A large electric utility's programme to acquire water for a new power plant in Utah's Sevier Basin in the 1980s provides an example of broadly engaging an initially suspicious and hostile irrigation community. Large numbers of irrigators participated, selling small proportions of their water allocation or selling their 'option to sell' to other growers. Each irrigation district grower-member received some form of benefit, and the revenues from the acquisition programme were spread broadly, helping to defuse local opposition (Saliba and Bush, 1987; Colby and Pittenger, 2006). Permanent retirement of irrigated lands also involves suspending irrigation, but this approach is not discussed here.

Forbearance programmes can differ in a number of other important respects, and it is useful to compare across programme design features. Some programmes use fixed offer prices (per acre fallowed, or per acre-foot of water made available) to induce irrigators to participate. Others use a process under which irrigators submit bids indicating the amount they will accept to fallow an acre of a specific crop. The latter, in general, is preferable from the perspective of programme cost-effectiveness, as measured in dollars paid per unit of water acquired.

Forbearance programmes must have an effective mechanism for measuring and monitoring the quantity of water made available through reduced irrigation. Thoughtful selection criteria to identify those irrigators and irrigated lands eligible to enrol in a forbearance programme can partially address this issue. The criteria specifically must screen out acreage that has not been recently or regularly irrigated. To avert disagreements about the amount of water 'conserved' per acre fallowed, some programmes cap the water credited per acre in a fallowing programme (IID, 2007a).

Yardas (1989, p1) warns that programmes that acquire and retire agricultural water that has not been consumptively used 'perpetuate the conflict' because no new water will be freed up to mitigate wetland losses or to improve water quality. Some relatively objective public agency must take on the politically onerous task of differentiating wet water from paper water. 'Inactive water rights account for nearly a quarter of the decreed irrigation total in the lower reaches of the Truckee and Carson Basins' (Yardas, 1989, p3).

Fragmentation of irrigation conveyance systems can create higher costs and system inefficiencies if some lands are in forbearance and other lands at the end of ditches are being irrigated. Conveyance costs per unit of land served and conveyance losses both can increase significantly. Programme design can be structured to attempt to attract lower-productivity lands and acreage at the end of canals into the programme. Sunding (1994) suggests that 'price discrimination based on location and past land productivity' and could be fruitful.

All forms of irrigation forbearance raise concern over local economic impacts. One-time programmes have only short-term impacts due to the temporary nature of the land fallowing. Criteria developed to choose among lands offered for forbearance can include factors to favour those lands that result in the least local economic effects. For instance, some programmes place a priority on enrolling lands where growers agree to implement dry-land crop production so that farming continues during irrigation suspension. A programme could be designed to favour crop acreage that is not labour-intensive to avoid displacing higher levels of farm labour.

Third-party economic impacts, while a concern in any programme, only have been monitored and analysed for a subset of western US forbearance programmes. Howitt (1994), McCarl et al (1997) and Keplinger and McCarl (1998) generally conclude that local economic impacts of land fallowing are minor compared to the benefits of the programmes, and compared to the stimulus of the economic activities in the region that benefit from the dry-year supplies produced by the programmes. The IID-SDCWA fallowing programme set up a fund to offset third-party economic effects, a fund that was intended to be allocated by 'a committee of economists appointed by the county supervisors and the selling and purchasing agencies' (Howitt and Hanak, 2005, p79).

The level of payment offered to growers, relative to typical net income per acre from crop production, is a key programme design issue. Most US programmes have paid growers many times higher than the net returns from irrigating their most marginal crops. The per-acre payment should lie between the net returns to mid-value and low-value crops, for several reasons. First, the programme's water acquisition costs will be lower and, second, growers will have a stronger incentive to more carefully target low net return crops and portions of their farm. Keeping higher value crops and land areas in production preserves a higher level of net farm income. Higher net returns per irrigated acre do not necessarily correspond to higher employment per acre or to higher local input purchases to support crop production. These local economic impact issues need to be compared on a crop-by-crop basis for specific localities.

Forbearance programmes may need to be revised each year of operation in their early years, to take advantage of what is learned and to function more effectively. For instance, the Klamath Bank in Oregon switched from a fixed offer price to a bidding system and was able to obtain water from irrigators more cost-effectively. The eligibility and selection criteria also have been refined over time. It is important to maintain the ability to modify programme features, an 'adaptive management' approach to successful long-term irrigation suspension programmes.

To summarize, forbearance programme selection criteria should encompass multiple concerns: (i) acquiring 'wet water' rather than 'paper water'; (ii) retaining higher value crops; (iii) maintaining profitable crop production; (iv) maintaining labour-intensive crops; (v) encouraging forbearance on acreage with less efficient irrigation systems; and (vi) prioritizing forbearance on lands that contribute most to water quality problems or environmental damage.

The parties needing water from agriculture need a mechanism to pay for the programme costs. Some cities have introduced drought surcharges on their customers to pay for programmes. In some programmes, concerns about long-term decline of agriculture are addressed by limiting temporary forbearance to a maximum amount in any given year and to only being executable a limited number of years each decade, three years out of each ten-year period for instance (UAWCD, 2004).

Some forbearance programmes require the municipality obtaining the water from growers to implement measures such as an Increasing Block Rate Structure or mandatory outdoor water restrictions on the city's water ratepayers. Clauses such as these are intended to ensure that water will not be used to support new growth and become a permanent draw of water away from agriculture. Programme provisions may also govern the timing of water transfers to reduce damage to native stream flow patterns and fish and wildlife.

Some programmes incorporate farm water conservation as well as fallowing. For instance, a programme involving a large irrigation district and the City of

San Diego relies on fallowing for the first 10 years and then gradually switches to agricultural water conservation projects, which are intended to completely replace land fallowing over the life of the long-term agreement (IID, 2007a).

Some environmental impacts of land fallowing may be addressed by requiring participating irrigators to engage in monitored land management measures such as weed control, erosion control and other remedial measures when necessary (MWD, 2004).

## Conclusions and policy implications

Arizona has experienced various eras with respect to water transactions. Early transfers involved purchases of large 'water farms' by cities seeking to increase the quantity and reliability of their supplies. While these purchases caused concern, intense statewide controversy over urban water transfers erupted later when it became clear that the Ground Water Management Act would limit pumping in the AMAs. After the passage of the GMA and the completion of the CAP, there was both incentive and conveyance infrastructure to move large quantities of water from rural areas of western Arizona to the central Arizona cities. Arizona moved into a new era of high rural–urban conflict over large-scale water transfers from western Arizona to the central municipalities. This round of transfers ended through state legislation that regulated such transfers and through recognition that new supplies delivered through the CAP would provide more than adequate supplies for many years to come.

From the early 1980s until the present time, there have been regular and relatively uncontroversial transfers of groundwater rights (within the AMAs), of CAP allocations, of effluent and of Colorado River water through the Arizona Water Bank. In the 1990s and continuing to the present time, Native American tribes have participated in water transactions both through leasing water to cities and by acquiring water to fulfil the terms of their water rights settlements. Now, in the face of climate change, innovative transactions are being discussed to improve supply reliability through dry-year fallowing, a strategy advanced in dialogue over the Yuma Desalting Plant. Such transactions can be temporary and leave irrigated farming intact in normal water supply years.

The Arizona experience provides some useful observations for other regions facing decreasing supply reliability, growth, aboriginal water claims and competition for water. Transferring water out of agriculture to improve the reliability of supplies for municipal, environmental, tribal and recreation needs is an important component in regional water management, especially in areas where climate change and other factors make water supplies more variable from year to year. Temporary transfers can ameliorate economic disruption during long-term drought, sustaining economic activity in those sectors to which water is being moved. Nevertheless, negative impacts create controversy and policies must be carefully developed to govern transactions. Arizona has had successful experiences with water transactions, despite uncertainty over water rights created by the slow and cumbersome water adjudication processes underway throughout much of the state. Well-defined entitlements to groundwater and CAP water have facilitated specialized markets for these types of water allocations.

Instantaneous, standardized transactions in water are not a desirable or realistic policy goal. Nevertheless, transactions and economic incentives should play a significant role in regional water policies, both inducing movement of water to alleviate dry-year hardships and giving water users incentives for more efficient water use. Water transactions are more commonly akin to complex diplomatic negotiations than to commodity exchanges. Public policies should not necessarily seek to minimize the cost of reallocating water because appropriately structured costs may facilitate efficient reallocation by giving transacting parties an incentive to account for social costs of transfers (Colby, 1995).

Transaction costs are part of the costs of voluntary transfers and include costs of searching for water supplies, negotiating price and obtaining public agency approval for the proposed transaction. Transactions' costs are an important issue in designing forbearance programmes. If the costs of water transactions are too high, beneficial transfers will not take place and water supplies will remain locked into outdated use patterns. However, some transaction costs are justified by the need to account for third-party impacts and the need for hydrologic, legal and economic data to address transfer impacts.

In summary, the ability to move water to new places and purposes is an essential water management tool, especially during drought and given climate change implications for supply reliability. Innovative transactions allow water to move out of agriculture to alleviate temporary shortages. Water scarcity creates tensions worldwide and voluntary transactions are one important strategy for addressing such conflicts.

## References

Arizona Public Service Co. v. John F. Long, 773 P.2d 988 (1989)
Arizona Revised Statutes, Groundwater Transportation Act of 1991
Arizona Water Settlements Act of 2004, US Statutes at Large 118 (2004): 3478
Bark, Rosalind (2006) 'Water reallocation by settlement: Who wins, who loses, who pays?' bepress Legal Series. Working Paper 1454. http://law.bepress.com/expresso/eps/1454
Colby, Bonnie G. (1995) 'Regulation, imperfect markets and transaction costs: The elusive quest for efficiency in water allocation,' in D. W. Bromley, (ed.), *Handbook of Environmental Economics*, Blackwell, Oxford, pp475–502
Colby, B. G. and Kathy Jacobs (2006) *Water Policy for Urbanizing Arid Regions*, Resources for the Future Press, Washington, DC
Colby, B. G. and K. Pittenger (2006) 'Structuring voluntary dry year transfers LasCruces', *New Mexico WRRI Conference Proceedings*, 25 January

Colby, B. G., Katie Pittenger and Dana Smith (2006) 'Water transactions and flexibility during drought', in B. G. Colby and Kathy Jacobs (eds), *Water Policy for Urbanizing Arid Regions*, Resources for the Future Press, Washington, DC

Colby, B. G., John Thorson and Sarah Britton (2005) *Negotiations Over Tribal Water Rights*, University of Arizona Press, Tuscon, AZ

Howitt, R. E. (1994) 'Empirical analysis of water market institutions: The 1991 California water market', *Resource and Energy Economics*, vol 16, no 4, pp357–371

Howitt, R. and E. Hanak (2005) 'Incremental water market development: The California water sector 1985–2004', *Canadian Water Resources Journal*, vol 30, no 1, pp73–82

IID (2007a) *Fallowing Programs*. Retrieved 9 November 2007 from www.iid.com/Water_Index.php?pid=267

IID (2007b) Participating Fields. Retrieved 16 November 2007 from www.iid.com/Water_Index.php?pid=285 then 2006_participants.pdf

IID (2007c) 2007–2008 Fallowing Program Solicitation Announcement. Retrieved 16 November 2007 from www.iid.com/Water_Index.php?pid=2797 then 2007-2008-Fallowing-Program-Solicitation-Announcement.pdf

IID (2007d) Proposals sought for 2007–08 On-Farm Fallowing Program. Retrieved 16 November 2007 from www.iid.com/Water_Index.php?pid=2797 then Proposals-sought-for-2007-08-On-Farm-Fallowing-Program(revised).pdf

Keplinger, Keith O. and Bruce McCarl (1998) *The 1997 Irrigation Suspension Program for the Edwards Aquifer: Evaluation and Alternatives* (Technical Report No. 178), Texas Water Resources Institute, College Station, TX

McCarl, B. A., L. L. Jones and R. D. Lacewell (1997) *Evaluation of 'Dry Year Option' Water Transfers from Agricultural to Urban Use* (Technical Report No. 175), Texas Water Resources Institute, Texas A&M University, College Station, TX

McGinnis, Mark A. and Jason P. Alberts (2005) 'Southwest water decisions', *The Water Report, 20* (15 October 2005), pp1–15

MWD (2004) MWD Board approves the principles of agreement for proposed Palo Verde Program, July 2001 press release retrieved 20 September 2005 from www.mwdh2o.com/mwdh2o/pages/news/features/11%5F01/paloverde01.htm

Reclamation (2007) *Lower Colorado River Multi-Species Conservation Program*, Final Implementation Report, Fiscal Year 2008 Work Plan, and Budget Fiscal Year 2006 Accomplishment Report. June 2007.

Saliba, Bonnie Colby and David B. Bush (1987) *Water Markets in Theory and Practice: Market Transfers, Water Values, and Public Policy*, Westview Press, Boulder, CO

Thorson, J., S. Britton and B. Colby (2007) *Tribal Water Conflicts: Essays in Law, Economics and Policy*, University of Arizona Press, Tuscon, AZ

UAWCD (2004) Water District Agreements Enhance River Flows. www.uawcd.com/documents/agreements_1-26-04.htm

*Water Strategist* (June 2003) Rodney Smith (ed.), Published by Strat Econ, Clarmenont, CA

Yardas, D. (1989, June) *Water Transfers and Paper Rights in the Truckee and Carson River Basins*. Prepared for the American Water Resources Association Symposium on 'Indian Water Rights and Water Resources Management', Missoula, MT

Zuni Indian Tribe Water Rights Settlement Act of 2003, US Statutes at Large 117 (2003) p782

# 5

# Groundwater Management Issues and Innovations in Arizona

*Katharine L. Jacobs*

Recent International Panel on Climate Change (IPCC, 2007) findings indicate that the arid subtropical regions of the world are likely to expand in the context of global warming, and the proportion of the human population facing water supply limitations is expected to increase. The combination of higher temperatures, causing increased aridity, and anticipated changes in atmospheric circulation mean that even if total precipitation increases on average across the globe, drought is likely to become an even greater problem than it is today. The IPCC findings also warn that the intensity of precipitation is likely to increase, so both extremes – floods and droughts – will increasingly challenge water managers.

Arizona is particularly concerned about the consensus findings of the IPCC, because increased warming, particularly in summer, will affect the quality of life and demand for water in the state and because projections for water supply look relatively bleak under even the most conservative of the IPCC scenarios. Downscaled information from the majority of the climate models indicate that flows in the Colorado River, which provides about 40 per cent of Arizona's total water supply, may reduce by up to 40 per cent in the next 50 years according to the worst case scenarios (Hoerling and Eischeid, 2007; Barnett et al, 2008) or by up to 11 per cent in the next 100 years under the most optimistic scenario (Christensen and Lettenmaier, 2007). These reductions are primarily a result of drying caused by higher temperatures (reduced soil moisture, increased evapotranspiration and reservoir losses), but also result from changes in snowpack and snowmelt regimes. Some models predict reductions in precipitation in this region as a result of changes in the jet stream. If precipitation reductions occur in addition to increased temperatures, impacts on water supplies will be even more dramatic. Seager et al (2007) refer to drought as the new 'normal' for the southwest.

The implications of these changes on groundwater will be substantial – both because the opportunities to replenish aquifers will reduce as surface flows are diminished and evapotranspiration increases, and because water users will turn increasingly to groundwater to offset reductions in surface water availability. Little is known about the specific implications of climate change on groundwater recharge in specific basins, but most signs point to reduced opportunities for recharge.

## Arizona's water supplies

Arizona uses close to 8 million acre-feet (9867mcm) of water annually, of which about 40 per cent is groundwater. An additional 40 per cent comes from imported Colorado River water; the final 20 per cent is other surface water supplies and effluent. The central Arizona metropolitan region has been overdrafting groundwater (withdrawing more water than is replenished annually) since the 1940s. A key objective of the 1980 Groundwater Management Act (discussed below) was to limit the overdrafting of groundwater to ensure a more reliable long-term water supply. Arizona's surface water supplies are relatively limited, so groundwater supports a large proportion of the total demand in the state – including 60 per cent of the domestic water supply.

With the exception of the central highlands of the state, groundwater is relatively plentiful and is contained in vast underground aquifers in alluvial basins throughout the state. Groundwater is particularly plentiful within the 'basin and range' province, which consists of a series of uplifted mountain ranges separated by broad, alluvial valleys. However, this groundwater is considered largely 'non-renewable' because the rate of replenishment is very low. The state's renewable surface water supply is largely developed, and few opportunities remain for further enhancement or augmentation of the supply through dam or reservoir building. Water flows in Arizona streams and rivers are variable and fully allocated to existing water users. The Salt River Project controls the majority of the water supplies in the Salt and Verde River watersheds, serving an average of a million acre-feet annually to the Phoenix metropolitan area. There are few surface water supplies of any significance available within most of the remainder of the state. Effluent, or municipal wastewater, is used for landscape irrigation, artificial recharge and power production to a significant extent within the urban areas of the state.

## The Arizona Groundwater Management Act

In the US, water supply allocation is primarily delegated to states, while water quality standards are generally set by the federal government. Although this formula sounds simple, it is in fact very complex, since many watersheds cross state (not to mention international) boundaries, and water quality and quantity issues are closely intertwined. Arizona is fortunate that significant efforts to manage the state's groundwater supplies were initiated more than 25 years ago, well before most states made serious efforts to manage groundwater and before the new challenges of climate change and decadal drought cycles were understood. The establishment of the Arizona Groundwater Management Act (the Act) in 1980 was truly a 'watershed' event for the state, as it established an entirely new regulatory structure. The Act established a new groundwater rights system within areas of the state called Active Management Areas (AMAs). The AMAs cover about one-fifth of the land area of the state, 80 per cent of its population, and about half the total state water use. The AMA boundaries generally follow groundwater aquifer boundaries.

Source: Arizona Department of Water Resources.

**Map 5.1** *Arizona groundwater basins and active management areas*

Management goals for the five AMAs are generally related to achievement of 'safe-yield', a balance between withdrawals of groundwater and the amount withdrawn from the aquifers on a long-term average basis. Safe-yield is to be attained by the year 2025 through a series of mechanisms, including both supply augmentation and mandatory reductions in groundwater use. No new irrigation for agricultural purposes is allowed within AMAs, and a limited permit system is available for new industrial uses. Municipal development is only allowed in AMAs if an 'assured water supply' for the proposed use can be proven. This means that water availability for 100 years must be demonstrated before new subdivisions can be developed, and the water supply must be primarily from renewable sources. There must also be a demonstration that the supply is of adequate quality and is physically and legally available, and that there is the financial capacity to build whatever delivery and treatment works are required. This is the most stringent land use/water supply demonstration in the US, although there are many who do not believe that it is sufficiently stringent to ensure a reliable water supply into the future. This issue is discussed later in this paper.

**Table 5.1** *A comparison of the five active management areas in Arizona*

| AMA | Management goal | Size (sq miles) | Acres irrigated agriculture | Groundwater overdraft (estimated) (acre-feet) | Total water use (acre-feet) |
|---|---|---|---|---|---|
| Phoenix | Safe-yield | 5386 | 287,000 | 300,000 | 2,000,000 |
| Pinal | Allow development of non-irrigation uses and preserve existing agricultural economies in the AMA for as long as feasible, with the necessity to preserve future water supplies for non-irrigation uses | 4096 | 275,000 | 100,000 | 800,000 |
| Tucson | Safe-yield | 3869 | 33,600 | 100,000 | 360,000 |
| Prescott | Safe-yield | 485 | | 11,300 | 20,000 |
| Santa Cruz | Maintain safe-yield and prevent local water tables from experiencing long-term declines | 716 | 5000 | Variable, depending on surface flows | 20,000 |

Within AMAs the conservation of water is mandatory for all large users (those who have wells with a pump capacity that exceeds 35 gallons per minute) and there are specific, enforceable water use requirements for the agricultural, municipal and industrial sectors. These conservation requirements are adopted within 'management plans' that have the effect of administrative law. A series of five management plans must be adopted within each AMA in the 45-year period between 1980 and 2025 – the fourth of the five plans is currently being developed to cover the period between 2010 and 2020. The management plans are designed to iteratively work towards safe-yield through a combination of demand management and supply enhancement strategies, with periodic adjustments in the regulations with each new management period.

Outside of AMAs water use restrictions are less stringent – in fact, there are few regulations at all – and renewable water supplies are not commonly available. Although recent statutory changes allow for a demonstration of 100-year water 'adequacy' before new subdivisions can be established outside of AMAs (but only if authority to do so is adopted by a unanimous vote of the county board of supervisors), the standards are much less strict than in AMAs and most of the other regulatory provisions of the AMAs do not apply.

Water supplies in Arizona are owned by the public, but the state allocates the rights to use the water. In contrast, water quality standards are generally set by the federal government, although responsibility for enforcement is sometimes delegated to state agencies.

Surface water is managed under a prior appropriation 'first in time, first in right' system within Arizona. Although the surface water rights of the state have not been quantified, except within certain watersheds where water rights are established by court decree, the implications of this failure to define surface water rights are not as important as they might appear. This is because there are very few free-flowing streams remaining in the state – the majority of the flows have either been diverted for human uses or have disappeared as a result of over-pumping of the groundwater aquifers. With the exception of headwater streams at higher elevations, there are only two free-flowing rivers remaining in the state, the San Pedro, which flows north from Mexico, and the Verde, which flows from the north-central part of the state southwest towards the reservoirs of the Salt River Project, which serves water to much of the metropolitan Phoenix region. Both of these rivers provide important habitat for migratory birds and have very high species diversity, and both support a number of threatened and endangered species.

In contrast, groundwater outside AMAs is managed under the 'reasonable use' doctrine, which essentially means that withdrawals are not managed, and the only standard is whether the water is used 'without waste'. There is little recourse for one groundwater user if his water supply is diminished by a subsequent user, because there is no priority system and no limitation on new uses. Although water users can sue in court for damages, there is very little legal protection. Because the surface waters and the groundwaters of the state are managed under separate legal doctrines, there are significant institutional limitations to water management, particularly as it relates to protecting instream flows and shallow groundwater levels that are important for habitat protection from depletion through groundwater pumpage. Even in AMAs there is no legal mechanism for limiting groundwater use to protect surface flows, with the exception of one of the AMAs (Santa Cruz), which has a management goal that allows local groundwater levels to be managed.

## Arizona water management innovations

### Active Management Areas (AMAs)

AMAs are regions within the state where there is enhanced regulatory authority over water use. The AMA concept includes state-mandated water management goals within hydrologic basins, with opportunities for local stakeholder input into the regulatory process and, as such, provides an interesting laboratory for evaluating regional water management within a larger state structure. There is some 'creative tension' between the regional AMA offices and the central offices of the Department of Water Resources in the development of these plans, but there are opportunities to tailor the plans in response to hydrologic conditions and public input. The desire for local self-determination (which is a very strong sentiment, especially in the rural parts of Arizona), versus the perception of many state regulators that centralized planning and implementation is more efficient, is a valuable source of balance in the system. Having centralized enforcement mechanisms provides some significant benefits, particularly if the state system is properly supported and administered (there are some issues in this regard, but the important point is that the system itself has significant advantages). A central enforcement authority does provide an equitable enforcement of statutes that can be difficult to achieve at the local level, due to the strong influence of politics and the emotive nature of water rights. Having some consistency in management across the AMAs is helpful in providing a level playing field for water using sectors and achieving economies of scale in developing expertise in water efficiency techniques. Standardized monitoring and reporting mechanisms are also useful because, in theory, this results in more consistent data gathering and the potential for trend analysis, leading to improvements in decision making.

### Groundwater rights and mandatory conservation

The groundwater rights system established by the Groundwater Management Act is also relatively innovative – all large groundwater users (over 35 gallons per minute pump capacity) established prior to 1980 have some form of 'grandfathered right' to groundwater, or right to continue to withdraw water. In most cases, access to groundwater is restricted for new users. The right to irrigate land is tied to specific parcels that were historically irrigated, while certain industrial categories of water use can be severed from the land and used in new locations. All grandfathered groundwater rights are subject to the conservation requirements in the management plans; in the case of agriculture, the amount of water each farmer can use is based on the water requirements associated with crops that are historically grown, multiplied by an increasing efficiency standard, unless the farmer opts into a 'best management practices' approach to conservation. Water companies serving municipal water users are regulated either on the basis of a gallons per capita per day target for their service area, or based on reasonable reductions in per capita use. This approach allows total water use to increase as the population served increases (an important consideration in a state that has an economy based primarily on urban growth) or through a best management practices approach that does not actually quantify the water use at all but uses performance requirements to encourage increases in efficiency. The assured water supply (AWS) programme provides the 'check' on the municipal water use, because of the requirement to use renewable supplies and demonstrate the availability of supplies 100 years in advance. There are some significant 'loopholes' in the water rights system, however. For example, it is possible to establish new industrial uses on groundwater relatively easily within AMAs and small, domestic wells (less than or equal to 35 gallons per minute pump capacity) are entirely unregulated.

## Assured water supply

The safe-yield goal within the three AMAs that are measured by this standard (Prescott, Phoenix and Tucson) is a basin-wide goal; the sum of all the groundwater pumping in the basin is compared to the total water supply inflows to calculate whether or not the goal is achieved. A basic tenet of the Act is that although many users, such as agricultural and industrial water rights-holders, are likely to continue to use groundwater in the future because their rights are 'grandfathered', the municipal sector is expected to move off groundwater and onto renewable supplies. There are several reasons why focusing on the municipal sector is critical and appropriate to achieving safe-yield. The municipal sector has always been expected to expand over time and ultimately to dominate the water use picture in the AMAs. Likewise, it has been noted that the municipal sector has the most resources to invest in renewable water supplies, and the most to gain from a secure water future. Therefore, the AWS programme establishes firm limits on how much groundwater can be used by municipal water providers. Although there are emergency provisions during times of shortage in the surface water supply system, the expectation is that, beyond the initial allocation of mined groundwater that was provided through the AWS rules, groundwater will not be a significant water supply for municipalities in AMAs, especially growing municipalities, in the future.

The concept of AWS is relatively unique to Arizona, although several states (including California) have subsequently adopted similar but less stringent statutes. The relationship between land use decisions and water supply is still less than ideal even with the AWS programme in place – in part because there are mechanisms that allow recharge to offset groundwater pumpage within the AMA but at a distance from where the groundwater itself is withdrawn. This saves money in transporting water from the Central Arizona Project (CAP) canal, which brings Colorado River Water to central Arizona. The CAP is the primary new source of renewable supplies used to prove availability of renewable supplies for AWS. However, allowing aquifer recharge to occur distant from the area where the groundwater is pumped, even if in the same AMA, does not prevent local impacts on the groundwater table, such as subsidence of the ground surface due to dewatering.

## Underground Storage and Recovery

The Underground Storage and Recovery programme was initiated in 1986, primarily to facilitate the storage of surplus water in aquifers and to protect the rights of the storers to recover that water in the future. The statutes related to this programme have been amended many times, but the basic approach to encouraging artificial recharge has proven to be a sound one that supports many components of Arizona's water supply programmes. It has also served as a model for similar statutes in other states. Using aquifers for the storage of water has several benefits, including a reduction in evaporative losses and opportunities for improving water quality through biological and chemical processes that occur as the water percolates into the vadose zone and blends with other water in the aquifer. An example of an improvement in the water quality that is of value for municipal entities is the removal of organic materials, which otherwise could lead to harmful by-products when mixed with chlorine for delivery to customers.

Credits for storing water underground can be generated at recharge facilities that are permitted for this purpose, and generally fall into two categories, underground storage facilities for 'direct' recharge and groundwater savings facilities. The former category generally involves constructed or managed recharge facilities that actually add water to the aquifer, while the latter provides credits to those who conserve groundwater in the aquifer by providing an alternative renewable supply to an entity that would otherwise have used the groundwater. The credits can be used to withdraw groundwater in the future that legally meets the AWS criteria as a 'renewable' supply. This 'indirect' approach to recharge is very efficient and inexpensive, although it may not always result in the same physical benefits as water stored in facilities that are deliberately located to benefit future municipal water uses.

## Arizona Water Banking Authority

The last of the innovations that we discuss is the Arizona Water Banking Authority (AWBA), which was created by statute in 1996 and authorized to function into 2016. The purpose of the AWBA is to store excess water underground to offset anticipated future shortages in CAP deliveries, particularly to 'firm' municipal water supplies that would otherwise be unreliable because of CAP's junior priority on the Colorado River system. The AWBA also supports Indian water rights settlements, facilitates achievement of the management goals for the AMAs and, last but not least, allows for interstate banking arrangements with other Lower Colorado River Basin states (California and Nevada). The primary incentive for its creation was the need to increase Arizona's use of its 2.8 million acre-foot Colorado River allocation to ensure that it would not be lost to California due to non-use. AWBA provided a mechanism for Arizona to greatly increase its water use relatively quickly and beneficially use its full CAP allocation – which was a major issue at the time given that Arizona's supply was underutilized and California was using Arizona's unused share.

AWBA is not like other water banks, in that it is not a broker for water rights and it does not establish a market for water. It exists primarily to store water to offset shortages in surface water deliveries for municipal users, and its functions are quite circumscribed. Nevertheless, it has been very effective in rapidly increasing Arizona's use of Colorado River water and in storing large quantities of water using existing recharge facilities at a relatively low cost. It has no water rights, and stands near the end of the line in priority to access water, so it is not in competition with other water users. It also does not own or operate its own storage facilities, and is not responsible for the recovery of the water it stores – the CAP itself recovers the water during times of shortage.

AWBA is partially funded by a pump tax on groundwater withdrawn within the three central AMAs that are part of the CAP service area. It also has access to a property tax levied by the CAP board if it is not needed for repayment obligations, and had general fund revenues in the early years of operation. AWBA is overseen by a five-member board of directors and is supported by staff who work within the auspices of the Arizona Department of Water Resources. More than 3 million acre-feet have been stored by the AWBA, the majority in the Phoenix and Pinal AMAs.

## Arizona's unresolved water issues

### Supply versus demand

From a big picture perspective, Arizona's water issues appear to be about whether the supply is adequate to meet the demand, especially in the context of the rapid pace of growth. Arizona is in an arid region and its total demand for water exceeds the renewable in-state supply. It is growing more rapidly than any other state in the US, and many doubt whether the water supply can keep pace with the rate of growth. However, on closer examination, the issues are much more nuanced. For example, the current water use for the state is close to 70 per cent agricultural, which means that municipal and industrial water use could triple by converting agricultural use to urban use. Agriculture appears to provide a good buffer for municipal use until it becomes apparent that significant new infrastructure would be required to bring high-priority agricultural Colorado River water to the central urban regions of the state.

To a great degree, the water supply issue in Arizona is not about total quantity of water, but rather about the way it is distributed geographically relative to the demand. In northern Arizona in particular, there is a need for additional water supply infrastructure, but the land area is very large and systems to deliver new supplies are far beyond the resource capabilities of the state. Additional infrastructure at a very large scale will also be required within the central part of the state within the next 50 years if growth continues as anticipated.

The CAP is a 330-mile canal system completed to Tucson in 1992. It currently imports about 1.5 million acre-feet (MAF) of water into the Phoenix–Tucson urban corridor, but its capacity to deliver water is nearly at maximum. Arizona's allocation of Colorado River water is 2.8MAF, but a second canal would be required to significantly increase deliveries. It took approximately 30 years from official authorization to completion of the $4 billion CAP, and it is highly unlikely that the federal government would be willing to subsidize another project of this magnitude. Furthermore, there is substantial resistance to 'ag-to-urban' transfers, especially in the case of permanently transferring water rights. Objections from the agricultural community include concerns about whether the agricultural economy could be sustained in a much-diminished state, impacts on rural economies and lifestyles, impacts on the tax base, and, significantly, public policy issues about food security in light of heightened international tensions. Even if all of these concerns are addressed through economic means, and it is clear in theory at least that the majority of them could be, at this time there is essentially no institutional mechanism to facilitate these transfers, although the dry-year options that Dr Colby describes in her chapter are clearly one of the paths of least resistance.

## Environmental sustainability

A second category of sustainability issues relates to environmental sustainability. The vast majority of surface water supplies in the state have already been diverted to serve human uses, and the remaining surface water supplies are dwindling in the context of drought and increased groundwater pumpage. At the present time, there are no institutional mechanisms in place to ensure that the water supplies for critical habitats for threatened and endangered species are protected, other than the Federal Endangered Species Act, which is a relatively blunt tool at best.

Although other western states, such as California and Washington, have state statutes that augment the federal requirements, Arizona has very limited ability to protect water for environmental purposes. This is partly because of the dichotomy between groundwater and surface water laws that was alluded to previously – unlike Colorado, for example, where judges allocate 'tributary groundwater' in the same way that surface water rights are allocated, Arizona's legal framework does not generally allow the two sources to be managed in tandem, despite the fact that they are hydrologically connected. (It should be noted that Arizona courts have determined that 'subflow' – which is groundwater that is hydrologically connected to the surface flows, especially within the 'younger alluvium' of a stream – does constitute surface water).

A limited number of 'instream flow' rights with very junior priority (all post-1980) have been issued, generally in headwater areas. These rights can be established for habitat protection and recreation purposes on public lands, but their junior priority and location in the headwaters means that the impacts of these rights will be limited. There are also state programmes such as the Water Protection Fund, which provides financial assistance to projects that protect waterways but cannot be used to purchase water for environmental purposes. In the two high-priority flowing rivers that remain, the San Pedro and the Verde Rivers, the ability to protect the flows against increased pumping from private wells is very limited. In the case of the San Pedro, local ordinances have been passed to enhance conservation efforts, increase recharge and water harvesting, and assist in requiring new water withdrawals further from the river (in part through transferable development rights), but these are at best delaying tactics for what many see as the inevitable drying up of the river – if new sources of supply are not found to augment the water budget. Importation options are actively being pursued, but there is no unallocated surface water available in the state, and there are concerns about groundwater importation from rural basins, even where such basins are not expected to develop in the near future.

## Adaptive capacity of water management institutions

A third category of issues facing Arizona is that the institutions were generally designed to meet the water supply needs of the 20th century, not the 21st. There are multiple ways in which the water management institutions are ill-prepared to meet the rapidly changing demand trends and the changing public values regarding water and land use. Although the Groundwater Management Act was very farsighted and virtually revolutionary in its day, the combination of incremental legal changes (many of which have weakened its provisions), inadequate resources within the state agency that administers it (the Arizona Department of Water Resources) and the unanticipated pace of growth have conspired to make its provisions less effective than they might otherwise have been.

The pace of change in the 'natural' environment has also exceeded the expectations of virtually everyone – scientists and water managers alike. Our understanding of the hydrologic implications of climate change is evolving rapidly, particularly our understanding of past variability (NRC, 2007), but is not keeping pace with the pace of change itself. Global change in general, including land cover change, significant changes in ecological systems, and globalization has resulted in a set of new challenges that our institutions are likely not well positioned to respond to. It is clear that they are not ideally designed for adaptive management to this constantly changing physical and economic context. Even though the Act does allow for incremental assessments of progress towards the management goals in each AMA as each successive management plan is written, it is not clear that the tools exist in the current law to ensure that safe-yield is achievable, let alone assured, in any of the AMAs.

## Institutional complexity and public engagement

Further, institutional complexity in the water management arena makes it very difficult to achieve even the simplest of objectives, and public perceptions are often very divergent from the perspectives of 'experts' who run the water management system. There are multiple layers of water management, unwritten rules of engagement and multiple regulatory agencies that participate in decision making processes. Public perceptions, and even those of elected officials, are often very strongly influenced by an incomplete understanding of the physical, economic, environmental and social realities of water supply and hydrology (Jacobs and Pulwarty, 2003). The media often exacerbate the misperceptions because they encourage alternative views of issues to be discussed, even in cases where there is a clear 'right' and 'wrong' technical answer to a question. Confusion is also exacerbated by the often conflicting roles that federal, tribal, state and municipal jurisdictions play in water management, and the multiple institutional layers mean that changes in policy are often well-nigh impossible to achieve.

Access to information within the highly balkanized and political water world is often very restricted, even in democracies, simply because so much knowledge is required to be effective in the 'inner circle' of water policy, leaving a very few 'empowered' individuals. The major players in the water arena – the cities, agricultural and industrial interest groups – are often most interested in protecting the status quo, because they have worked hard to optimize their water rights and positions in the context of the current institutional arrangements. Although most water rights information is in theory public information, so much sophistication is required to get access to water data and to understand how to use it properly that the water knowledge system is really controlled by a very few experts. Particularly outside of AMAs, there is a perception that inadequate data exist to determine the sustainable yield of the surface and groundwater supplies, yet in many cases there are studies and data that are not widely known. However, the reality is that there is much information available, but it is not easily accessible in a common format.

The Arizona Department of Water Resources has been working on a publicly available 'Water Atlas' in part to address this issue.

## Land use and water supply issues: Does AWS work?

A related issue is the fact that although the AWS programme does provide a way to connect sustainable water supply and land use in a more direct way than it is connected in virtually any other part of the US (with the exception of mandatory replenishment districts such as the Orange County Water District in California) there are still major problems with ensuring meaningful, integrated land use and water supply planning. This is due to a number of factors, not all of which relate to the AWS rules themselves. One of the land use–water supply AWS programme issues is related to the Central Arizona Groundwater Replenishment District, which was created in 1993 as a mechanism to help meet the consistency with management goal criterion in the Phoenix, Pinal and Tucson AMAs. The Central Arizona Groundwater Replenishment District (CAGRD) was created to provide landowners, developers and water providers who do not have direct access to CAP water, effluent or other renewable supplies with an alternative mechanism to help demonstrate a 100-year assured water supply as required under the AWS rules. The CAGRD replenishes the aquifers within AMAs on behalf of its members, which are 'member lands' or subdivisions, as well as 'member service areas' that are water companies. Within three years after groundwater is pumped by its members, the CAGRD must replenish the groundwater that its members use in excess of the amount that is allowable under the AWS rules. The water that is used for replenishment may be CAP water or water from any other lawfully available source, except groundwater withdrawn from within an AMA.

There are several issues associated with the CAGRD, including whether the renewable supplies that are available to it will continue to be reliably available in the future. The CAGRD is intended to use excess water as its primary source, but there is a constant reduction anticipated in the amount of 'excess' water available, both because of increasing demand for water in both the Lower and the Upper Basin and because of climate change. Whether the CAGRD can really provide reliable supplies to its customers in perpetuity without a significant proportion of its supplies coming from permanent water rights (and in light of climate change and Arizona's junior priority on the Colorado, whether even the permanent water rights will be of much value) is commonly cited as a major concern.

More directly related to the land use–water supply nexus, the CAGRD has been accused of facilitating sprawl because it allows land to be developed in outlying areas where physical access to renewable water supplies is generally nonexistent. By allowing development of outlying areas based on 'mined' groundwater, the accusation is that the CAGRD facilitates lower-cost, less sustainable land use and water supplies. Ultimately, many argue, the local groundwater supplies in such areas will no longer meet the physical availability criteria of the AWS rules, and expensive infrastructure will be required, possibly at great cost to society in general (while the initial developer did not bear any of those costs and is no longer responsible for them). Many are concerned that the AWS programme is a 'paper' rather than a 'wet water' solution to the AMA's water management issues.

## Achievability of the safe-yield goal

Another major water issue is whether the safe-yield goal in the AMAs is achievable. The rate of groundwater use exceeds the rate of aquifer replenishment in the Phoenix, Pinal, Prescott and Tucson AMAs. Despite expanded use of renewable supplies from the CAP and effluent, this situation is expected to continue past the statutory safe-yield goal date of 2025. Renewable supplies are not universally available, and financing to build infrastructure to transport renewable supplies to locations where they are needed is inadequate, despite the establishment of a revolving fund for infrastructure and other state financing mechanisms.

## The energy–water nexus

An emerging issue for the state (and the entire western US) is the nexus between energy and water. It has been estimated by some that the water requirements associated with energy production could dramatically increase in the coming decades, in part because energy production itself is increasing rapidly to meet demand, particularly in urban areas and in the US–Mexico border region. A separate reason for concern is that the types of energy production that seem most likely in the coming decades use a lot of water. Nuclear power production, for example, uses more water for cooling than other power production technologies, although improvements in design may reduce the water requirements. Similarly, the next increment of water available to the west seems likely to have even higher energy requirements than our current supplies. Desalination, for example, is a high energy user, even though technologies are improving regularly and reducing energy requirements on a per unit basis. Although many believe that there are huge opportunities for conserving both energy and water by reducing high water uses both indoors and out, there are also significant growth factors that seem to foretell an increasing spiral in water and energy use.

## Value of water versus price

Another critical issue for Arizona is that the value of water is not reflected in the prices that most users pay, which is one of the reasons that a regulatory solution was devised to address the state's groundwater overdraft problems. Although scarcity is starting to drive up the costs in some regions (a current water supply proposal for the Chino Valley area north of Prescott is that new development would pay $45,000 per acre-foot of imported supplies), in general, water rates are astonishingly low. Some users pay as little as $5 per acre-foot. Obviously, this huge spread in the price of water causes massive inequities and inefficiencies in water use and management, as well as incentives to develop property in areas with less expensive supplies. As prices and regulation increase, there are also more unregulated domestic groundwater wells drilled that are not subject to high costs. Water rates are far lower than electricity rates or other less essential items such as cable TV or phone service. The value of water for environmental and economic purposes far exceeds what people are currently paying, but there is significant social and political pressure to keep the costs of water low. This relates in part to the perception (shared in many other regions of the world) that access to inexpensive water is a basic human right. Yet, Arizonans are not ready to recognize the environment's 'right' to water as has been recognized elsewhere.

A key impetus for the Act was to encourage water use efficiency and the use of renewable water supplies in the state before groundwater supply problems reached critical proportions, to ensure that the economy was not damaged by perceptions that Arizona's water supply was not secure. Some have argued that the Act creates a 'regulatory drought' that is intended to avoid a future real or perceived crisis by restricting water use to ensure that there is sufficient supply for future use, and to facilitate a transition from agricultural to urban water uses. This regulatory drought was first enforced during a time when Arizona's access to excess water supplies was very high, and when perceptions of shortage were almost nonexistent. This may have turned out to be a strategic error (although it was not something that could have been foreseen) because support for the conservation programmes in the Act has actually eroded over time. Clearly, if the initial management plan requirements were just now being enforced, it might have been perceived quite differently by the public and decision makers because a decade of drought and the prospect of climate change is now of great concern to a large proportion of Arizona citizens.

## Tribal issues

Tribal issues are another source of concern for water managers in the state. There are 26 tribal entities within Arizona; tribal lands account for 28 per cent of the land in the state. Using the calculations based on 'practicably irrigable land' established in Arizona versus California in 1963, the water rights of the tribal entities are said to exceed the total amount of surface water in the state. The 1908 *Winters* case held that tribal reserved water rights date to the establishment of the reservation, so the water rights for tribes are of higher priority than most non-Indian water rights. The prospect of having such large unresolved claims for water looming over the water supply picture in Arizona has been a significant incentive to settle the water rights of the tribal entities. Arizona now has more Indian settlements than any other state, and the still-pending Arizona Water Settlements Act of 2004 is the largest settlement in the history of the US. It provides more than 653,000 acre-feet of water to the Gila River Indian Community, along with substantial monetary compensation. Although this is an enormous accomplishment, there are still major unresolved settlements, including the San Carlos Apache, the Navajo and the Hopi, and resources to settle these cases are shrinking on a regular basis.

## Water quality issues: Arsenic, salinity and emerging contaminants

Arizona also faces increasing challenges associated with water quality. There are significant groundwater contamination issues in the urban areas, many associated with industrial solvents. There are also multiple areas of the state where nitrates associated with fertilizer and wastewater disposal are elevated. However, the most critical water quality issues relate to arsenic, salinity and emerging contaminants. Arizona's water quality protection programme is considered to be very advanced, and is generally quite proactive. However, recently adopted federal standards for arsenic of 5 micrograms per litre pose a substantial problem – Arizona has more water companies in violation of this standard than any other state, and there is a significant effort to provide support to water providers from both a financial and technical perspective.

The salinity problem is generated in part by the tons of salt that are imported into Arizona every day from the Colorado River through the CAP, and in part from naturally occurring salts in both surface and water supplies within the state. The salts imported from the Colorado will eventually build up either on the land (because of applications to agriculture and landscapes, and wastewater or sludge disposal) or in the water supply. Particularly as a larger and larger component of the water supply is recycled, the salinity of the water will increase over time. Municipal effluent reuse tends to concentrate salinity significantly with each use cycle. Naturally occurring saline groundwater is also a problem.

Other water quality issues include newly discovered pathogens and various kinds of emerging contaminants, particularly pharmaceuticals and estrogenic compounds that survive the wastewater treatment process and negatively affect fish and amphibians in streams. These compounds have also been identified in drinking water supplies at low levels, but the effects on human health are not known. These compounds are not currently removed or destroyed in the context of traditional wastewater treatment, although more advanced treatment processes do provide higher removal rates. As wastewater is used more directly for drinking water purposes and for aquifer recharge, these concerns will continue to be magnified. There is significant research ongoing within the state focused on ways to manage these compounds.

## Transboundary issues

There are three major river systems and several groundwater aquifers that cross the Arizona–Mexico boundary; all of the aquifers are closely linked to surface flows. Managing cross-boundary supplies results in challenges to both water quality and water quantity, with particular concerns about untreated wastewater flowing north across the border, the potential for contaminated groundwater to enter the US, and the US's obligation to provide Colorado River water with acceptable salinity levels to Mexico. There are also habitat concerns related to the border, including recent disputes about the impact of the border fence on riparian habitat and animal migration. Multiple international, federal–state, and interstate conflicts relate to management of the Colorado River, not the least of which is how to manage water deliveries during drought. These conflicts are similar to conflicts on the Rio Grande River and multiple other locations throughout the world where rivers and lakes cross national boundaries.

## Conclusions and policy implications

In Arizona, a decade-long drought and the likelihood of climate change-induced reductions in water supply availability have already intensified competition for water. Rapid population growth, interstate and international water supply issues, and unmet environmental water needs are also increasing concerns about water sustainability. There is evidence of increasing conflict over access to renewable water supplies to meet the Assured Water Supply programme requirements within urban areas. Water supplies to meet future tribal settlement needs are very limited. Rural parts of the state would like to import water to offset current shortages, but their water supply options are very limited.

Arizona's water management system is unique, but many of the issues that remain unresolved are common to other arid regions of the world. Arizona's relatively sophisticated water management system is focused almost entirely on the urban portions of the state (within the AMAs), leaving the rural (non-AMA) areas with few tools to address water shortages that are periodically quite intense.

Values related to water and perceptions of threats to sustainability in the desert are changing as a larger proportion of the population is exposed to evidence of drought, climate change and rapid urbanization. There is a disconnection between the very high value the average Arizonan places on scenic beauty, recreation and habitat protection, and the virtual lack of protection of environmental values, particularly riparian areas. Arizona's bifurcated legal system – which manages surface water and groundwater through separate statutes – essentially prevents any meaningful protection of surface water flows and riparian habitats in areas where groundwater base flow provides a significant contribution to these habitats. The political climate is such that environmental flows are not likely to be protected any time soon – in many ways, Arizona is still the 'Wild West'.

Although the Groundwater Management Act was very far-sighted in acknowledging that there would be an agriculture-to-urban transition in water use, this transition is going relatively slowly in some parts of the state. Meanwhile, there is ongoing concern among the agricultural community about the viability of the agricultural sector in light of rapid urbanization, and in light of globalization, food security and economic opportunities associated with urban development. Roughly 70 per cent of the water use in the state is agricultural, indicating that agriculture is still a major force in the economy. Agricultural interests continue to wield considerable power in the legislature and they are often reluctant to facilitate retirement of agriculture or agricultural conservation efforts although it is economically advantageous to individual farmers, in part because they wish to protect agricultural lifestyles and communities as well as open space.

Even within the AMAs, where the water management tools and water supply infrastructure are most concentrated, it is likely that the mechanisms currently in place are inadequate to ensure renewable water supply availability in the long term for some categories of municipal users. Two of the five AMAs (Prescott and Santa Cruz) do not have access to the Central Arizona Project, and therefore have a much more limited ability to prove an AWS. An area of particular concern within the other three AMAs (Phoenix, Pinal and Tucson) is the Central Arizona Groundwater Replenishment District (CAGRD) discussed previously, which is designed exclusively to replenish overdrafted groundwater associated with new housing subdivisions in the AMAs. Concerns focus on the reliability of the water supply for its members and whether it is encouraging sprawl, exacerbating air quality and transportation issues.

There are currently difficulties in funding and providing technical support for effective long-range planning and drought planning throughout the state, even within the AMAs. There is rising concern about whether the AMAs can serve their original purpose adequately, or whether regional entities need to take on a larger role in water supply planning. Many communities, including Tucson, are engaging in conversations about how best to build a collective vision of how water should be used, and what sort of management entity will be most able to make the necessary tough decisions about water supply and wastewater management in the future.

## Looking to the future: Addressing increasing management challenges

Although Arizona's leaders have been remarkably far-sighted in investing in infrastructure and in building institutions to address water supply challenges, it is clear that significantly more investment and innovation will be required to ensure reliable water supplies for the burgeoning metropolitan regions as well as the rural parts of the state. Arizona does have significant water resources, and is particularly well endowed with water supplies by comparison to its neighbour to the northwest, Nevada. The critical differences are (i) Arizona's Colorado River allocation is an order of magnitude larger; and (ii) the substantial water supplies currently used by agriculture serve as a buffer against long-term shortages in the municipal sector. Still, there will be difficult times ahead.

## Increasing pressure on finite water supplies

In the past, water supply issues were addressed through engineering, supply-side solutions that moved water from one location to another, with or without storage features. Although dams are still considered as part of the solution set in the American west, despite the water management experts who have claimed that 'the era of big dams is over', they will definitely be more difficult and expensive to build in the future. In addition, the federal government is much less likely to subsidize western water projects than it was in the past. We are in an era of limits, where redeploying existing water supplies is far more likely than generating new ones. The role of the marketplace will increase, and there is likely to be more regulation as well – to the extent that economic signals do not always address public trust values and equity issues. Demand-side solutions, including conservation and increases in cost will move water to higher economic uses, but we are challenged to find ways to make these solutions more acceptable to the public. Water leasing, water banking and water transfers will become more common, and more public investment in restoring and protecting environmental values will be required. Increasing conflict over water supplies will be common, so a focus on collaborative and innovative solutions to water supply problems will be needed.

## Management goals: The need for a broader perspective

A balance between groundwater recharge and withdrawals, or 'safe-yield', is the management goal of four of the five AMAs. Although this goal provides a target for water management programmes, it is not currently an enforceable goal. The language in the Act requires that the AMAs 'attempt to achieve and thereafter maintain' the goal by 2025. Furthermore, many are concerned that since the concept of safe-yield focuses only on groundwater in aquifers, it cannot be used to protect surface water or riparian water uses. Under the current system, safe-yield can be accomplished while drying up streams either by direct diversion or by pumping groundwater that would otherwise have supported base flows. However, major changes in the Act that would move towards 'sustainable yield' for surface water supplies while also encouraging 'safe-yield' of the aquifers seem very unlikely. A broader management goal that requires a watershed-based balance as well as recognition of the need to manage local areas to limit damage due to subsidence, migration of poor quality water, protection of water-based critical habitat and to ensure physical availability of water supplies could be an improvement over the current situation.

The need for a broader perspective is also illustrated by issues of scale. For example, from the perspective of Arizona, Colorado River water is a renewable resource and it is appropriate to divert the full legal entitlement to offset non-renewable groundwater mining. Yet as anyone who has visited the Colorado River Delta can attest, looking at these diversions from the perspective of a healthy watershed at the scale of the Colorado itself changes the conclusion about whether the diversions of all of the basin states are truly renewable and appropriate, given the implications for the environment and for Mexico. The Colorado River now rarely flows to the sea, and the once-wondrous sea of grass in the Delta has been virtually eliminated.

## Sustainable development

It is unlikely that current concerns about the reliability of water supplies in Arizona will ever be entirely resolved, but it is clear that land use decisions need to be made in the context of long-term water supply availability. As in many other countries, the issue in the US is not the total quantity of water required, it is where the water supplies are located relative to the demand centres. Importation projects such as the Central Arizona Project have substantially increased the water supply options for the state. However, as has been demonstrated elsewhere, there is no 'free lunch' – importation projects have multiple costs beyond construction and operations, including environmental and social impacts. As energy costs increase, these options will become less feasible.

Reuse of municipal water and reductions in per capita use will also be needed to help balance the supply and demand equation, but both of these options require some behaviour modification and have water quality implications. As water supplies become more scarce, the trade-offs between water and energy become more intense, since the marginal supplies tend to require more treatment and/or more energy to produce.

Arizona's Assured Water Supply programme, which is focused within AMAs, has addressed the land use and water supply planning equation more directly than any other state programme, since it requires 100 years of reliable, renewable supplies to be secured before subdivisions can be approved. The Assured Water Supply mechanism is not perfect, but it provides a regulatory base from which even more significant requirements can be built by individual municipalities and water utilities. There are multiple lessons associated with Arizona's experience in this area – particularly in assessing the relative genius of those who have found ways to circumvent some of the basic requirements in order to minimize the costs of new development.

## Price versus value

Groundwater is often the cheapest source of water in Arizona, but it is also the most valuable because it is generally of higher quality than surface water, requires less treatment before delivery, is virtually always available (unless overdrafted), is stored for free in the aquifer and has few of the reliability issues associated with surface water. Thus, there is a disconnection between price and value, which is one of the reasons that a regulatory mechanism has been required to encourage reductions in groundwater use. This disconnection is also acute in the context of maintaining ecological flows, because there is currently no mechanism to quantify the aesthetic and environmental values associated with them and to translate that into an economic incentive to preserve these values. Until an economic mechanism is developed, it is clear that some form of additional regulation may be needed to protect instream flows in times of scarcity. The only mechanism that now exists, instream flow rights, has numerous limitations, but the greatest limitation is the low priority of these rights since they were established relatively recently. Clearly a hybrid mechanism that includes funding and other incentives for protecting instream flows, the groundwater that supports them and related habitats is needed. However, incentives alone may not be sufficient in light of the intense pressure that the few remaining streams are under in the context of expanding demand for water and decreasing supply.

## Climate change and supply variability: Adapting to faster pace of change

Climate variability has always provided a challenge to water managers, and indeed floods and droughts frame every management system. However, overlaying climate change on top of existing variability reframes both the means and the extremes – in ways that are not yet completely understood. It is clear that the pace of global change is escalating, including evidence of land use changes, precipitation and temperature and social conditions, especially those resulting from globalization. Significant uncertainties exist about future water supply availability, and attempts to reduce uncertainty are ongoing – but water managers are notably uncomfortable about making decisions in the context of uncertainty, and want, at the very least, to understand the best and worst case scenarios (Garrick et al, 2008). The IPCC has concluded that global warming will increase the severity of droughts in the mid-latitudes, including northern Mexico and the southern parts of the southwestern US. This increase in dryness is expected even in the context of the potential for more intense precipitation events, primarily because of the effects of temperature on evaporation and transpiration. Significant reductions in flow and changes in the seasonality of flow are expected in snowpack-dependent river systems. A range of studies (Christensen and Lettenmaier, 2007; Barnett et al, 2008) indicate that reductions in flow could be anywhere from 10 to 50 per cent in the next 50 years – even the low end of this prediction will be a significant problem for the over-allocated Colorado. Responding to a faster rate of change, while preparing for the possibility of non-linear changes or crossing of 'thresholds' requires improved observations and faster integration of scientific information into the water management process.

## Need for more adaptive management

In addition to physical changes in water supply availability, social changes are likely to affect water availability. A more flexible water rights system may be required to respond to changes in social values as well as changes in demand due to growth. Changing expectations regarding water quality also have significant impacts on water availability and cost, creating tensions between affordability and quality as well as between affordability and environmental health. Changing social values have already impacted the priority of water used for tribal settlements, protecting endangered species and recreational needs. These new demands are further stressing the already over-allocated water systems of the western US. Addressing these competing demands will require both economic and regulatory solutions that are not currently in place. However, regardless of good intentions, changes in priorities and management approaches result in economic and political dislocations, so careful assessment of the costs and benefits of alternatives is required. Existing institutions are not well constituted to support adaptive approaches, because adaptive approaches provide less certainty, are more costly (in part due to increasing monitoring and more proactive solutions), and require more professional judgement to manage.

Preparing for rapidly changing conditions requires close scrutiny of the ability of water management institutions to adapt and innovate – can they adapt as quickly as they need to, considering the rate of change in the physical and social environments? Will they be able to move away from the notion that climate conditions are relatively stationary, and adjust to the new reality that 'stationarity is dead' (Milly et al, 2008)? Water management institutions at all levels will need to become more proactive and strategic in order to prepare for a future that is inevitably different from the past. Although Arizona has proven to be quite adept and strategic in responding to hydrologic and political imperatives in the past, the pace of innovation will need to increase (Colby and Jacobs, 2006).

## References

Barnett, T. P., D. W. Pierce, H. G. Hidalgo, C. Bonfils, B. D. Santer, T. Das, G. Bala, A. W. Wood, T. Nazawa, A. A. Mirin, D. R. Cayan and M. D. Dettinger (2008) 'Human-induced changes in the hydrology of the Western United States', *Science*, no 319, pp1080–1083

Christensen, N. and D. P. Lettenmaier (2007) 'A multimodel ensemble approach to assessment of climate change impacts on the hydrology and water resources of the Colorado River Basin', *Hydrology and Earth System Sciences*, vol 11, pp1417–1434

Colby, B. and K. L. Jacobs (eds) (2006) *Arizona Water Policy: Management Innovations in an Urbanizing, Arid Region*, Resources for the Future Press, Washington, DC

Garrick, D., K. L. Jacobs and G. M. Garfin (2008) 'Models, assumptions and stakeholders: Planning for water supply variability in the Colorado River Basin', *Journal of the American Water Resources Association*, vol 44, pp381–398

Hoerling, M. and J. Eischeid (2007) 'Past peak water in the southwest', *Southwest Hydrology*, January/February, 18–19 and 35

Intergovernmental Panel on Climate Change (IPCC) (2007) Fourth Assessment: The Physcial Science Basis, Working Group I, Geneva, IPCC

Jacobs, K. and R. Pulwarty (2003) 'Water resource management: Science, planning and decision-making', in R. Lawford, D. Fort, H. Hartmann and S. Eden (eds), *Water: Science, Policy, and Management*, American Geophysical Union, Washington, DC

Milly, P. C. D., J. Betancourt, M. Falkenmark, R. M. Hirsch, Z. W. Kundzewicz, D. P. Lettenmaier and R. J. Stouffer (2008) 'Stationarity is dead: Whither water management?' *Science*, vol 319, pp573–574

National Academy of Sciences (2007) *Colorado River Basin Water Management: Evaluating and Adjusting to Hydroclimatic Variability*, National Academy of Sciences Press, Washington, DC.

Seager, R., M. Ting, I. Held, Y. Kishnir, J. Lu, G. Vecchi, H. Huang, N. Harnick, A. Leetmaa, N. Lau, C. Li, J. Velez and N. Naik (2007) 'Model projections of an imminent transition to a more arid climate in Southwestern North America', *Science*, vol 316, no 5828, pp1181–1184

# 6

# Water Policy in Australia: The Impact of Change and Uncertainty

*Lin Crase*

In November 2007 the Australian people went to the polls to elect a Federal government. For those with an interest in water policy formulation in Australia, these events attracted unprecedented interest. After all, water had been controlled by the various state jurisdictions in Australia for over 100 years. Moreover, whilst the Federal government has played a more active part in water policy recently, Section 100 of the Australian Constitution remains unchanged and resolutely proclaims the sovereignty of the states over water:

> *The Commonwealth shall not, by any law or regulation of trade or commerce, abridge the rights of the States or of the residents therein to the reasonable use of waters of rivers from conservation or irrigation.*

As would happen, the outcome of the federal poll had the potential to significantly modify water policy in Australia. The incumbent conservative government, led by then Prime Minister John Howard, was convincingly defeated making way for a new Labour government.

This casts some doubt over the future direction of the ambitious *National Plan for Water Security* (*the Plan*), which was hurriedly announced by Prime Minister Howard in late January 2007. *The Plan* was to have seen the Federal government extend its control over water resource management, particularly in the Murray-Darling Basin, and invest significant sums of public money ($A10 billion) in the process.

On the other hand, the changing political landscape at the federal level may well deliver no discernable impacts on water policy. Proponents of this school of thought could point to past struggles to modify even the most trivial parts of water legislation and attendant regulations. Moreover, the most recent announcements by the Labour government would suggest that, even with a substantial majority, radical alterations to the policy setting are unlikely. This policy inertia derives from the political minefield that characterizes efforts to reallocate water in a mature water economy. In the Australian case, some of these forces exert strong influence to stave off reform generally, whilst others drive policy in opposing directions, thereby limiting progress.

Critical forces included in this context emanate from six main areas. First, there is a long-standing vested interest from irrigated farming to maintain the status quo. Accompanied by a wider decline in the competitive advantage of agriculture, this motivation has seen the agricultural lobby desperately mobilize political resources to cushion its constituents from the impacts of any reallocation of the resource. Second, there is evidence of growing environmental ambitions amongst the general community, heightened by mounting concern about the potentially deleterious impacts of climate change (see, for instance, CSIRO, 2005). Third, social observers involved in water policy bemoan the emerging social and economic divide between metropolitan and rural Australia. This group is prone to casting water reform as another genre of 'economic rationalism' destined to exacerbate social discord. Fourth, over the past two decades there has been a fixation by governments with fiscal conservatism. This has resulted in a determination to invoke (or be seen to invoke) versions of the user pay principles, which often result in perverse incentives and outcomes (see Dwyer, 2006). The fascination with 'user pay' has also provided fuel for agricultural lobbyists who continue to call for more holistic measures to account for the impact of (quite often small) reallocation decisions (Benson, 2004, px). Fifth, there has been an undisputed and sometimes unquestioning enthusiasm for the use of markets as a vehicle for allocating resources. At times, this has occurred without the necessary groundwork to make the most of market-based reforms (Brennan, 2006). Arguably, there has been limited analysis outside the neoclassical economic tradition, say in the area of property rights and transaction costs, which, were they considered more fully, may have delivered alternative solutions (Pagan, 2007). Finally, and almost ironically, there has been a resurgence of interest in engineering solutions to address water shortages. This movement is clearly at odds with some of the aforementioned influences, but has nevertheless captured attention in policy circles.

Notwithstanding the vibrant contemporary backdrop that these forces provide for policy analysis, any understanding of water policy formulation in Australia requires consideration of path dependencies and history. It is the purpose of this chapter to sketch the various suasive forces that have historically driven water policy and review the current suite of responses on offer. The chapter is purposely broad, since a detailed examination of sectoral policies, jurisdictional differences and community responses is not feasible within the allowable space. The chapter itself is organized into four parts. In the next section a synoptic overview is provided of the historical, legal and institutional influences over water policy, to give substance to the aforementioned spheres of influence. The subsequent section is then used to articulate the most recent policy episodes, with a particular emphasis on national policy trends. A cautious prediction of future directions is provided in the following section before brief concluding remarks are offered in the final section.

## An overview of the influences in the Australian water policy debate

Australia is frequently cited as being the driest inhabited continent on earth (see, for instance, National Archives of Australia, n.d.). In reality, Australia's water resources would be more accurately described as highly variable in both spatial and temporal terms (Letcher and Powell, 2008, p17). Moreover, many have pointed to the fact that Australia actually has relatively abundant water resources when considered on a per capita basis. For instance, UNESCO (2002) ranks Australia as having the 40th highest annual water availability per person, well in front of nations like Switzerland (77th), France (104th) and Spain (117th). Thus, the water resource problem in Australia is not so much about the lack of water per se, but rather hinges on the mismatch between present use and the episodic and stochastic character of Australia's rainfall. To understand how a fundamental discord between use and resource capacity of this magnitude could emerge, it is necessary to briefly trace some of the important historical, legal and institutional influences over Australian water policy.

Indigenous Australians held values and undertook practices that acknowledged the variable character of Australia's water resources. In this regard, Australian Aborigines lived within the limits of water resources for at least 40,000 years (Magarey, 1894–1895). Indigenous production systems were thus intrinsically linked to ecosystems that were unique to the Australian climate. These systems had the capacity to expand rapidly during times of plenty and retained sufficient resilience to withstand long periods of drought. In the case of those indigenous communities occupying the Murray-Darling Basin, this meant that the river systems upon which they relied would frequently flood during spring before retreating to modest streams, which sometimes ceased to flow altogether in summer.

The arrival of European settlers radically transformed this balance in as little as 200 years. Colonial settlement commenced in the late 1700s but serious 'development' of water resources did not commence until the mid-1800s. Tasmania was the first colony to play an active part in irrigation development following the 1840–1843 drought. Other colonies followed a similar trend, seeking to shore up what were perceived as unreliable water supplies (Hallows and Thompson, 1998). However, shoring up Australian water resources for the purpose of irrigation is no easy feat. Smith (1998) observes that to achieve an equivalent level of water supply security an Australian storage needs to be twice as large as that of an average dam in the world and six times as large as those in Europe. Regardless of changes in technology, these basic parameters remain unchanged and add a significant cost to the maintenance of irrigated agriculture in Australia.

Notwithstanding the relative cost disadvantage of irrigated agriculture, all Australian colonies set about building the infrastructure required to promote irrigation during the late 19th and early 20th centuries. Musgrave (2008) describes this first phase of irrigation development in detail but notes, in particular, the influence of Alfred Deakin and the Chaffey Brothers in Victoria. Both Deakin and the Chaffeys were determined to replicate the social and economic transformation they had witnessed in California. Put simply, there was a strong belief that bringing irrigation to the inland of Australia would be the vehicle for raising land values and establishing a capable yeomanry willing to raise families in the hinterland – this was the stuff of nation building!

Arguably, this original conceptualization of irrigation (and agriculture generally) has proved more enduring than one might think. High value agriculture is still described in policy circles as one typified by orchards and vineyards (Brennan, 2007) rather than one which is in tune with the climatic and economic realities of Australia in the 21st century.

The development of irrigation also brought with it important institutional and legal considerations that pervade current policy development. First, each colony (and then the states) developed largely autonomous administrations, although some general consistencies emerged. Amongst these was the general abandonment of the riparian doctrine and its replacement with state control accompanied by rights to use water under licence. The state also became an active participant in allocation and use decisions in most jurisdictions, partly as a consequence of the policy to absolve the debts of irrigation trusts using general revenue and partly as a result of the establishment of water bureaucracies that took it upon themselves to continue this national building exercise. Enthusiasm for irrigation development as an expression of nation building continued for most of the 20th century and it was not until the 1960s that policy makers and the community more generally began to question the benefits of closer settlement, driven in the first instance by increased concern about the fiscal impost of such endeavours (Musgrave, 2008, p40).

Outside of irrigation, additional powerful bureaucracies had emerged in the context of hydroelectricity. This occurred primarily in Tasmania and in the Snowy Mountains area, which straddles New South Wales (NSW) and Victoria. The mantra of these bureaucracies was peculiarly similar to that associated with irrigation – the engineering and administrative capability of the state was to be used to 'harness nature' and build prosperity.

The experience of the Snowy also demonstrated the capacity of the Commonwealth government to bring competing states' interests into line. Following the Second World War both NSW and Victoria had offered competing proposals for

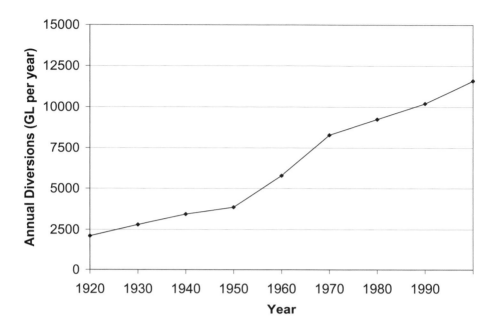

*Source:* State of the Environment Advisory Council (1996)

**Figure 6.1** *Diversions for consumptive uses in the Murray-Darling Basin*

the development of the Snowy and it was only as a result of intervention by the Commonwealth that a compromise was reached. This continued a tradition that dated back to the early 1900s when Commonwealth influence assisted in the formulation of the Murray-Darling Basin Agreement, which sought to resolve inter state conflict over water sharing in 1914.

In sum, by the end of the 1960s, 'development' was only beginning to be questioned at all levels of government, and this was principally on financial grounds. Fiscal concerns were joined by another powerful force in the subsequent two decades, as the level of extraction from Australia's river systems began to manifest in concerns about the environment. This was particularly the case in the Murray-Darling Basin where enthusiasm for water resource development had been most pronounced (Figure 6.1).

Thus, by the 1980s, water policy makers faced several pressing challenges. First, they confronted a farming population that had been encouraged by decades of state policy to expand water use in the belief that this would bring prosperity to inland communities and the nation generally. Arguably, this had bred a form of agrarian fundamentalism that extended beyond agricultural communities – many metropolitan constituents remained convinced of the nobility of agriculture and struggled to conceptualize the finite nature and fragility of water resources. Second, the economic rationale for irrigation and state sponsorship thereof was increasingly difficult to maintain. There existed a general thrust towards smaller government and this was clearly inconsistent with continued subsidization and control of agricultural activities. Third, a significant portion of the community was becoming aware of the environmental ills that beset the most developed of Australia's river basins. Prediction of increased salinity (MDBC, 1998) and loss of biodiversity were accentuated by newsreels showing large tracts of the Darling River filled with blue-green algae. Fourth, a powerful water bureaucracy with an emphasis on engineering achievement would need to be transformed in order to deal with the problems of the later half of the 20th century.

## Recent reform episodes

### The Murray-Darling Basin cap

The role of the Commonwealth government in achieving consensus in the context of water development was briefly described in the previous section. However, the prominence of the Federal government in water affairs has grown markedly over the past 25 years, particularly in the context of environmental demands. In 1978 the Commonwealth passed the National Water Resources Financial Assistance Act. The Act had arisen from the need to fund water conservation and salinity mitigation measures in the Murray-Darling Basin. The trend for the Commonwealth to involve itself in water policy in response to environmental issues was again evidenced in 1995. The Murray-Darling Basin Ministerial Council, comprising representation from the Commonwealth and state jurisdictions with an interest in the Murray-Darling, invoked a 'cap' on water extractions from the Basin in 1995. Initially an interim measure, the cap was moved to a permanent footing for NSW, Victoria and South Australia (SA) in 1997. The cap aims to limit extractive use to the level that attended 1993–1994 development (MDBC, 1998). Notwithstanding the challenges experienced in monitoring and maintaining the cap in all jurisdictions (see, for instance, MDBC IAG, 2005), the cap represented one of the most overt policy reforms aimed at addressing the environmental status of rivers and markedly contrasts with the approach of earlier decades. Other significant legislation to support the cap had been implemented, or was being considered, at the state level at about this time. This included the move towards the specification of licences in volumetric terms and the introduction of transferable water entitlements.

### The Council of Australian Governments

The next phase of reform in which the Commonwealth played a major part occurred under the auspices of the Council of Australian Governments' (CoAG) micro-economic reforms. CoAG gained prominence as part of the national agenda to enhance competitiveness within the Australian economy. In broad terms, the sale of a number of Commonwealth assets created a pool of funds that was used to convince state governments of the benefits of reform. Put simply, if states did not meet the targets set as part of a competition policy framework, they would not share in the spoils of privatization.

In the context of water, two major phases of CoAG reform occurred – one in 1994 and a second in 2004. The original Water Reform Framework comprised five main elements covering:

1 the introduction of pricing practices aimed at recovering costs, being consumption based and removing (or at least making overt) cross-subsidies;
2 the development and implementation of a system of volumetric and tradeable water allocations that were separable from land and which recognized the needs of the environment;
3 the separation of regulation, water service delivery and resource management functions;
4 two-part tariffs were adopted for urban water users, where practicable;
5 all future investments in water infrastructure were to meet both economic and environmental sustainability criteria.

The precise definitions to be applied to these criteria were not provided and in the context of discussion pertaining to current policy tends this is worth noting. It is also worth noting that much of the original enthusiasm for using markets was relatively myopic and was not always accompanied by careful specification of property rights or adequate investments in monitoring.

Somewhat disgruntled by the perceived slow progress of the early CoAG reforms and in response to failings within the design of the Water Reform Framework, the Commonwealth pressed for additional reforms in 2004. These occurred under the guise of the National Water Initiative (NWI) and set about creating yet another water bureaucracy – the National Water Commission. Importantly, the NWI sought to address several of the property rights deficiencies that became apparent with the stimulation of trade under earlier reforms. This amounted to insisting that water access rights should be specified as a perpetual share of the consumptive pool of water resources. A bifurcation between the consumptive and environmental/public uses of water was also introduced, with the latter given statutory recognition with superior claims to consumptive demands.

In addition to specifying water rights in this manner, the NWI also endeavoured to assign more clearly the risks between the parties to a water access right. More specifically, any reduction in a share of the consumptive pool due to climate change, drought, bushfires or improvements in bona fide knowledge about the capacity of systems to sustain extractions was to be borne by water users. The latter is limited until 2014, at which point a proportion of this risk moves to the different levels of government. Risks of reductions in access arising from government policy are to be carried by government. Quiggin (2008, p71) observes that whilst this approach may prima facie have some appeal, it is likely to be beset by several problems. These range from the short-term difficulty of reaching consensus amongst the state jurisdictions to medium- and longer-term problems related to distinguishing between the different forms of risk and the related issue of returning over-allocated systems to a more sustainable footing.

## The National Plan for Water Security

The latter of these problems received particular attention in 2007 with another foray by the Commonwealth into water policy – The National Plan for Water Security (the Plan). As noted in the introduction, the Plan was both hurriedly prepared and ambitious, perhaps reflecting the political desperation of a government who had been 'on watch' for over a decade and in whom the public was beginning to express its discontent. As Watson (2007, p1) noted, the authors of the Plan were 'not claiming spurious accuracy for their major proposals. As subsequently emerged, the ten-point Plan to spend $A10 billion over ten years was prepared in haste, well away from the troublesome gaze of Treasury and Finance officials and the experienced eye of the Murray-Darling Basin Commission.'

There were two main elements to the Plan that deserve special mention in the context of this chapter. First, the largest portion of the funding ($A6 billion) was assigned to elaborate engineering solutions to enhance irrigated agriculture. This 'modernization' of irrigation was claimed to deliver 'water savings' that could then be used to underpin environmental sustainability. By way of contrast, a mere $A3 billion was earmarked for buying back water from farmers in over-allocated systems. Second, the Plan foresaw an increased role of the Commonwealth in the administration of water generally and the Murray-Darling Basin in particular.

## Water for the Future

At the time of writing the policy approach to be pursued by the national Labour government is only beginning to be formalized. In April 2008 the Minister for Climate Change and Water released a broad outline of the government's water policy in the form of 'Water for the Future'. This document generally mirrors the Howard government's approach inasmuch as non-trivial public funds have been earmarked for the purpose of 'modernizing irrigation' whilst a lesser but significant emphasis has been placed on restoring balance by buying back water access rights. In the context of the former of these policy thrusts, the federal government has undertaken to co-sponsor the renovation of irrigation infrastructure in Victoria to the tune of about $A1 billion. This generosity was largely driven by Victoria's resistance to earlier calls by the Howard government to cede its powers over water resources to the Commonwealth. There seems little doubt that other state jurisdictions will be queuing up for similar assistance based on the naive view that such measures will generate fungible 'water savings'.

## Lessons and challenges

The upshot of the NWI, the Plan and Water for the Future provides some salient lessons in water policy formulation in a mature water economy where path dependencies arising from previous policies constrain the choices of the present and today's choices apply caveats to future options. Even a cursory reading of the historical context provided in the second section of this chapter might leave readers with the impression that the current national policies broadly comprise a roadmap for a trip 'back to the future'. The deficiencies of exuberant investments in irrigation that occurred between the mid-1800s and the later 1900s seem likely to be revisited if the thrust of the Plan and Water for the Future remains unaltered by the new Federal government. Moreover, the expansion of the water bureaucracy into farm-level decisions is reminiscent of the intrusive regimes of past eras, when engineers knew best and governments were driven by a sense of nation building.

On a more promising note, the introduction of buy-back (albeit a more modest financial component of the Plan) as part of the national water agenda represents a refreshing departure from the usual rhetoric, which proclaims the sovereign right of irrigation communities to 'their water'. Buying back water from willing sellers has regrettably been demonized by both agricultural lobbyists and politicians who lack the capacity or will to explain the costs of alternatives. Buy-back is widely acknowledged as the cheapest and most feasible mechanism for dealing with over-allocation problems (Quiggin, 2006; Crase et al, 2007; Watson, 2007). And yet the debate frequently degenerates into discussion on pecuniary externalities, in the form of stranded irrigation assets for example.

The stranded assets argument hinges on the notion that the withdrawal of water from some irrigation uses or the exit of irrigators from the industry will leave those remaining in the industry with an unreasonable financial burden due to the less-intense use of irrigation infrastructure. Regrettably, some of the earlier pricing reforms have combined to add weight to this argument. First, by shifting much of the cost of infrastructure on to the volumetric component of water tariffs, any movement of water rights to other locations now disproportionately impacts on the water tariff faced by those in a communal irrigation network. Second, by deliberately breaking the nexus between water and land, such that the latter is no longer charged for the benefit bestowed by water infrastructure, remaining water users will carry more of the burden than would otherwise be the case. Third, the proclivity of governments to adopt charging regimes that include a rate of return on sunk investments, many of which were previously fully funded by taxpayers, results in higher than optimal volumetric water charges. The latter of these perverse outcomes is not limited to irrigation users and also applies to urban water customers (Dwyer, 2006).

The persuasiveness of the stranded assets argument and the accompanying hysteria about water leaving agricultural districts undoubtedly explains the return to favour of engineering solutions in policy circles. After all, renovating irrigation districts and subsidizing on-farm capital investments is hardly likely to draw criticism from the agricultural sector. In addition, and as was noted earlier, the iconic status of agricultural pursuits also continues to resonate with much of the electorate, regardless of the abandonment of closer settlement schemes several decades ago. Importantly, the costs of these investments can also be more easily disguised than market purchases of water. This is assisted by the proclivity of the press to resort to quantifying mystical 'water savings' with crude metrics such as swimming pools, the number of hot showers or the like (Stanhope, 2007).

# A cautious look to the future of Australian water policy

So, what is the future direction of Australian water policy? On the one hand the CoAG agenda and subsequent Plan and Water for the Future pointed to an expanded role for market mechanisms and the use of price and markets to exert influence over allocation decisions. On the other hand, the state appears reluctant to withdraw from investments in irrigation infrastructure, which under conventional economic scrutiny would likely not pass muster. This schizophrenic policy stance is also evident in matters relating to urban water policy. Here, higher prices are often advocated by economists as a means of bringing dwindling supplies into balance with demand (Grafton and Kompas, 2007), although water bureaucrats simultaneously favour restriction regimes that dictate the desirability of one household or industrial use over another.

## Inter-sectoral considerations

Core to understanding the future direction of water policy is the hydrological fundamentals described earlier. These imperatives are clearly at odds with the current distribution of water resources in this nation. The present water allocation is also at odds with the current relative economic might of sectors that ultimately compete for the resource. To illustrate this point, data from the Australian Bureau of Statistics is provided in Figure 6.2.

Figure 6.2a shows that agriculture remains the most expansive user of Australia's water resources, standing at about two-thirds of all water use nationally. Figure 6.2b reveals the relative economic contribution of different sectors within the Australian economy. Here, agriculture plays a relatively minor part and changes in the terms of trade for this sector point to a continued contraction on this front. Whilst similar trends may be evident elsewhere in the world, there are few nations with the hydrological character of Australia. In addition, many of Australia's agricultural systems are predicted to be severely impacted by anthropogenic climate change (CSIRO, 2005).

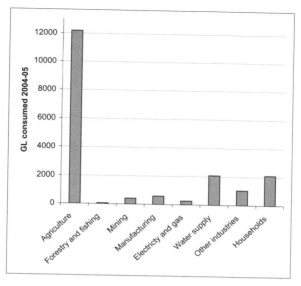

*Note:* 1 GL = 1 million cubic metres.
*Source:* Australian Bureau of Statistics Water Account 2004–2005

**Figure 6.2a** *Water consumption by sector (GL per year)*

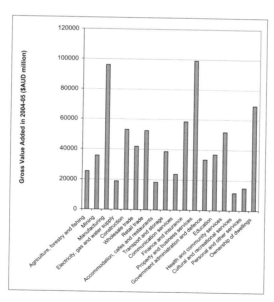

*Note:* $1 AUD = $0.74 to 0.79 US for 2004–2005.
*Source:* Australian Bureau of Statistics National Accounts 2004–2005

**Figure 6.2b** *Industry gross value added ($AUD million)*

Against this background it is difficult to see how irrigated agriculture can persist in its current form in Australia, at least in the longer term. Notwithstanding that the enduring affinity of urban voters with their rural cousins will result in political compromises that see public funds expended on 'adjustment programmes' and (arguably futile) attempts to reinvigorate irrigation districts, the water shortage itself seems likely to force change. In this context it is worth noting that runoff in the southern portion of the Murray-Darling Basin is predicted to decline by at least 15 per cent by 2030 as a result of climate change. Similarly, the capacity of rainfall to supply urban water in Western Australian is falling, with a 50 per cent reduction in reservoir flows observed between the 1970s and early in the 2000s (Pittock, 2003). Currently, almost all major metropolitan centres face some form of water rationing and in the case of a number of these (Adelaide, Melbourne, Canberra and perhaps to a lesser extent Perth) this could be easily alleviated by the transfer of modest amounts of water from agriculture (Quiggin, 2006).

In the interim, urban water users are beseeched to 'save every last drop' (Watson, 2006). To date this argument has been generally well received, not for the veracity of the cause itself, but for two other reasons. First, the Australian population remains broadly accepting of the arguments pertaining to the sanctity of agriculture. More generally this attitude has arisen from more than a century of water policy that saw the expansion of irrigation and sponsorship of agriculture as noble expressions of nation building. Second, the fixation with Australia's perceived water scarcity persists, regardless of the facts and with little regard to the costs carried by most Australians as a consequence of the current resource allocation.

In the minds of some (the author included) policy change occurs only when the transaction costs of change are surpassed by the costs of tolerating the status quo (Saleth and Dinar, 2004). Clearly, the costs of the status quo are on the rise in the context of the present setting of Australian water policy.

## Broader trends in water policy

Rather than endeavour to predict the future with precision, it is worth considering some of the emerging trends to deal with these dilemmas. In this instance, the focus is on initiatives that go beyond the standard political responses described earlier (i.e. markets for some and not others, public intervention to direct behaviour and state investment in various forms of infrastructure). Moreover, in some cases these initiatives arise from the actions of individuals in an effort to subvert the intentions of the state.

First it is worth turning attention to the urban communities that presently confront restriction regimes, in part as a consequence of the Balkanization of rural and urban water users (Freebairn, 2005). In this regard, legislators have already become aware of the deleterious social consequences of extended restriction regimes. For instance, Cooper (2007) describes the symptoms of water restriction fatigue and the link between persistent urban water restrictions and civil unrest, some of which is being reported in the press (e.g. ABC, 2007). In this context imaginative solutions are already emerging. Some urban businesses have entered the market where irrigation allocations have been historically sold and purchased water to alleviate the costs of restrictions. Others have exploited the government subsidies that attend industry and household water-saving investments. For example, households can install elaborate recycling technologies and then expand indoor water use to maintain outdoor amenity without attracting penalties. Regrettably, the distributional consequences of these actions inevitably favour the rich over the poor. Other urban and industrial users have invested in groundwater extraction to avoid the gaze of government.

In the irrigation sector a similar trend has been evident. This has been driven by substantial gaps in water legislation that fail to acknowledge the conjunctive nature of the resource. Coupled with the limited capacity of the state to meter groundwater extraction (or its reluctance to adequately fund metering and enforcement), the progressive tightening of access to surface water supplies has shifted demand from surface water to groundwater. Regrettably, this leaves policy makers constantly playing 'catch up' as they endeavour to rein in excessive extractions of surface water and simultaneously monitor and control groundwater use.

Perhaps the most encouraging trend is the move towards reconciling the needs between competing demands via a range of imaginative market instruments. More specifically, options contracts between various users are only now emerging and demonstrate considerable promise. On the environmental front, option contracts are presently in place to bolster environmental flow regimes in the Murrumbidgee Valley (Hafi et al, 2005) with the intent of increasing the environmental gain from what would otherwise be a modest flood event.[1] In situations where urban communities are already physically connected to irrigation communities, market instruments such as options contracts also offer a remedy to a particularly thorny problem – the movement of water resources from agriculture to urban users. As has already been shown elsewhere (e.g. Michelsen and Young, 1993) options contracts avoid the politically untenable issue of agricultural interests having to forgo water access rights to accommodate the needs of urban users. This is achieved by guaranteeing ongoing financial support to agricultural producers in return for a right to exercise an option over water access rights in times of acute scarcity. Not only do such market innovations offer considerable promise, inasmuch as they potentially defray the political costs of change, they provide a mechanism for more accurately balancing the climatic, ecological and economic reality of the Australian context within the growing demand for water resources.

Without wishing to overstate the benefits of this approach, it is worth reflecting on the nature of Australia's water resources as described in earlier sections of this chapter. Much of the agricultural activity that is euphemistically described as 'high value' in this country mimics the high-value production of European nations. And yet, as we have already established, Australian hydrology is not like other nations and it is spurious to conceptualize agricultural value using the same framework. Rather, what is required is a framework that acknowledges the extant constraints and one that takes account of likely future conditions, in the form of the predictions of climate change, for example.

## Redefining high-value water use

When these variables are taken into account, the future distribution of Australia's water resources takes on a new dimension. Agriculture, which currently dominates the use of water resources, does not pale into obscurity per se. Rather, the agricultural systems that use much of Australia's water and typify conventional thinking as 'high value' and 'sustainable' begin to be reshaped. In this light an agricultural system that exploits water when it is plentiful and gains resilience during drought, say by transferring water to urban and industrial users, becomes the dominant regime. Obviously, there are many social and political obstacles to be overcome before this is realized. For example, in Victoria there is an enduring perception that perennial horticulture backed by a conservative allocation regime is the key to future agricultural prosperity in the state. This is partly promoted by interstate rivalries and the proclivity of NSW agriculture to be dominated by annual cropping. And yet simultaneously, the Victorian government is investing large sums of public money to connect agricultural and metropolitan water users via what is euphemistically referred to as the 'north–south' pipeline.

At the time of writing, the benefits to be bestowed on Victoria's agricultural water users in the Goulburn Valley (i.e. in the north) are in the form of increased public investment in irrigation infrastructure. Regrettably, these investments would appear to be at odds with the necessity for water to be available to metropolitan Melbourne during dry years and the corollary that agricultural activities should have the capacity to contract at that time. In addition, such investments would appear to ignore the predictions of climate change inasmuch as 'more efficient' irrigation infrastructure counts for little when stream flows fall by 15 per cent and urban users bid water away. Alternatively, agricultural systems geared to exploit the benefits of a relatively plentiful year and with the capacity to contract as water becomes scarce would be better synchronized to these realities. Put simply, it is difficult to see how the present configuration of high-value agriculture will manage to balance current constraints, sectoral demands and future trends in the longer term.

## Conclusion and policy implications

To appreciate Australian water policy in 2008, it is necessary to have some understanding of the innate hydrology that attends the continent and the history of resource policy in this country. This chapter has endeavoured to highlight significant forces on both these fronts. First, Australian hydrology is variable, both spatially and temporally. This has led to the misguided perception that Australia is a dry continent. In fact, Australia has ample water resources for most human needs but the historical allocation of the resource to achieve social and economic ends has resulted in a difficult policy landscape. Second, the determination to manufacture a European agricultural outcome in a non-European landscape has left a legacy of institutions, beliefs and infrastructure that are costly to maintain and troublesome to change.

Nevertheless, change has occurred and seems destined to continue. Many of the significant policy modifications began towards the latter part of the 20th century and, unlike earlier episodes, the Commonwealth played an expanded role relative to the states. Initially spurred by concerns about the financial viability of irrigation and later accompanied by apprehension about the environmental impacts of excessive extractions, policy makers sought to rein in water use, and simultaneously make users more accountable for their actions. Some of these reforms achieved progress but it would be premature to proclaim the job as complete. Policy failures have emerged and to some extent these have been addressed by subsequent reforms. However, at a national level at least, policy schizophrenia has increasingly become the norm in recent times.

In this chapter I have rationalized this response as a balance between the costs of policy change and the costs that attend the status quo. In addition, I have argued that the future environment is unlikely to resemble the present, particularly given the predictions of climate change and the evident economic and social trends. Importantly, policy makers need to take care when formulating policy in this environment and resist the temptation to 'pick winners' on the basis of values formed in a different time.

## Note

1   Hillman in this volume provides a compelling explanation of the necessity for such flow regimes.

## References

ABC. (2007) Man charged with murder after lawn watering dispute. Online. Available from: www.abc.net.au/news/stories/2007/11/01/2078076.htm?section=justin (accessed 25 November 2007)

Benson, L. (2004) *The Science Behind the Living Murray: Part 2*, Murray Irrigation Limited, Deniliquin

Brennan, D. (2006) 'Current and future demand for irrigation water in Western Australia', *Resource Management Technical Report*, Department of Agriculture and Food, Perth, Western Australia

Brennan, D. (2007) 'Water policy reform in Australia: Lessons from the Victorian seasonal water market', *Australian Journal of Agricultural and Resource Economics*, vol 50, no 4, pp403–423

Cooper, B. (2007) 'Water restrictions in Southern Australia', Unpublished paper presented at Water Pricing and Equity Conference, La Trobe University, Wodonga

Crase, L., J. Byrnes and B. Dollery (2007) 'The political economy of urban–rural water trade', *Public Policy*, vol 2, no 2, pp130–140

CSIRO (2005) 'Climate change is real', *CSIRO Fact Sheet*, CSIRO, Canberra

Dwyer, T. (2006) 'Urban water policy: In need of economics', *Agenda*, vol 13, no 1, pp3–16

Freebairn, J. (2005) 'Early days with water markets', Unpublished paper presented at Industry Economic Conference, Melbourne, September

Grafton, Q. and T. Kompas (2007) 'Pricing Sydney water', *Australian Journal of Agricultural and Resource Economics*, vol 51, no 3, pp227–242

Hafi, A., S. Beare, A. Heaney and S. Page (2005) 'Derivative options for environmental water', 8th Annual AARES Symposium: Markets for Water – Prospects for WA, Office of Water Strategy, CSIRO, Swan River Trust, WA Water Corporation, Perth

Hallows, P. and D. Thompson (1998) *The History of Irrigation in Australia*, Australian National Committee on Irrigation and Drainage, Mildura

Letcher, R. and S. Powell (2008) 'The hyrdological setting', in L. Crase (ed.), *Water Policy in Australia: The Impact of Change and Uncertainty*, RFF Press, Washington, DC, pp17–27

Magarey, A. (1894–1895) 'Aboriginal water quest', *Proceedings of the Royal Geographical Society of Australasia, South Australian Branch*, pp1–15

Michelsen, A. and R. Young (1993) 'Optioning agricultural water rights for urban water supplies during drought', *American Journal of Agricultural Economics*, vol 75, pp1010–1020

Murray-Darling Basin Commission (MDBC) (1998) *Managing the Water Resources of the Murray-Darling Basin*, Murray-Darling Basin Commission, Canberra

Murray Darling Basin Commission IAG (2005) *Reports on CAP Implementation*. Online. Available from: www.mdbc.gov.au/nrm (accessed 25 November 2006)

Musgrave, W. (2008) 'Historical "development" of water resources in Australia: Irrigation in the Murray-Darling Basin', in. L. Crase (ed.), *Water Policy in Australia: The Impact of Change and Uncertainty*, RFF Press, Washington, DC, pp28–43

National Archives of Australia (NAA) (nd) 'A sunburnt country?' www.naa.gov.au/learning/schools/online/just-add-water/a-sunburnt-country.aspx (accessed on 15 April 2008)

Pagan, P. (2007) 'Evaluation of institutions for interstate water trading involving the ACT', Master of Philosophy Thesis, CRESS, Australian National University, Canberra

Pittock, B. (2003) *Climate Change – An Australian Guide to the Science and Potential Impacts*, Australian Department of Environment and Water Resources, Australian Greenhouse Office, Canberra

Quiggin, J. (2006) 'Urban water supply in Australia: The option of diverting water from irrigation', *Public Policy*, vol 1, no 1, pp14–22

Quiggin, J. (2008) 'Uncertainty, risk and water management', in L. Crase (ed.), *Water Policy in Australia: The Impact of Change and Uncertainty*, RFF Press, Washington, DC, pp61–73

Saleth, R. M. and A. Dinar (2004) *The Institutional Economics of Water: A Cross-country Analysis of Institutions and Performance*, Edward Elgar, Cheltenham and the World Bank

Smith, D. (1998) *Water in Australia: Resources and Management*, Oxford University Press, Melbourne

Stanhope, J. (2007) 'Government doing its part to reduce water use', Press release by the Chief Minister, 25 October, ACT Government, Canberra

UNESCO (2002) 'Water availability per person per year', www.unesco.org/bpi/wwdr/WWDR_chart1_eng.pdf (accessed 15 October 2007)

Watson, A. (2006) 'Contemporary issues in water policy', *Connections – Farm, Food and Resources Issues*, vol 6, no 1

Watson, A. (2007) 'A national plan for water security: Pluses and minuses', *Connections – Farm, Food and Resources Issues*, vol 7, no 1

# 7

# The Policy Challenge of Matching Environmental Water to Ecological Need

*Terry Hillman*

In recent years there has been almost universal acceptance of the principle that, if we wish to ensure the long-term utility of our water resources, we need to support the ecosystem that sustains those resources. We also generally acknowledge that these ecosystems are threatened (to varying degrees) by the consumptive use of water and the management regimes and infrastructure established to support that use. A balance needs to be struck that supports the human use of water but manages the risk to the ecosystem. The risk can be reduced through changes to infrastructure, the pattern of demand, and the return of some water from human use to sustaining the ecosystem. This chapter concentrates mainly on the last of these.

Extensive understanding of the multiple links between river flow and the ecosystem would permit a seamless and optimally efficient link between the provision of water for environmental purposes ('environmental water') and supplying the needs of the ecosystem. This knowledge is not currently available, although many of the general principles are understood. As a consequence it is necessary to develop management responses that deal with water allocation and environmental application at a range of spatial and temporal scales that accommodates both the heterogeneous needs of the ecosystem and the requirement for reliability and predictability of the human water resource.

This chapter uses examples from the Murray-Darling Basin to examine the links between a range of management responses and the ecological needs they seek to satisfy. The Murray-Darling Basin occupies 14 per cent of Australia's land mass and supports about 40 per cent of its rural production, including the majority of its irrigated agriculture. Detailed descriptions of the system are provided elsewhere by Walker (1986) and Crabb (1997). The Basin lies in four state jurisdictions, New South Wales (NSW), Victoria, South Australia (SA) and Queensland plus the Australian Capital Territory. It is also the concern of the Australian (Federal) Government so that its management (and exploitation) is in the hands of six separate governments. Whilst there appears to be general agreement that the Basin's water resources are over-allocated, with a consequent risk to its long-term health and security, it is likely that an array of management solutions will be developed by the various jurisdictions. This creates special difficulties for the management of a return to more balanced conditions and the development of policy that will ensure that. Flexibility will need to be an important ingredient of the policy response. But flexibility is not encouraged where the management of a single shared resource is in the hands of state governments that may see their primary responsibility to be in ensuring that their constituent water users gain maximum access to the resource.

## Ecological needs

'How much water does the ecosystem need?' was a common question in early debates about environmental water. Because the river ecosystem evolved in the absence of human intervention, it is adapted to using all of the water. Whilst maintaining a wild river in pristine condition may need all of the water, this is neither a helpful answer nor a reasonable aim. The task is to manage a working river: rehabilitation rather than total restoration. Thus, the ecological needs we have to account for are those that sustain the system indefinitely – and, in the present context, that subset influenced by human river management. Aquatic organisms have, by definition, to be in water for some or all of their lives. Water is their natural medium as air is ours. The ecological role of water is much more complex than that, however. The ecological needs supplied by aspects of hydrology can be grouped as follows:

- *Habitat extent and complexity.* The extent of water defines the extent of habitat. As well as providing more space, an increase in volume makes available additional habitat such as snags (coarse woody debris, e.g. fallen trees), instream benches and plant beds. This process also tends to increase the heterogeneity of available habitat as it incorporates complex in-channel features, backwaters and so on. Increased habitat complexity supports greater biodiversity and provides refuge from predation (Balcombe and Closs, 1996).
- *Habitat quality.* The biota of much of the Murray-Darling Basin is adapted to a wide range of water quality. Aspects relating to flow include occasional incursions of highly saline groundwater during periods of low flow and short-term inputs of low oxygen/high organic load water following floods. Many of the native species are able to tolerate short periods of high salinity but, in many cases, some life-stages such as eggs or immature individuals are intolerant.

- *Transport.* Downstream flow is the main means of distributing nutrients in rivers. Water returning from the floodplain as over-bank flows decline also carries terrestrial litter (an important long-term food supply for macroinvertebrates) and an array of food items, from dissolved organic carbon to small organisms produced in floodplain wetlands, back to the river. Flow is also a dispersal mechanism for macroinvertebrates (drift) and some large-bodied fish species native to the Murray-Darling Basin have pelagic eggs and early larval stages that also disperse with flowing water.
- *Connectivity.* Increased flow volume provides connection between otherwise isolated components of the riverine ecosystem. An example is the periodic connection between floodplain wetlands and the main river channel which allow the refreshing (and recharging) of the floodplain bodies and temporary access to their high productivity for the channel system (Boon et al, 1990). Also, as part of their reproductive cycle, a number of native fish migrate upstream (either seasonally or on other signals) (Humphries et al, 1999). This movement is inhibited by channel blockages, insufficient depth of water between instream pools, or insufficient density of snags to provide shelter during migration.
- *Signals.* Cues for a number of important ecological processes are often drawn from flow events. This is particularly important in the southern two-thirds of Australia, including the Murray-Darling Basin, where rainfall events are unpredictable with very large inter-annual variation. Some important native fish species are cued to migrate and breed by changes in flow rather than by seasonal signals (Briggs, 1990). Others tend to breed seasonally but only complete the process successfully under appropriate flow conditions. Successful recruitment amongst many water bird species is also linked to hydrological signals. These birds are stimulated to nest and breed in flooded wetlands and/or will abandon their nests before the young are successfully fledged if the water level recedes too quickly (Crome, 1988).

For all of these important ecological processes the seasonal cycle is replaced partially or entirely by a facultative response to information embedded in the rivers' flow patterns.

## The human footprint

A number of flow-modifying factors resulting from human activities in the Murray-Darling Basin (and common to most highly managed river systems) can be identified that, whilst aimed at managing the river for human benefit, incidentally damage its ecological function – in modern parlance 'collateral damage'. These are examined more fully elsewhere (Close, 1990; Hillman, 2007) and can be summarized here as follows:

- *Total volume change.* The volume of flow through the system is modified in two ways: inter-valley transfer and abstraction. Inter-valley transfer describes the redirection of water from streams outside the Murray-Darling Basin to add to its total water resource. Most notable is the Snowy Mountain Scheme in which water is redirected from east-flowing rivers into the upper reaches of the Murray and Murrumbidgee (with the substantial generation of hydro-electricity). Abstraction refers to the removal of water from the system (for rural, urban and industrial use). In recent years this has resulted in a reduction of about 75–80 per cent in the amount of water that might otherwise have flowed from the Murray-Darling Basin system to the sea (Jones et al, 2001).
- *Physical barriers.* A large number of structures intersect flow in the Murray-Darling system, ranging in size from major storage dams in the upper catchments through large diversion and navigation structures to small block dams that maintain head or pool levels for pumping. All break the upstream–downstream connection (unless over-topped by flow). In addition the body of non-flowing water held behind them represents a barrier to flow-adapted organisms and those depending on drift for dispersal. Barriers in the form of levee banks flank the lowland reaches of most of the larger rivers in the Basin, interfering to a varying extent with lateral connectivity between floodplain and main stream.
- *Seasonal flow inversion.* The primary object of flow management in the Murray-Darling Basin is to ensure the timely (and reliable) supply of water for irrigation. This results in a tendency for flows in the winter/spring to be captured for later release in summer/autumn, leading to an inversion of the normal seasonal flow pattern particularly in those catchments that derive the majority of their inputs from snow and winter rainfall (in terms of volume, the majority in the Murray-Darling Basin). The consequent seasonal shift is maintained in the reaches between the major water storages and reaches downstream from which major irrigation supplies are diverted (Close, 1990). Further downstream the 'natural' seasonal pattern is restored albeit with lower volumes of water.
- *Loss of flow classes.* For any river the long-term hydrograph can be divided into flow classes ranging from 'zero flow' to 'extreme flood' (e.g. high flow events with a return frequency of 1 in 50 years) and any number of intermediate classes (Puckridge et al, 1998). In regulated rivers the relative frequency of occurrence of these classes may be significantly changed or they may disappear completely (e.g. in the absence of regulation, the Murray may have ceased to flow on some occasions during the past century but management has ensured that at least some flow occurred at all times) (Close, 1990).
- *Change in frequency of significant flow events.* In line with changes in the temporal distribution and frequency of flow classes is the risk that flow events of particular ecological significance may become less frequent or less reliable. Hillman (2007) gives the example of an increased risk that, under extreme circumstances, the period between conditions suitable for successful water bird

recruitment in the Murrumbidgee River (tributary of the Murray) could exceed the life expectancy of the birds. Such an outcome is catastrophically different from a mere extension of the period between recruiting events – undesirable as that may be.

- *Reduced short-term variability.* Day-to-day variation in water level is a natural characteristic of rivers. Managing flow on the basis of downstream demand generally results in extended periods of little change in stage height, often interspersed with sudden and substantial changes. This pattern of flow can threaten bank stability and may disrupt the hydraulic balance between groundwater and the river. Short-term variability in water level is a contributing factor to maintaining productivity and biodiversity in littoral communities.
- *Depressed summer temperatures.* Water in large reservoirs tends to stratify during warmer months, with cold and anoxic water trapped at the bottom. Release of water via low-level offtakes at this time (the irrigation season) severely reduces temperature and oxygen levels downstream – and risks other pollution problems. Water downstream of several large dams in the Murray-Darling Basin is regularly too cold to support the successful breeding of native fish species.

## Recovery of water for the ecosystem

The negative effects of the human footprint on the ecosystem of the Murray-Darling Basin has been analysed in detail elsewhere (Jensen, 1998; Thoms et al, 1998; Kingsford, 2000). A significant number of these can be redressed, at least in part, by changes to management that do not involve the provision of more water (Hillman, 2007). However, an independent assessment of the Murray-Darling Basin, in light of international experience, suggested that 1500GL/year should be moved from human use back to the river system to avoid significant risk of continued environmental decline and collapse (Jones et al, 2001).

Since that time a number of schemes have been devised to recover water for the environment. There is no intention to analyse these schemes in depth other than to observe the links between the method of acquisition and the constraints that this places on deploying the water for ecological benefit. Water is made available for environmental use through four basic mechanisms: statutory regulation, acquisition of water use entitlements, changes to infrastructure and water management, and catchment management.

Statutory regulation refers to rules controlling the timing and quantity of water diversions from a stream. In Australia this is mostly the province of state governments – within overall collaborative management arrangements in the case of the Murray-Darling. It is most prominent in rivers without major storage facilities.

In the Murray-Darling Basin, acquisition of water use entitlements is a market-based process. The purchases may be funded directly or indirectly by governments (Federal or state) and not-for-profit organizations. Government-funded modification of infrastructure is also an important means of gaining water for the environment in the Murray-Darling Basin. To date these include:

- Efficiency gains in water distribution infrastructure. (Note, however, that additional water gained through preventing 'leakage' from the distribution system is acquired at the ultimate expense of some other (probably unaudited) component of the system such as groundwater or seepage-fed wetlands (Watson, 2008) and that there is a high risk that water recovered in this way will move to human use through further allocation or increased reliability of supply.)
- Regulation/restriction of water entering (and subsequently evaporating from) large wetlands artificially filled by high irrigation flows (and weir pools) in summer.
- Rationalization of artificially high flows in natural channels as a means of satisfying small, isolated consumptive demands.

To be an acceptable means of providing water for the environment, mechanisms must be in place to move the water from its point of recovery (either physically or, more usually, by substitution) in a way that avoids concomitant socio-economic or ecological damage – a triple-bottom-line approach.

Currently catchments are not managed with the express purpose of influencing water yield in the Murray-Darling, but commercial forestry and farm dams have recently been identified as potential threats to the shared water resource by the Murray-Darling Basin Commission and their future development is being monitored.

In part the mix of recovery methods is dependent on the degree to which a stream is regulated (the size of artificial water storage available in its catchment) and the extent and nature of the irrigation industry it supports. Importantly, water obtained through any of these mechanisms is likely to be available to contribute to environmental flows – that is, is not 'subverted' by the granting of additional diversion entitlements – because total diversions are 'capped' in the Murray-Darling Basin by intergovernmental agreement (see Crase in this volume). The task remains to apply that water to the maximum benefit of the Murray-Darling ecosystem.

## Meeting ecological needs

Given the acquisition of water for the environment, the task remains to use it to the greatest benefit of the ecosystem – perhaps to seek to achieve the same levels of efficiency in deploying environmental water that we would wish to see in agricultural applications. Sadly, our understanding of the complex relationships between the river and aquatic ecosystem is slight compared to the knowledge of crop–water relationships. Our current approach is a mixture of supporting identifiable and valued components of the ecosystem and attempting to supply general environmental support on the basis of first principles. The following discussion uses examples of water recovery programmes in the Murray-Darling Basin to assess possible approaches to determining the best uses of environmental water. In discussing these programmes it is important to note the extreme drought conditions that have prevailed in the Murray-Darling Basin over the past decade. Three of the six driest years on record have occurred in the current drought since 2002 (D. Dreverman, MDBC, pers. comm.). These conditions have seriously compromised the actual recovery of water, although not the development of policy and protocols.

Measures to obtain and supply environmental water fall into one of two categories that might be termed 'conservation of *hydraulic habitat*' and 'satisfying *identified needs*'.

Hydraulic habitat conservation is based on the so called Field-of-Dreams hypothesis – 'Build it and they will come'. It seeks to restore important elements of the flow pattern of a stream where their frequency has been reduced significantly, or timing modified, by diversion of water for human use. The aim is, without necessarily identifying all the beneficiaries, to provide hydraulic habitat of sufficient diversity, and in a seasonal pattern, able to support a functioning aquatic ecosystem appropriate to the stream. This is generally achieved through rules dictating flow conditions under which water may not be taken from the stream – a flow-dependent protocol for sharing the water resource.

This is the principle behind the development of Stream-flow Management Plans (SFMPs) in streams in Victoria including those in the Murray-Darling Basin. They are developed through applying the 'FLOWS method' (SKM, 2002), a detailed protocol for developing objectives and hydrological evaluations through a system of community consultation and expert panel assessment. For this discussion the salient feature of the FLOWS method is the review of (modelled) hydrographs describing historic (without diversions) flows and flows under current (or planned) levels of diversion. This comparison is assessed with a view to the preservation of a proportion of ecologically significant hydrological characteristics such as those summarized in Figure 7.1.

The Living Murray (TLM) programme (MDBC, 2007) is an example of the second approach to deploying environmental water: satisfying identified needs. TLM is a collaborative programme aimed at the rehabilitation of specific ecological components at six so-called 'icon sites' on the floodplain of the Murray and much of the main channel of the river. On 25 June 2004, First Ministers of the Murray-Darling Basin Ministerial Council from New South Wales, Victoria, South Australia, the Australian Capital Territory and the Australian Government signed the 'Intergovernmental Agreement on Addressing Water Overallocation and Achieving Environmental Objectives in the Murray-Darling Basin'. This

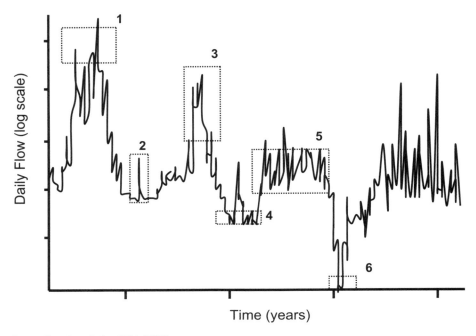

Source: Developed after SKM (2002)

1 *Over-bank flow* – recharge wetlands, connect mainstream and floodplain;
2 *Summer fresh* – support riparian vegetation, re-suspend fine sediment;
3 *Bank-full flow* – connect stream with backwaters and floodplain channels;
4 *Low flow* – trigger recruitment in some fish species;
5 *High flow* – trigger migration/reproduction in some fish species, maximize habitat complexity;
6 *Zero flow* – reset system, disadvantage some introduced species.

**Figure 7.1** *Hypothetical hydrograph with examples of ecologically significant hydrological characteristics and their role in river ecosystems*

made $A500 million available for the recovery of water by direct purchase of entitlements and 'works and measures' (infrastructure development) with the aim of making 500GL/year available for ecological purposes at the icon sites by 2009. An additional $A200 million was contributed by the Australian Government in 2006. TLM can be seen primarily as a pilot study concentrating on achieving some identified outcomes at sites for which a significant amount of data already exist. There has been wide consultation and considerable scientific effort has gone into assessing the icon sites, developing watering plans and, importantly, the design of monitoring programmes to quantify planned outcomes; all extensively reported elsewhere (MDBC, 2007). As of July 2007 approximately 370GL of water had been identified for recovery. The important features of TLM in terms of this chapter are:

# The Policy Challenge of Matching Environmental Water to Ecological Need 117

- environmental water is targeted at specific ecosystem components (at the various sites these include floodplain woodlands, waterbirds and native fish);
- specific outcomes are hypothesized and monitoring programmes established to test those hypotheses;
- deployment of environmental water is planned by an inter-jurisdictional expert group.

TLM further differs from FLOWS methodology in the previous example in that all of the water recovered can be converted to an equivalent volume stored in major reservoirs. This provides a level of independence in deciding when and where to apply the water and the possibility, if storage capacity allows, of accumulating the environmental reserve.

Both forms of providing water for riverine ecosystems, the conservation of hydraulic habitat and the redress of specific ecosystem needs, are legitimate vehicles for developing water management policy. To a large extent their value depends on the nature of the rivers to which they are applied and the management and water use regimes associated with those rivers. Both also have intrinsic strengths and weaknesses. Hydraulic habitat conservation has advantages in that:

- it can be built into sets of management rules that apply indefinitely;
- its underlying concepts are easily understood and discussed in community fora;
- the risk of missing important ecological components or unforeseen specific needs is low since the process seeks to restore key elements of the 'natural' flow regime; and, as a result,
- there is a limited need for specific ecological knowledge of each river.

The disadvantages include:

- little flexibility in addressing individual environmental goals;
- often limited opportunity for targeted monitoring (and associated adaptive management) other than longer-term assessment of river condition;
- risk of relatively inefficient use of environmental water and limited opportunity to refine that use.

Conversely, the identified ecological needs approach provides:

- clear hypothetical links between applied environmental water and ecological outcomes, which can be tested (monitored) and the watering process refined – leading to increased efficiency in using environmental allocations over time;

- flexibility in using the water: the potential to match the ecological task to the available water and to make choices between competing demands.

However, clear disadvantages include:

- significant risk that critical components of the ecosystem will be overlooked or unsupported leading to undesired outcomes in total;
- continued need for high-level technical input.

In essence, the management of identified ecological needs can be viewed as the high risk/high potential benefit approach. To maximize the ecological benefit from environmental water allocations, particular care with management and policy development is essential.

## Optimizing deployment of environmental water

In much of the work on environmental flows in the Murray-Darling Basin to date, the emphasis has been on processes to reserve or acquire environmental water. Its deployment in support of the riverine ecosystem has received less attention – a situation probably exacerbated by the prevailing extended drought conditions and the consequent lack of water (as distinct from entitlements to water) with which to work. Scarcity and cost of the resource would indicate that optimum deployment will be the key to successful, long-term, management of the river system, however. Management frameworks and the policies that guide them must lead to maximum water use efficiency in supporting the river ecosystem. To achieve this, the mechanisms for delivering environmental water must:

- have a sound *ecological basis*;
- be positioned in a '*learning framework*' to support future refinement;
- have sufficient *flexibility* to accommodate temporal changes in ecological need and/or contingencies.

*Ecological Basis.* Much of the scientific support for the development of environmental water plans in the Murray-Darling Basin has been based on assessments by expert panels. The FLOWS methodology also includes consultation with regional community members in order to incorporate 'local knowledge' into the process. Expert panels provide a convenient means of accessing current knowledge and, particularly where hydraulic habitat rehabilitation is the objective of the project, constitute a useful means of designing/assessing flow management plans – particularly where scheduled reviews are built into the process. They are also dependent, of course, on the amount and quality of research (their own and others) that has provided the knowledge base underlying their advice. Expert panel assessments have been improved in recent times with the combination of targeted field visits with sophisticated hydrological models that incorporate probabilistic forecasts and spell analysis of identified key hydrological features and extreme events (see e.g. Cottingham et al, 2007).

The expert panel approach may also be used in developing environmental flow programmes based on identified ecological needs. As part of the TLM programme Zukowski and Meredith (2007) assembled a team of ecologists with expertise in a range of biota and ecological processes thought to be potentially at risk from current management practices in the Murray and to be capable of responding positively to environmental water allocations. Having visited the sites involved, each scientist was given the task of developing hypotheses linking components of the ecosystem to flow characteristics (in the range likely to be achievable in TLM) that could be tested as part of a pilot application of environmental flows. Zukowski and Meredith (2007) then integrated a selection of these hypotheses to create an environmental flow – and outcome monitoring – programme for icon sites as part of TLM. Extended drought conditions have precluded the trial of this process to date.

Environmental flow recommendations must be based on a clear understanding of the riverine ecosystem, and the social, economic and cultural costs (as well as those to the ecosystem) of failing to apply them must be clearly articulated.

*Learning Framework.* Zukowski and Meredith (2007) provide an example of how environmental flow programmes can (and should) be set into a learning framework. Basing the programme on a set of hypotheses, stated a priori, ensures that the assumed knowledge linking flow to ecological outcome can be tested through an appropriate monitoring protocol and the programme thereby refined through the application of adaptive management principles. By properly constructing and testing management hypotheses in this way learning is gained in the immediate application and the resultant understanding can be transferred to other situations with reasonable confidence.

*Flexibility.* Although providing a suite of flow/ecology hypotheses provides a sound basis for designing a flexible environmental flow programme – in that different subsets of outcomes can be sought at different times – in practice flexibility is difficult to achieve in a programme as large and expensive as TLM because of the complexity of the decision making process it requires. (In this case a degree of inflexibility has some advantages as it protects a level of continuity needed to test the basic assumptions inherent in this *pilot* programme.) In the long term, flexibility can be introduced into large, complex and inherently bureaucratic programmes such as TLM through the development of decision support mechanisms (e.g. 'if–then' trees), which, in this case, might be based on the full suite of possible hypotheses assembled by Zukowski and Meredith (2007). Inevitably, however, the success of such an approach would depend on identifying and accounting for all contingencies – a difficult task at our current level of knowledge. There is probably value in smaller schemes that can respond more tactically and at shorter notice.

Currently several small water recovery schemes are being developed – for example RiverBank is funded by the NSW Government (DECC, 2008), and a not-for-profit conservation group, the Nature Conservancy Council, is setting up a trust account based on philanthropic donations. These and others are established to purchase small parcels of diversion entitlements as they come onto the market with the objective of moving the water from rural applications to the environment. Planning the mechanism by which this is achieved is not yet complete particularly in the latter case (again partly because entitlements do not closely equate to water under the current extreme conditions). A small-scale scheme in an early stage of planning is being developed by a large irrigation company with the support of Australian Government funding. The 'River Reach' programme is looking to develop a system of purchasing options on parts of individual diversion licences in the Murrumbidgee valley to be activated in years in which allocations of water are high. This enables water users to keep the water to which they are entitled in dry years but to forgo some of their allocation in wet years, resulting in additional water (about 44GL on current estimates) to the environment in wet years but none in dry years. Such a pattern lends itself to 'top-up' allocations – adding water to already-occurring flow events (including other environmental releases) to enhance their ecological efficacy. Theoretically this is a valuable addition to the environmental flow 'armory' but, on two counts, it requires a sophisticated arrangement for making decisions about its deployment. First, as its efficacy is dependent on flows currently in the system, release needs to be triggered by current or imminent hydrological events. Second, its value can be maximized only if it is deployed tactically in response to the most critical ecological demand for which that quantity of water will be beneficial. Hillman (2007) identified several ecological requirements in the Murrumbidgee system that could, potentially, benefit from such 'top-up' flows. It was noted, however, that the relative urgency of these requirements varied substantially over time and were dependent on antecedent conditions – as an extreme example, the urgency of recreating flow conditions conducive to the breeding of waterbirds or native fish increases dramatically as the time since the last successful recruitment event approaches the lifespan of the fish or bird species concerned.[1] It is clear that the effective tactical use of these relatively modest parcels of water is dependent on a combination of decisions reached in 'real time'. The framework for such decisions has not been set out in this case.

One means of approaching the need to make tactical decisions about deploying environmental water at the regional scale is to establish small local groups dedicated to the task. This is one of the bases for creating expert and community groups in the FLOWS protocol. It is refined further in the New South Wales Murray Wetland Working Group (MWWG), set up in 1992 initially as a community group charged with developing and implementing wetland management programmes that were well researched, technically sound and community-endorsed. Several of these programmes involved the exclusion of artificially high summer flows as part of the rehabilitation of wetlands, which led to the accumulation of water available for environmental purposes. The group now manages 30–40GL of environmental water per year that is used in rehabilitation projects on public and private lands (in situations in which water is obtained too late in a season for useful application, it may be sold to support the group's work) (Nias, 2005). Members of the group are volunteers from the region selected on the basis of skills. They include scientists, engineers and irrigation farmers. Currently, they use environmental allocations for the rehabilitation of wetlands in the NSW section of the Murray Valley but, increasingly, their skills are being called upon for other projects.

The ability of MWWG to make tactical decisions is sometimes limited by constraints relating to its government-based funding, but in principle, and recognizing its achievements to date, it provides a useful model for a machinery to manage moderate quantities of water at the regional (intra-valley) level – particularly if those allocations are to furnish a tactical response to a variety of contingency issues.

## Conclusions and policy implications

It has been acknowledged, at least by the Federal and state governments involved, that the water of the Murray-Darling system is over-allocated – that is, too little is left over after consumptive use to dependably support the riverine ecosystem, thus creating an unacceptable risk that the health of the river system will collapse (and with it the quality and utility of the water resource). Policy makers have recognized this situation but large-scale policy development has been hindered in the past by jurisdictional differences amongst the states that make up the Murray-Darling Basin and current realignment of responsibilities between those states and the national government is again increasing the complexity of policy development.

At a finer spatial/temporal scale there is sometimes a gulf between ecologists and policy makers to the extent that ecologists tend to be impressed by the complexity of the systems they deal with and attempt to accommodate and account for it. Policy developers on the other hand tend to favour generalized approaches and seek simpler rules that are suitable for legislation and/or application over a broad scale. In reality the management of environmental water requires attention at all of these scales and any approach likely to succeed in the Murray-Darling Basin will need to accommodate them all.

The ways in which the riverine ecosystem can be supported through management of water allocation is not uniform across the Basin. It depends significantly on the hydrological nature of the stream, the degree to which artificial storages have been developed, and the types of water-dependent agriculture (and other human activities) it supports. Different combinations of these factors demand different environmental water management responses. These differences extend to the 'philosophy' underlying water recovery (hydraulic habitat rehabilitation versus addressing identified ecological needs) and the ways in which water can be deployed. They will be reflected in the available management responses and the policies that support them.

Our knowledge of the links between river flow and the aquatic ecosystem is growing rapidly but is still meagre. Policies that govern environmental flow programmes and the management regimes that put them in place should support – in fact should ensure – the adoption of adaptive management principles in environmental flow programmes. All management actions or interventions should be linked to predicted outcomes with rigorously designed monitoring programmes that test those predictions and the knowledge on which they are based. This adds to the expense of the programme but failure to learn from management actions is a waste of money, time and water.

Flexibility in responding to new information or contingencies is important but difficult in large, long-term projects. It may be necessary to develop environmental watering operations at a range of scales – regional to Basin-wide – if large-scale rehabilitation and agile response to contingency are to be encompassed by environmental flow provisions.

Finally, despite the fact that the on-coming climate change is generally acknowledged and the prediction of significant reduction in catchment runoff (perhaps 40 per cent) accepted, planning and management is still based on historic data and/or current conditions. For instance the cap on diversions from the Murray-Darling system is based on an inter-government agreement to limit diversions to a level equivalent to diversions in 1993–1994. In the future that level could easily exceed the total volume in the system (as it has in the past few years). Likewise ecosystem rehabilitation activities tend to be benchmarked against some estimate of conditions in the absence of irrigation and other human activity. Whilst these are reference points and certainly not goals, it is still true that the system with 40 per cent less water might certainly be quite different even in the (hypothetical) absence of human activity. The significance of this to restoration is yet to be explored (do our efforts go towards achieving some new state or do we attempt to create 'zoos'?), although it may be argued that the ecosystem of the Murray-Darling Basin is sufficiently modified by past management that the distinction might be academic. Nonetheless adjustments to climate change are likely to be profound both for consumptive water use and for the ecosystem. Ecologists, economists and policy developers (inter alia) need to address the issue in concert.

## Note

1  This example is relevant. Flow models indicate that, had current river management practices been in place throughout the 20th century, the period between successful bird-breeding events would have exceeded the lifespan of the waterbirds on three occasions.

## References

Balcombe S. R. and G. P. Closs (1996) 'Macrophyte use by fish assemblages in the littoral zone of a Murray River billabong', Paper presented at the Australian Society for Limnology 35th Congress, Berri, South Australia

Boon, P. I., J. Frankenberg, T. Hillman, R. Oliver and R. Shiel (1990) 'Billabongs', in N. Mackay and D. Eastburn (eds), *The Murray*, Murray-Darling Basin Commission, Canberra

Briggs, S. (1990) 'Waterbirds', in N. Mackay and D. Eastburn (eds), *The Murray*, Murray-Darling Basin Commission, Canberra

Close, A. (1990) 'The impact of man on the natural flow regime', in N. Mackay, and D. Eastburn (eds), *The Murray*, Murray-Darling Basin Commission, Canberra

Cottingham, P., M. Stewardson, D. Crook, T. Hillman, R. Oliver, J. Roberts and I. Rutherfurd (2007) *Evaluation of Summer Inter-valley Water Transfers from the Goulburn River*, Peter Cottingham & Associates report to the Goulburn Broken Catchment Management Authority, Melbourne

Crabb, P. (1997) *Murray-Darling Basin Resources*, Murray-Darling Basin Commission, Canberra

Crome, F. H. J. (1988) 'To drain or not to drain? Intermittent swamp drainage and water-bird breeding', *Emu*, vol 88, pp243–248

Department of Environment and Climate Change (DECC) (2008) Documents and reports relating to the Riverbank programme. www.epa.nsw.gov.au/education/nswriverbank.htm (accessed 12 February 2008)

Hillman, T. (2007) 'Ecological requirements: Creating a working river in the Murray-Darling Basin', in L. Crase (ed.), *Water Policy in Australia: The Impact of Change and Uncertainty*, RFF Press, Washington, DC

Humphries, P., A. J. King and J. D. Koehn (1999) 'Fish, flows and flood plains; Links between freshwater fishes and their environment in the Murray-Darling River system, Australia', *Environmental Biology of Fishes*, vol 56, pp129–151

Jensen, A. (1998) 'Rehabilitation of the River Murray, Australia: Identifying causes of degradation and options for bringing the environment into the management equation', in L. C. de Waal, A. R. C. Large and P. M. Wade (eds), *Rehabilitation of Rivers: Principles and Implementation*, Wiley, New York

Jones, G., A. Arthington, T. Hillman, R. Kingsford, T. McMahon, K. Walker, J. Whittington and S. Cartwright (2001) *Independent Report of the Expert Reference Panel on Environmental Flows and Water Quality Requirements for the River Murray*, Report to the MDBC, Canberra

Kingsford, R. (2000) 'Ecological impacts of dams, water diversions and river management on floodplain wetlands in Australia', *Australian Journal of Ecology*, vol 25, pp109–127

Murray-Darling Basin Commission (MDBC) (2007) Documents and reports relating to the Living Murray Initiative. http://thelivingmurray.mdbc.gov.au/ (accessed 4 December 2007)

Nias, D. (2005) *Adaptive Environmental Water in the Murray Valley, NSW, 2000–2003*, NSW Murray Wetlands Working Group, Albury

Puckridge, J. T., F. Sheldon, K. F. Walker and A. J. Boulton (1998) 'Flow variability and the ecology of large rivers', *Marine and Freshwater Research*, vol 49, pp55–72

Sinclair Knight Merz (SKM) (2002) *FLOWS – A Method Determining Environmental Water Requirements in Victoria*, Report prepared for the Department of Natural Resources and Environment, Sinclair Knight Merz, Melbourne

Thoms, M. C., P. Suter, J. Roberts, J. Koehn, G. Jones, T. Hillman and A. Close (1998) *River Murray Scientific Panel on Environmental Flows: River Murray from Dartmouth to Wellington and the Lower Darling*, Murray-Darling Basin Commission, Canberra

Walker, K. F. (1986) 'The Murray-Darling River System', in B. R. Davies and K. F. Walker (eds), *The Ecology of River Systems*, Dr. W. Junk, Dordrecht

Watson, A. (2008) 'Water policy in Australia: Old dilemmas and new directions', Paper prepared for International Workshop on Sustainable Water Management Policy in Agriculture: Options and Issues, 15–16 May 2008, Seoul, Republic of Korea

Zukowski, S. and S. Meredith (2007) *Lindsay and Wallpolla Islands Structure Specific Water Management Plan*, Draft report to Malley Catchment Management Authority, Murray-Darling Freshwater Research Centre, Mildura

# 8

# Water Management in Spain: An Example of Changing Paradigms

*Alberto Garrido and M. Ramón Llamas*

## Major water policy landmarks

Spain's water policy has undergone a rapid process of piecemeal reforms, beginning in 1985, experiencing fundamental amendment in 1999 and ending in 2007. In this section we review these reforms and summarize their main implications.

### The 1985 Water Law

In many respects the 1985 Water Law (WL) forms the core of water legislation in present day Spain. At the time it was enacted, it replaced the 1879 water law and its amended version in 1896. The 1985 WL opened a new era for water policy for a number of reasons: (i) water resources were considered to be public domain, saving a few exceptions of groundwater use (which are part of the root of the problems related to groundwater use that this chapter also reviews); (ii) it laid down the water planning principles that eventually would materialize in three failed attempts at establishing national hydrological plans; (iii) it consolidated a financial regime for water users that delivered important benefits, the irrigators being the most favoured group; (iv) it consolidated the institutional role of the basin agencies, granting them autonomy, financial resources and personnel to become the actual decision makers in all water issues within the basin boundaries; and lastly (v) it defined a model of co-decision making, in which direct water users and interested adminstrations have taken an active role in all water planning and management at basin level.

## The 1999 Water Law reform

The reform of 1999 amended the 1985 Water Law, changing three fundamental issues (Garrido, 2006). First was the regulation of the exchange of water rights, permitting right-holders to engage in voluntary water transfers and the Basin Authorities (*Organismos de Cuenca*) to set up water banks or trading centres in cases of drought or of severe scarcity. The second aspect focused on public corporations building water works and recovering the costs by means of sounder financial arrangements. The third was a subtle, but crucial, consideration of desalinized and reused water as belonging to the public domain, on an equal footing with other water sources, and the issuance of special water rights granted to its users. The first issue was clearly the most controversial and, in retrospect, the most relevant, based on the initiative, of the 2001 and 2004 Laws of the National Hydrological Plans, reviewed below.

## The European Union's Water Framework Directive (2000)

The European Union's (EU) Water Framework Directive (2000) (WFD) is the most relevant water policy initiative of the last 20 years, perhaps the most advanced international initiative based on world standards. Its mandates include significant changes of focus in areas such as water pricing, ecological objectives, political processes, public participation and a new approach to water planning. It also includes transition waters (estuaries) and coastal waters, a fact that has created serious jurisdiction problems in Spain and the recognition of noteworthy scientific gaps in understanding. For Spain, as well as most other EU countries, the WFD implies a rebalancing of priorities from ensuring water supplies to all economic users to improving the ecological status of all water bodies. To achieve this overarching objective, a programme of measures, included in new water planning documents, that passes the test of cost/effectiveness (not cost/benefit) must be approved for all European water demarcations (main watersheds) by 2009. The general goal is that all surface and groundwater bodies should achieve a good ecological status by 2015. Countries facing unsurmountable difficulties to meeting the quality standards of heavily modified water bodies must petition the European Commission (EC) to obtain derogations in the time schedule (two potential extensions to 2021 and 2027) or even downgrade the targets of good ecological health, possibly to the point of not improving it at all. They must provide cost–benefit analyses demonstrating that meeting the normal standards would entail disproportionate costs. Once the WFD entered into force, no single issue related to water resources would remain unaffected by one or another provision of the WFD.

## The 2001 and 2004 Laws of the National Hydrological Plans

The 2001 and 2004 Laws of the National Hydrological Plans (NHP) approved and repealed a major inter-basin water transfer project, the so-called Ebro water transfer (Arrojo, 2001; Albiac et al, 2003). While many other initiatives approved in the 2001 NHP were maintained in the 2004 NHP and have already been partially implemented (for example, the NHP still includes the construction of about 100 new large dams), the Ebro transfer epitomizes the breakdown of consensus of a century-old mode of thinking, planning and executing water policies. By any measure, the Ebro transfer was a flawed and extremely expensive project. And yet the scarcity problems along the Mediterranean coast from Catalonia to the eastern coast of Andalusia have not been solved to the extent most studies indicate. In Catalonia, there are calls to reactivate the project of transferring water from the Rhône in response to the severe drought at the beginning of 2008. The Ebro transfer is still demanded by politicians and users along the Mediterranean arc. However, the implications of the approval and subsequent repeal of such a big project go beyond the discussion of alternative plans to solve water problems, however important the beneficiary regions may be. It is an indicator of the inability to create bipartisan agreements on issues that transcend four-year political periods. Furthermore, the Ebro transfer paved the way to devolve competences to the Autonomous Commmunities on inter-community basins that had previously been granted to the Central Government (Spanish Government) in the 1985 Water Law, and originated in the creation of the Ebro basin agency in 1926.

As recently as 2007, approval for the reform of the Autonmous Statutes of Catalonia, Andalusia, Aragon and Valence consolidated the power of the regional governments on water affairs. One consequence of this devolution process is the transfer of competences from Madrid to Seville (the Andalusian capital) for the management of the Guadalquivir basin, even though this basin includes territory from two other Autonomous Communities. It should be noted that some of these provisions have been brought to the Constitutional Court (the Spanish equivalent of the American Supreme Court) for being in potential breach of the Constitutional consideration of inter-community basins as being a national jurisdiction. It is ironic that some of these appeals brought to the Constitutional Court have been filed by socialist regional (autonomous) governments, against the Statutes of Autonomous regions that are also controlled by the socialist party. In other words, water issues override the limits of political affiliations.

In 2004, the government that brought the repeal of the Ebro water transfer to the legislative quickly approved programme AGUA[1] (the acronym in Spanish of the Initiative for Water Management and Utilization). AGUA was meant to replace future supplies from the transfer arrangement with 20 large seawater desalination and wastewater reclamation plants (see Downward and Taylor, 2007). By the final months of the government's term, very few of these plants have been built and become operative. In total only 214 million cubic metres of desalination capacity of the 700 million cubic metres planned for 2004–2008 have become operative. Some of the planned plants are struggling to sign firm contracts with future customers, totalling a demand that justifies size and capacity So, if history repeats itself, the March 2008 election will dictate whether AGUA continues or whether the Ebro transfer is rescued.

## Miscellaneous initiatives: The Guadiana programme, water banks, new planning criteria and programmes of measures

Less important initiatives, such as the Guadiana programme, the establishment of water banks and new criteria for drafting programmes of measures issued by the Ministry of the Environment will be discussed below, where we consider in detail four case sudies that look at different aspects of water policy in Spain in 2007.

## Drivers of change

Four main drivers of change are giving rationale and momentum to the most recent policy initiatives. First is the widespread recognition that many water bodies have suffered severe deterioration. It is beyond dispute that restoring water quality is a formidable task that requires large investment, a better administration and a great deal of participation and education. Second, water demand is still growing insatiably, especially where resources are scarce. Economic development and growth, the construction boom, the tourism sector and a competitive export-oriented agricultural sector jointly contribute to worsen already polluted water environments. Third is the increasingly indisputable fact that climate change poses a serious challenge for the Iberian peninsula. Most models predict larger evapotranspiration, lower and more unstable precipitation regimes and lower river runoff. Agricultural demand is likely to grow, adding further pressure to the catchments and supply systems. And fourth, the Common Agricultural Policy (CAP) has shifted the support measures from production incentives and specific sectorial programmes to completely decoupled support. Farmers are now free to grow the crops they want. Associated with the influence of agricultural policy is the final result of the WTO trade round in order to decrease import barriers which today see most developing countries exporting their agricultural products to the EU (and to Spain). The results of the WTO agreements may have a significant impact on the ecnomic feasibility of a good number of current Spanish crops that today are mainly exported to the EU. Finally, the EC mentioned in its report, 'Health check of the CAP', the objective of ensuring the sustainable use of water resources (EC, 2007). As we will review below, none of these drivers lacks factual and scientific support.

The reports of Article 5 submitted by Spain to the EC in 2007 (MMA, 2007a) contain numerous and updated data proving support to the first two drivers. This report identifies the reasons behind the bad ecological quality of the main river basins. For decades, industries, animal feedlots and cities have spilled untreated water into rivers and natural waterways, or let it filter to aquifers. Furthermore, MMA (2007a) projects that by 2015 most basins will see their main parameters worsening or stabilizing at best. Groundwater quality is experiencing similar trends. The quality of drinking water is diminishing at alarming rates, while at the same time we see two-digit growth rates in the consumption of bottled water.

Water demand projections are equally worrying. Iglesias (this volume) estimates that agricultural water demand will increase by 10 points to 30 per cent because of global warming. A recent study of crops' evapotranspiration in the Guadalquivir basin[2] (with 880,000 hectares of irrigated land) show that water demand of crops may range from 3.45 to 5.3 billion cubic metres, depending on whether the spring and summer are wet or dry (Aquavir, 2006). However, the economic feasibility of this demand will depend on factors such as the above-mentioned future WTO agreements and on the implementation of the WFD principle of full-cost recovery. Spanish irrigated agriculture has been heavily subsidized in the past. The range of variation of crop demand in the Guadalquivir is equivalent to the urban consumption of 30 million people in one year. However, most analyses show that per capita consumption is stable in Spain (MMA, 2007a), and the economy's growth is increasingly becoming decoupled from water use growth.

Compounding the growth of water demand, the MMA (2007c) projects that runoff in most basins will be lower and more unstable. The impact on the mountain areas and the snow regime will be severely modified, if the findings from the Rhône (Bravard, 2008) can be applied to the Iberian basins. In addition, according to MMA (2007c), runoff regimes will become more unstable and prone to extremes. The consequences for the managing of reservoirs are that security levels for containing floods may need to be increased, reducing in turn the storage capacity. The recognition of these processes and implications is appearing in official documents and political statements alike, becoming a motto for raising awareness and a rationale for numerous initiatives. As dubious as the MMA reports may be, they indicate the major trends and convey information that before the reports were compiled was dispersed or simply ignored.

## Changes in the agricultural water demand

The fourth driver is the reform of the EU agricultural programmes, and its indirect implication on agricultural water demand. Up until 2003, support granted to the farm sector by the CAP was based on price support mechanisms or per hectare direct payments. As a result of both, farmers' incentives to grow certain crops (virtually all crops except fruits and vegatables) were driven by relative subsidy differences as well as quotas and other acreage limits. Examples of these distortions are numerous and telling. Since 2003, farmers have been less restricted and may grow the crops they wish; as a result, their decisions are far more influenced by prices and food demand. Furthermore, crops which were rarely irrigated ten years ago, like olive oils and vineyards, now occupy 800,000 hectareas of irrigated acreage. The interesting feature of these crops is that they require less water application and can endure tougher conditions of water stress than the crops experiencing decline, such as sugar beets, cotton, corn and tobacco.

This shift of cropping patterns has huge implications for many water stressed basins. One is that the opportunity cost of water is now more transparent and is connected to farms' different profitability. As a result of this, farmers are more open to market signals and less relunctant to exchange water rights than they were ten years ago (Garrido et al, 1996). Second, in many areas farm water demand is now more flexible in order to accommodate actual hydrological conditions. Flexible allocation and drought contingent programmes can find more room within the farming sector to face water scarcity periods. While water exchanges so far have moved small amounts among different users, they represent a qualitative difference with profound consequences for the future. Third, the water footprint of olives and vineyards altogether is 3.6 billion cubic metres, whereas both crops occupy 3.6 million ha; whereas cereals' internal footprint is 6.3 billion m$^3$ and acreage is 6.8 million ha (Rodríguez Casado, 2008). Garrido and Varela-Ortega (2008) show that the irrigated acreage of corn and other field crops, like cotton and sugar beet, are losing importance in favour of crops better adapted to the Spanish climate.

Trade in farms' products in Spain is also becoming more integrated. Novo et al (2009) have evaluated the volume of water and its economic value when 'virtually' traded just in the commerce of grains and cereals in Spain. They showed that the net import of virtual water with cereals was 5 billion m$^3$ and grew steadily from 1997 to 2005, by which time it totalled 9 billion m$^3$.

The technological and engineering factors connected to farmers' water use are also becoming crucial. At the irrigation district level, the government has completed modernization and rehabilitation projects in old districts totalling 1.3 million hectares (Barbero, 2005). In most cases, farmers have been requested to pay up to 50 per cent of the cost, although they were given preferential treatment in that they could borrow it back in the form of 50-year loans. These projects entailed, in many cases, a complete refurbishment of the irrigated districts, converting 19th-century design into 21st-century infrastructures. At the farm level, drip irrigation technology is now the commonest in Spain, occupying more than 1.3 million hectares in 2005.

In terms of labour use, agriculture has shown a stable downward trend as Figure 8.1 attests. In terms of macroeconomic profitability, Spanish agriculture has experienced a marked process of capitalization that has been followed by reduced margins and tighter economic conditions. While the index of animal and plant

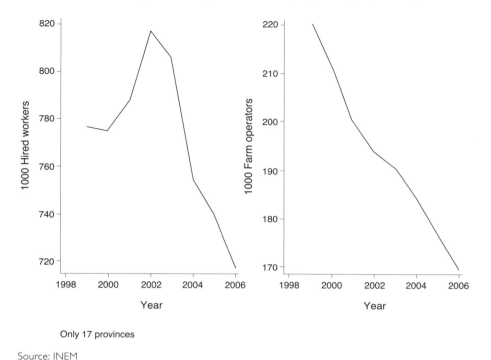

Only 17 provinces

Source: INEM

**Figure 8.1** *Farm employment trends (hired and farm operators in 1000s)*

prices at the farm gate reached 107.6 and 106.2 in 2006 (with 100 in 2000), the indexes of farm input prices have grown to 133 (fuel and energy), 145 (nitrogen fertilizer) and 123 (farm capital goods) (MAPA, 2007).

The value of food products obtained on irrigated land has kept growing in terms of constant prices since 2000, as shown in Figure 8.2. The figure plots total agricultural output obtained in irrigated land and dry land (evaluated in billion € of 2000), as well as irrigated and dry-land acreage (in million hectares).

## Changes in social discourse and the breakdown of consensus

One of the strongest forces underlying water policy reform in Spain, and yet one which is poorly understood and analysed, has been the breakdown of a century-long consensus. Up until 1994, when the first attempt to pass a Law of National Hydrological Plan failed, civil engineers had provided the intellectual leadership and technical capacity to design and execute water plans. In the last ten years, many other professional and scientific fields have become as much, if not more, influential in the most controversial discussions. In particular, hydrogeologists,

132  *Issues in Water Resource Policy*

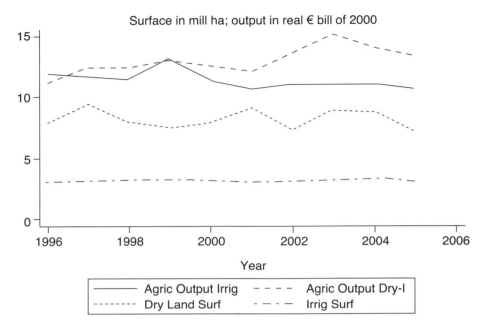

Source: Anuarios MAPA

**Figure 8.2** *Total agricultural output and surface (separating rainfed and irrigated crops)*

agronomists, chemists, ecologists, economists and other social scientists are now more prevalent than civil engineers, and are increasingly filling the vacancies in basin agencies and top management positions in the environmental departments of both regional and national governments. In this respect, the Spanish situation is similar to the one described by Dooge (1999) and by Allan (1999) in many other countries.

The consequences of opening the 'water resources' agenda to numerous professions cannot be sufficiently stressed. First, while civil engineers focused almost exclusively on water quantities and flows, the importance of water quality and river systems' ecological status gained prevalence with the enforcement of the WFD. Droughts and floods were soon joined by reports of ecosystem destruction and water pollution in the media, changing the view of the general public and redirecting the discourse of many politicians.

The discussions and debate about the 2001 NHP gave rise to another equally important breakdown of consensus. In this case, regional disputes over transboundary rivers became explicit and turned into political ammunition. Although the management of inter-community water resources is, according to the Spanish Constitution, a national jurisdiction, some Autonomous Communities claimed area-of-origin rights in order to question the grand Ebro transfer scheme.

The beneficiary regions, in turn, claimed that inter-community basins were a national jurisdiction and inter-basin transfers were strategy projects for the whole country. While the 2001 NHP was stopped soon after the Socialist Administration came into office in 2004, the conflicts subsided but did not disappear. For one thing, the region of Castille-La Mancha demanded that the Tagus–Segura transfer should eventually be phased out, on the basis that the region itself needs the water resources that are transferred annually to the Segura basin. Furthermore, the 2004 political term opened a period of political discussions in Catalonia, Andalusia, Valencia, Castille-La Mancha, Aragón and Basque Country, among others, to draft and approve new Autonomous Statutes. These statutes represent the cornerstone of the political autonomy of the Autonomous Communities (ACs)and mark the dividing line between the competencies of the central administration and those of each AC. The Catalonian Autonomous Statute was the first to be established, but it was soon followed by a number of other ACs. The implications of the redefinition of the Autonomies' regimes for water and the management of inter-community river basins are doubtful. On the one hand, all new statutes define to a larger or smaller extent new competencies over inter-community basins; the Andalusian going so far as to declare in article 51 that the region 'has exclusive competencies over the Guadalquivir resources that flow within its territory and do not affect other Autonomous Community', adding that '[those competencies] should not affect the National Planning of the hydrological cycle, … nor be in breach with article 149 of the Constitution', which establishes the exclusive competencies of inter-basin river basins. On the other hand, the Andalusian Statute has been brought to the Constitutional Court on the grounds, among others, that the Guadalquivir provisions of her Statute breach the constitutional principles. While the court has not pronounced on this issue, the Andalusian regional government has already been given competencies on the Guadalquivir and set up a regional office to manage it.

While it is still too soon to ascertain the impacts of this process of devolution, a prudent judgement would indicate that the role of the central government in inter-community basins has been diminished. Water policy is increasingly a regional policy, and regions, with the eventual support of their Autonomous Statutes, will surely develop their own legislative initiatives.

## Case studies

Against the dynamic process of institutional, environmental and economic changes summarized above, there are processes ocurring at a lower scale that perhaps better exemplify the profound transformation of Spanish water policy. In the first case, we review the way economics has recently permeated many facets that not long ago were totally devoid of an economic dimension. In the second case, we look at the way decades-long problems of groundwater overdrafting have been approached.

The following case studies are offered to provide a complementary view of the major trends discussed above. In the first case, focusing on the increasing role of economic instruments, we wish to illustrate how distant water allocation and management in Spain were from any sense of economic rationality. In this we integrate notions such as scarcity values, cost recovery rates, externalities and non-market values, together with rents and profit accruable from productive uses. We wish to show with this example that little progress had been made since 1989 in the economic area until the 1999 WL reform and the WFD of 2000 recognized that water policy could not progress without the support of economic instruments. With the second case study, looking at a succession of attempts to tackle the most pressing problems related to groundwater use, we wish to illustrate how statutes, however clear and sound, fail in the absence of economic compensation and water rights redefnition. The last case sudy, looking at the economic rationale of integrated water management, is proposed against the devolution process in the area of water management among the Autonomous Communities. It shows that cooperative behaviour along the entire watershed is the most cost-effective means to achieve the objectives of the Ministerio de Medio Ambiente (DMA), and provides a rationale to maintain the basin perspective that Spain has had since 1926 and that the DMA extends even to internationally shared river basins.

## Changes in the economics of water resources, including flexible allocation instruments, voluntary arrangements, and water prices

### An economic analysis and evaluation of the Ebro transfer

The project of the Ebro transfer has been thoroughly documented (Arrojo, 2001; Albiac et al, 2003). A grand scheme of inter-basin connections from the Ebro delta, northeast to Barcelona (with about 200 million m$^3$ of capacity and 150km long) and southwest to Almeria (800 million m$^3$ capacity and almost 800km long). The project was made public by the government in 2000, giving rise to five intense years of discussions, debate, street demonstrations and political battles. According to most analysts, including those contracted by the government itself (Hanemann, 2003), the project had three major flaws. First, it disregarded the balance and tides and sedimentation in the lower reaches of the Ebro, including its delta. Second, it was based on shaky evaluations of the demands it was meant to supply, primarily farmers relying on overexploited aquifers or insufficient water sources. And third, its cost–benefit analysis (CBA) was fatally wrong. Different teams reached very negative CBA results (San Martín González and Pérez Zabaleta, 2002; Garrido, 2003; Hanemann, 2003; Albiac et al, 2006). Linked to this was the fact that the option to add additional supplies in the most remote locations using desalination was not considered in the analysis of alternatives. The project's costs evaluation was flawed also, according to all external reviewers (Arrojo, 2001): the project's costs would be shared equally by all users, irrespective of their location and distance from the headwaters. Marginal cost pricing was disregarded, so cross-subsidization effectively kept the price at the end of the project cheap at the cost of the remaining customers. In terms of financing and designing grandwater works, the Ebro project still represents a landmark in wrongdoing and poor design.

## The Article 5 Spanish report to the European Commission

WFD's Article 5 establishes that each Member State should carry out, for all its river basins: (i) an analysis of its characteristics; (ii) a review of the impact of human activity on the status of surface waters and on groundwater; and (iii) an economic analysis of water use (see MMA, 2007a). This represents a massive study for the whole country, and a completely new approach to the inherited criteria with which water statistics were previously collected and recorded. Spain submitted its report and was given a good mark by the EC (72 points, ranking 6th out of 27 Member States; EC, 2007). The findings of these reports cannot be sufficiently stressed. They pertain to the evaluation of cost-recovery rates in the agricultural sector – very close to 100 per cent, simply because the costs are evaluated using inadequate rates for the amortization of the infrastructure. They show that about 50 per cent of agricultural water use has a profitability of less than €0.02/m$^3$. But groundwater users incur costs that are five to ten times the tariff paid by farmers using surface resources. The reports also illustrate how cheap urban water is in most cities in comparison to other EU countries (a factor of 2 with respect to the mean, and 3 with respect to Germany, Denmark or Sweeden, for example). At present industrial and urban water rates (see MMA, 2007b), sewage treatment can only ensure filtering, oxygenation and decantation. In the metropolitan city of Seville only 20 per cent of urban wastewater undergoes tertiary treatment, and most other medium to large Spanish cities do no better. These figures are taken to indicate that by 2015, on most water quality parameters and with no change in behaviour, the situation will either stabilize or worsen, giving little hope for improvement.

The Article 5 report has three main political implications. First, water prices will need to be raised significantly for all water services. This is because the pressure and impact from water services still have appreciable deleterious effects on the ecological status of most water bodies that will need to be addressed by more expensive water treatment and pollution abatement. Second, out of all agricultural water uses, about 30–40 per cent is still uncompetitive, despite significant growth in the adoption of technology and the intense pace of rehabilitation at the district level in the last ten years. As we review below, with the growth of demands and tighter water balances, incentives to initiate water exchanges will increase, exacerbated by the enforcement of programmes of measures (in pursuant to WFD article 11). Despite its drawbacks, resulting mainly from the fact that it was compiled from information that was not specifically collected for Article 5, the report provides a clear picture of all surface water uses, with the pressures and impacts mapped for the whole country; however, it does not address groundwater resources. The value of this information is still dubious, but, in the medium term, it is likely to help redefine notions such as 'uses of general interest' or 'structural deficits' in the most arid or semi-arid basins. Although perhaps more subtle than the others, this reveals the third major implication: the fact that so much information – properly organized and readily accessible– has been generated. Policy actions can now be easily judged on all accounts by the general public, media and the academic community.

## The application of article 9 of WFD, regarding the implementation of 'full-cost recovery prices'

Drawing up water tarifs is one of the cornerstones of the WFD (see Article 9). And yet, little is known about the extent to which water charges will 'take into account the environmental and resource costs' in addition to the financial costs. The EC seems to follow the principle of averted environmental costs, which in general generates very narrow and limited definitions of environmental costs. Even more difficult is the notion of resource cost, a concept that needs functioning water markets to become apparent and self-evident. Ironically, if water trading becomes a common practice, there will be no need to incorporate the resource cost component in the charges.

Irrigation is by far the largest consumer of water in Spain and is perhaps the sector that is most vulnerable to higher water prices. It remains to be seen whether Article 9 is applied in its fullest extent to irrigation. In a book edited by Molle and Berkoff (2007), *Irrigation Water Pricing Policy: The Gap Between Theory and Practice*, the contributors came to the conclusion that the role of water pricing in the agricultural sector should be downgraded. In Spain, most studies concur that water charge increases (within the range of political feasibility) results in severe income effects and little reduction in water use (Berbel et al, 2007).

## The creation of 'water banks' and the increasing occurrence of voluntary water exchanges

It was stated in the first section that, although the Water Law reform opening the era of water makets was enacted in 1999, the first effects were not seen for almost seven years. The law opened two ways for right-holders to lease out their rights, either to the basin authorities or to another user. The simplest way is an agreement between two right-holders and their decision to file a permission to formally exchange the right. The basin agency has 30 days to respond and, unless major technical, environmental or third-party difficulties are encountered, the petition will be granted. Very few, albeit significant, exchanges have been reported.

Consider the case of a big commercial farm in Almería (southeast) that purchases rice fields in the marshes of the Guadalquivir basin, 300km away from Almería and in a different basin. As a water right-holder, the commercial farm files a request to transfer its water rights linked to the rice paddies to Almería, using an inter-basin water transfer that connects the headwaters of the Guadalquivir with another basin (the Negratin–Almanzora aqueduct). This sale was approved despite the potential third-party effects of water resouces that, in the absence of the transfer, would have flowed to the Atlantic ocean 300km along the Guadalquivir river.

In another case, an irrigation district in the Tagus basin leases out all its water rights to a set of users in the Segura basin, using again another inter-basin aqueduct (the Tagus–Segura aqueduct). The revenue generated for the farmers by the contract is larger than the value of the crop the farmers would have produced in a normal year (Garrido, 2006). The agreement was especially profitable for two reasons: first, the district was undergoing a rehabilitation project to reduce the extremely large water allotments, which were transferred in full in the sale; second, during the season for which the rights were transferred and the rehabilitation project was being implemented, farmers would have had difficulties irrigating their fields. The farmers leased out their full allotments from headwater resources that had been very inefficiently used for years to users located in another basin.

The last case involves a subtler exchange that entailed no water transfer at all, but the obligation to maintain the minimum levels of key reservoirs. These levels are statutorily connected to the management of the Tagus–Segura aqueduct, so that the amount of resources that can be transferred in each year is conditioned by the state of the reservoirs at given dates. Through the purchase of the water rights of users serviced from these reservoirs, the purchasers could effectively increase their rights to transfer resources across the basin, simply keeping the levels above the minimum thresholds.

These three large-scale transfers illustrate the type of exchanges that will be more frequently requested. In general, they serve the purpose of moving water from the south central plateau to the southeast. For the moment, the basin authorities and the Ministry of the Environment have been granted these transfer requests. But once the third-party impacts are identified and evaluated, such transfers will perhaps become more difficult (see Colby (1990), in her seminal work on water trading and its institutional impediments as proxies of environmental taxes). Colby's thinking also fits with the fact that the government of Castille-La Mancha, the main area of origin in most exchanges, is erecting institutional barriers to prevent users located in their territories from selling water to others in adjacent Autonomous Communities.

The second route to enable water exchanges is by means of the so-called water banks or exchange centres. Not strictly an office or agency, these centres are hosted, run and located in the basin agencies themselves. Garrido (2007a, 2007b) shows that centres are a much more efficient means for promoting water exchanges, for a number of reasons, including transparency, control, avoidance of third-party effects and market activity and scope. And yet, the experience so far has been limited to the Jucar, Segura and Guadiana basins. Since these water centres have been primariy used to tackle the severe problems of the overexploitation of groundwater resources, we review them in the next section.

## Tackling the most pressing problems associated with intensive use of groundwater resources

Since the enactment of the 1985 Water Law, which included special provisions to tackle the problem of overexploited aquifers, there have been at least four major initiatives to manage groundwater resources. In short, these were (i) the declaration of overexploited aquifers and the mandate to enforce regulations and implement management plans; (ii) an EU agri-environmental programme, only applicable to Aquifer 23 in the Guadiana Basin, with subsidies to farmers curtailing their water consumption; (iii) the use of inter-basin transfers, both in the case of the southeast coastal areas and in the Upper Guadiana; and lastly, (iv) The Especial Plan of the Upper Guadiana (PEAG, Spanish acronym), and the creation of exchanging centres in the Segura, Jucar and Guadiana basins (Llamas and Custodio, 2002; Varela-Ortega, 2007).

Varela-Ortega (2007) traces the history of the emblematic Aquifer 23 in the Southern Castillian plateau, linking the ups and downs of its piezometric levels with the first three rounds of initiatives just mentioned. Clearly, option (i) failed; option (ii) succeeded, but the financial cost was very high; and option (iii) failed because option (ii) was not sustainable. In the end, the PEAG was approved in 2007 with a total budget for 20 years of €5 billion (equivalent to the proposed Ebro transfer) and part of its subprogrammes are now operating, although under PEAG the basin would reduce to a meagre 200 million m$^3$.

Underlying these initiatives, but undermining them too, was the recognition that tens of thousands of users in virtually all basins had no legal rights or concessions to the groundwater resources they had been tapping for years. Any effort to reduce total extractions in the overdrafted hydrogeological units had to be accompanied by the closure of the 'alegal' or 'illegal' uses. So far all attempts have failed, and any reduction of total extractions has come from the efforts made by both legal and illegal users.

In 2005 it was clear to all managers, analysts and users that something new had to be given a chance. The option to use buyouts of water rights, permanent or temporary, gave a rationale to the establishment of exchanges centres (*centros de intercambio* in Spanish). We will review the different approaches taken in the Jucar and Guadiana. In the Jucar basin, the Offer of Public Purchase (*Oferta pública de adquisición de derechos*, OPA) was targeted to farmers tapping groundwater resources near the Jucar's headwaters. Its objective was to increase the water table in Castille-La Manche to ensure that Jucar flows to the Valencia region increased from historical lows. Farmers were given the option to lease-out their rights for one year in return for a compensation ranging from 0.13 to 0.19 cents per m$^3$, the variation depending on the distance of the farmer's location to associated wetlands or to the river alluvial plain. The OPA was launched in two rounds, the first with disappointing results in terms of farmers' response while the second had more success. The purchased waters served the unique purpose of increasing the flows, enabling more use downstream in Valencia. But the OPA did not have not any specific beneficiaries dowstream, other than the increase of flows.

The OPAs of the Guadiana followed a completely different approach and were meant to address serious problems of overexploitation in the Upper Guadiana. As stated before, the OPA formed part of a more ambitious programme of aquifer recovery, called the PEAG. The Guadiana's OPA made offers to purchase permanent water rights to groundwater, paying farmers €6000–10,000 per hectare of irrigated land. Note that, since these farmers had seen their allotments reduced in preceeding years, what the Guadiana basin was truly purchasing from the farmers was about 1500–2500m$^3$/ha, effectively €2–4 per m$^3$. The Guadiana basin agency has the objective of 'purchasing' the water rights to 50,000 hectares of irrigated land, and is budgeting €500 million for the whole plan. A marked difference to the Jucar's OPA is that the Guadiana exchange centre will transfer part of these rights to other farmers (growing vegetables) and to the Autonomous Community of Castille-La Mancha. The Guadiana basin will grant less rights than it has purchased, allocating the difference to wetlands and to increasing the piezometric levels of the aquifers. One subtlety of the Guadiana scheme is the fact that, while farmers entering the programme must surrender their private rights (honoured because they were in the catalogue of private waters before the 1985 water law was enacted), those that gain access to them will be granted 30-year 'concession' rights (which is more attenuated property than the others). So the Guadiana operation had this other dimension that in the long term will imply that the basin agency has more users with 'concessions' than with private rights.

Livingston and Garrido (2004), reviewing US and Spanish experiences with overexploited aquifers, hypothesized that OPAs such as those of the Jucar and Guadiana would be the only feasible solution. What these authors overlooked was that OPAs would also serve the purpose of water reallocation, entitling government agencies with water rights, that in turn would allot to other users. A question that has not been addressed in the Guadiana case is the price that will be asked of the new users, and whether the exchange centre will incur losses or be able to recoup the costs of the purchase.

## The cost-effectiveness rationale of programmes of measures

We will now review the main breakthrough of the Cidacos Pilot project from the third perspective of cost-effectiveness.[3] This project, completed in 2003, was promoted by Spanish institutions to develop a conceptual framework for the application of WFD's Article 11 definitions of 'programmes of measures'. Gómez and Garrido (2007) summarized the rationale of the use of cost-effectiveness

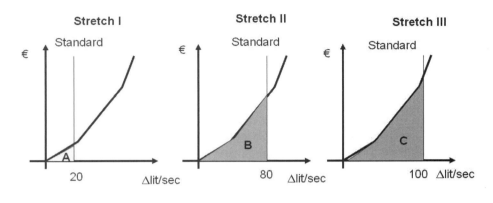

Overall costs = A + B+ C

Source: Gómez and Garrido (2007)

**Figure 8.3** *Cost-effective programme with three independent water bodies*

in the selection of the programmes of measures that are least costly. Consider the parameter of water flow in a given river that is divided into three stretches. Obtaining good ecological status (GES) implies that rates of flow must be increased by, say, 20, 80 and 100 litres/second respectively in the upper, middle and lower stretches of the river. In Figure 8.3, marginal costs curves are represented against rates of flow on the horizontal axis, for the three stretches of the river.

One is prompted to ask whether this approach is cost-effective. Since stretch I is upstream of stretches II and III, it would perhaps be reasonable to go beyond the required level in stretch I (20l/s) and perhaps move the marginal cost curves in stretches II and II to the right (a reduction of costs). Figure 8.4 represents the option to increase the standards in stretches I and II, and the resulting cost reduction in stretch III. If the overall cost can be reduced by going beyond the standards in some stretches, the most cost-effective programmes of measures will focus more on the upper than downstream reaches.

What the Cidacos project showed and put into practice goes beyond this simple reasoning. The project designed a cost-effective programme of measures, mapped them in the Cidacos basin, linked them with the different agents (users and pollutants) and, in a final stage, put the programme out to discussions in hearings, following the WFD mandate about public participation. The general public of Navarre (the region of the Cidacos) participating in the discussions understood the whole rationale of the programmes and accepted the differential treatment of pollutants along the basin. They even agreed on a financial scheme and on criteria to share the costs. This case study was taken by the EC and integrated into the WATECO guidelines, jointly with other pilot studies, that were met

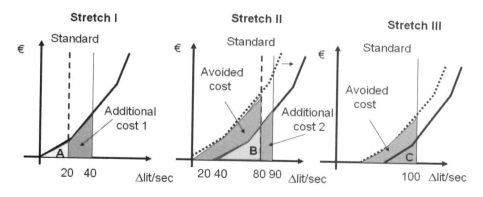

**Figure 8.4** *Least cost programme, integrating the standards of three connected water bodies*

to help Member States conduct the economic analyses mandated by the WFD, including the selection of a cost-effectiveness analysis of programmes of measures. In Spain, the Cidacos project inspired tens of tenders put out by the basin agencies to conduct similar studies.

## Drawing useful lessons from the Spanish example

This paper has summarized the major developments and challenges of the recent history of water policy in Spain. The following lessons can be drawn:

1 Large water projects are not the solution to unsustainable water uses or enhancing water supply reliability. More flexible alternatives (with and without technologies), that ensure some screening of the beneficiaries and a sound financial scheme are prerequisites for giving the green light to grand water works.
2 Flexible and adaptable solutions, which rely on technologies, infrastructure and demand management instruments are more complex and require multiple standpoints and longer approval periods. The context must be clear before innovative schemes get through. In general, once crises, major landmarks or groundbreaking progress occur, it is easier to plan and implement complex solutions.
3 The actual costs of supplying water at subsidized prices multiply, spilling over on to other users, the taxpayer and the environment, especially when

scarcity becomes acute. Cheap water granted in the form of concessions create perceptions in their holders of being 'entitled' to water resources. When trading systems are established, extraordinary rents will be created by those selling the water. While many would find this offensive, a continuous functioning of the market will tend to erode the rents.
4   Rigid, hierarchical and top-down planning models fails when water hegemonic thinking and political coalitions break down; all the more so if there are also regional disputes.
5   Accessible information, science-based decision making and public participation are key elements in breaking through entrenched and adversarial positions.
6   Innovative water policies require strong budgets, sound finance and equitable burden distribution.

## Notes

1   'Agua' means water in Spanish.
2   It includes Guadalete and Barbate Andalusian basins.
3   This section borrows from Gómez and Garrido (2007).

## References

Albiac, J., J. Uche, A. Valero, L. Serra, A. Meyer and J. Tapia (2003) 'The economic unsustainability of the Spanish national hydrological plan', *International Journal of Water Resources Development*, vol 19, no 3, pp437–458

Albiac, J., M. Hanemann, J. Calatrava, J. Uche and J. Tapia (2006) 'The rise and fall of the Ebro water transfer', *Natural Resources Journal*, vol 46, no 3, pp727–757

Allan, J. A. (1999) 'The Nile Basin: Evolving approaches to Nile water management', Paper presented at the Conference on Water Quality, convened by the International Institute of Water Quality Sciences, Jerusalem

Aquavir (2006) *Superficie de los Cultivos de Regadío y sus Necesidades de Riego, en la Demarcación de la Confederación Hidrográfica del Guadalquivir*, Sociedad Estatal Aguas de la Cuenca del Guadalquivir, SA. y Empresa Pública Desarrollo Agrario y Pesquero, SA

Arrojo, P. (ed) (2001) *El Plan Hidrológico Nacional a dabate*, Bakeaz, Bilbao

Barbero, A. (2005) 'The Spanish National Irrigation Plan', Paper presented at OECD Workshop on Agriculture and Water: Sustainability, Markets and Policies, 14–16 November, Adelaide, South Australia

Berbel, J., J. Calatrava and A. Garrido (2007) 'Water pricing and irrigation: A review of the European experience', in F. Molle and J. Berkoff (eds), *Irrigation Water Pricing Policy: The Gap Between Theory and Practice*, CABI, IWMI, Wallingford, UK, pp295–327

Bravard, J.-P. (2008) 'Combined impacts of development and climate change on the Rhône river (Switzerland, France)', in A. Garrido and A. Dinar (eds), *Managing Water Resources in a Time of Global Change: Mountains, Valleys and Flood Plains*, Routledge, London

Colby, B. G. (1990) 'Transactions cost and efficiency in Western water allocation', *American Journal of Agricultural Economics*, vol 72, pp1184–1192

Dooge, J. C. (1999) 'Hydrological science and social problems', *Arbor*, vol 646, pp191–202

Downward, S. R. and R. Taylor (2007) 'An assessment of Spain's Programa AGUA and its implications for sustainable water management in the province of Almería, southeast Spain', *Journal of Environmental Management*, vol 82, pp277–289

European Commission (EC) (2007) 'Towards sustainable water management in the European Union – first stage in the implementation of the Water Framework Directive 2000/60/EC', Communication from the Commission to the European Parliament and the Council, Brussels, 22 March 2007. COM(2007) 128 final

EC (2008) 'Health check of the Common Agricultural Policy', http://ec.europa.eu/agriculture/healthcheck/index_en.htm

Garrido, A. (2003) 'An economic appraisal of the Spanish National Hydrological Plan', *International Journal of Water Resources Development*, vol 19, no 3, pp459–470

Garrido, A. (2006) 'Analysis of Spanish water law reform', in B. R. Bruns, C. Ringler and R. Meinzen-Dick (eds), *Water Rights Reform: Lessons for Institutional Design*, International Food Policy Research Institute, Washington, DC, pp219–236

Garrido, A. (2007a) 'Water markets design and experimental economics evidence', *Environmental and Resource Economics*, vol 38, no 3, pp313–330

Garrido, A. (2007b) 'Designing water markets for unstable climatic conditions: Learning from experimental economics', *Review of Agricultural Economics*, vol 29, no 3, pp520–530

Garrido, A. and C. Varela-Ortega (2008) 'Economía del agua en la agricultura e integración de políticas sectoriales', Panel de Estudios, Universidad de Sevilla-Ministerio de Medio Ambiente

Garrido, A., E. Iglesias and M. Blanco (1996) 'Análisis de la actitud de los regantes ante el establecimiento de precios públicos y de mercados de agua', *Revista Española de Economía Agraria*, vol 178, pp139–162

Gómez, C. M. and A. Garrido (2007) 'Cost effectiveness analysis for the WFD', in Pulido (ed), *Economics in Water Management Models. Applications to the EU Water Framework Directive*, Springer

Hanemann, W. M. (2003) 'c. Economics: Findings and recommendations', in A. J. Horne et al (eds), *A Technical Review of the Spanish National Hydrological Plan (Ebro River Out-Of-Basin Diversion)*, Fundación Universidad Politécnica de Cartagena, Spain

Livingston, M. L. and A. Garrido (2004) 'Entering the policy debate: An economic evaluation of groundwater policy in flux', *Water Resources Research*, vol 40 no 12, WS12S02, doi:10.1029/2003WR002737

Llamas, R. and E. Custodio (eds) (2002) *Intensive Use of Groundwater: Challenges and Opportunities*, Balkema Publishing Company, Amsterdam

MAPA (2007) Anuario de Estadística Agroalimentaria. Ministerio de Agricultura, Pesca y Alimentación. (www.mapa.es/es/estadistica/pags/anuario/introduccion.htm). Last accessed 17 August 2008

Messerli, B., R. Weingartner and D. Viviroli (2008) 'Mountains of the world – Water towers for the 21st Century', in A. Garrido and A. Dinar (eds), *Managing Water Resources in a Time of Global Change: Mountains, Valleys and Flood Plains*, Routledge, London, pp11–31

Ministerio Medio Ambiente (MMA) (2007a) *El Agua en la Economía Española: Situación y Perspectivas. Informe Integrado del Análisis Económico de los Usos del Agua. Artículo 5 y Anejos II y III de la Directiva Marco del Agua*, Ministerio Medio Ambiente, Madrid

Ministerio Medio Ambiente (MMA) (2007b) *Precios y costes de los servicios de agua en España. Informe integrado de recuperación de costes de los servicios de agua en España. Artículo 5 y Anejo III de la Directiva Marco de Agua*, Ministerio de Medio Ambiente, Madrid

Ministerio Medio Ambiente (MMA) (2007c). *Evaluación de los Impactos del Cambio Climático en España*, Ministerio de Medio Ambiente, Madrid

Molle, F. and J. Berkoff (2007) *Irrigation Water Pricing Policy: The Gap Between Theory and Practice*, CABI, IWMI, Wallingford, UK

Novo, P. (2008) 'Are virtual water "flows" in Spanish grain trade consistent with relative water scarcity?' *Ecological Economics*, in press

Rodríguez Casado, R. (2008) *La huella hidrológica de la agricultura española*, Trabajo Fin de Carrera, UPM

San Martín González, E. and A. Pérez Zabaleta (2002) 'Una evaluación económica del trasvase del Ebro según la Directiva Marco del Agua', in L. Del Moral (ed), *III Congreso Ibérico sobre Gestión y Planificación del Agua*, Universita de Sevilla and Pablo Olavide

Sauchyn, D., M. Demuth and A. Pietroniro (2008) 'Upland watershed management and global change – Canada's Rocky Mountains and Western Plain', in A. Garrido and A. Dinar (eds), *Managing Water Resources in a Time of Global Change: Mountains, Valleys and Flood Plains*, Routledge, London, pp32–47

Varela-Ortega, C. (2007) 'Policy-driven determinants of irrigation development and environmental sustainability: A case study in Spain', in F. Molle and J. Berkoff (eds), *Irrigation Water Pricing Policy: The Gap Between Theory and Practice*, CABI, IWMI, Wallingford, UK, pp328–346

# 9
# Policy Issues Related to Climate Change in Spain

*Ana Iglesias*

## The reality of climate change

Climate change is already happening. *The Fourth Assessment Report of the Intergovernmental Pannel on Climate Change* (IPCC, 2007) clearly shows that the climatic variations over recent decades have had noticeable direct consequences in natural ecosystems, glaciers and agricultural systems in many regions. Many areas of the world are already struggling today with the adverse impacts of an increase in global average temperature. The scientific literature also suggests that observed changes in climate have affected the frequency and intensity of extremes (drought, floods and heatwaves). The alarming number of extreme weather events that have occurred during the last five years may be the consequence of climate change and suggest that climate change is resulting in the increase in natural climate disasters, at least in some regions. The IPCC defines climate change as a statistically significant variation in the state variables that define the climate of a region (such as temperature or precipitation) or in its variability persistent over an extended period of time (typically decades or longer periods).

The Kyoto Protocol to the United Nations Framework Convention on Climate Change (UNFCCC, 1992) imposed certain reductions of greenhouse gases (GHGs) production on ratifying countries, since the accumulation of GHGs in the atmosphere was found to increase global temperatures and changes in the climate (IPCC, 2007). Two main policy interventions have been identified for combating climate change – mitigation and adaptation. According to the UNFCCC (1992), there is a clear difference between mitigation (reduction of greenhouse gas emissions and carbon sequestration) and adaptation (ways and means of reducing the impacts of, and vulnerability to, climate change). Until recently, UNFCCC negotiations have focused primarily on mitigation; however, it is now clear that the objectives of human well-being in the future should be addressed, stressing the importance of adaptation. Regardless of international progress to reduce emissions of the greenhouse gases that cause climate change (mitigation policies), the climate system will continue to adjust for the next few decades to past and present emissions. This will bring unavoidable impacts on natural and human systems, presenting the challenge of a second response to climate change – adaptation – to prepare for and cope with these impacts.

In contrast to this clear understanding of the concepts of 'climate change' and 'mitigation', the concepts of impacts, vulnerability, risk and adaptation are not defined in either the UNFCCC or the Kyoto Protocol; the terms are used loosely by many scientific and policy communities and they also have a meaning in common usage. It has been observed that interpretation of some of these key terms by scientific groups or policy makers can be quite different, which may lead to varied or false expectations and responses (Levina and Adapmas, 2006). Nevertheless, understanding and quantifying the adaptation responses to climate change is a key issue, since they are key determinants to the economic impacts to society. Stern et al (2006) argue that 'the overall costs and risks of climate change will be equivalent to losing at least 5 per cent of global GDP each year'. Although this has been challenged by many economists with wide-ranging experience working in climate change (Tol, 2006) since it ignores and contradicts numerous unquestionable results (Sachs, 2001; Fankhauser and Tol, 2005; Nordhaus, 2006; Nicholls and Tol, 2006), the analysis in Stern et al (2006) contributed to an open discussion about the cost that society is willing to undertake, therefore eliminating any doubt about the need to adapt in order to avoid unwanted damage.

## Addressing the adaptation challenge

There is a general consensus about the unsustainability of the present model of development and about the need to reach a balance among equity, economic security and the environment. Climate change will likely affect people inside society, creating or reinforcing new forms of social and economic discrimination. In particular, the sustainable management of freshwater resources – especially focusing on the availability of safe drinking water – is one of the main challenges to our present social model of development.

The management of water resources needs to incorporate the principles of sustainable development in order to deal with the increasing pressure on freshwater resources. This pressure arises mainly from the following factors:

1   *Population.* Over the 20th century, global population has tripled while water withdrawals have increased by a factor of about 7.
2   *Pollution.* The effects of industry and agriculture intensification have resulted in major pollution problems in many regions of the world; this is linked (together with scarcity) to the degradation of aquatic ecosystems.
3   *Governance.* Poor governance has been a result of fragmented and uncoordinated management, top-down institutions and increased competition for finite resources.
4   *Climate change.* The impacts of climate change on freshwater resources affect all sectors of society.

There is a high degree of social and scientific awareness about the potential impacts of climate change and the need to adapt water management to hotter and more extreme conditions. It is certain that the need for increased spending as a result of intensified damage caused by extreme weather events will lead to a loss of rural income and economic imbalances between the more and less prosperous parts of Europe and also to environmental damage. Nevertheless, adaptation policies, strategies and concrete measures are fragmented and uncoordinated in most cases. This is in part due to the diverse perception and value that different groups in society place on the issue of climate change and in part is due to the difficulty encountered in evaluating the potential cost of inaction.

Societies have shown, throughout history, a great ability to adapt to changing conditions, with or without a conscious response by citizens and government (Mendelsohn et al, 2004). However, it is likely that the changes imposed by climate change in the future will exceed the limits of autonomous-endogenous adaptation, and that policies will be required to support and enable different sectors of society to cope with similar changes.

The European Commission (EC) has recently adopted a Green Paper entitled 'Adapting to climate change in Europe – options for EU action' (EC, 2007a). This sets out options to help the adaptation process and focuses on four priority areas, including early action to avoid damage and reduce overall costs. Adaptation efforts may have to be stepped up at all levels and in all sectors, and may benefit from coordination across the European Union (EU). The Commission will publish a White Paper containing more concrete policy proposals in 2008.

The present chapter aims to provide an understanding of the potential implications of climate change relevant to policy development in the EU leading to the formulation of measures to reduce the vulnerability of the water sector to climate change. In the following section the chapter assesses the risks of climate change to water resources, providing concrete examples from Spain. The next section explores the challenges and opportunities for developing adaptation policy options aiming to reduce the social vulnerability to the projected impacts of climate change. In the following section, the chapter evaluates how current policy instruments – especially agricultural and water resources policies – work towards adaptation, and potential options for integrating adaptation into them. Finally, in the conclusion and policy implication section, the chapter draws on the results presented in the previous sections to suggest some policy implications related to climate change and water resources elsewhere.

# Climate change risk to water resources

## The European context

There have been several thousand studies into the potential impacts of climate change on water resources, with many different approaches (e.g. physical modelling, econometric analysis) and definitions (e.g. impacts, vulnerability, risk, adaptation). Studies have focused on particular issues (e.g. agricultural water pressure, ecosystem services), time frames (e.g. 2020s, 2050s and 2100), scenarios (e.g. IPCC SRES, 2001) and spatial scales (with a focus on national and global scales). Consequently, our knowledge of the potential impacts is diverse and fragmented. Nevertheless, the projected impacts pose challenges for many water-dependent activities and magnify the regional differences in Europe's natural resources and assets. Although there is a large variation in projected impacts in each EU region, overall the studies are consistent in the direction of change and the spatial distribution of effects. In general in the northern areas, most sectors of the economy are benefited by climate change, providing that projected extremes do not become catastrophic events. However, these potential opportunities will only be possible if water requirements are met. In most of the central and southern areas of Europe water availability is projected to decrease under all scenarios considered. In addition, concurrent altered carbon and nitrogen cycles may have significant implications for soil erosion and water quality.

The effects of climate changes on major water management determinants and expected social and ecological consequences are summarized in Table 9.1. Most studies agree that climate change is likely to have the following common consequences across Europe (EEA, 2007):

- Increased demand for agricultural water in all regions due to expected increases in crop evapotranspiration in response to increased temperatures. The potential for decreasing water demand due to the direct effects of $CO_2$ on the crop have been challenged (Long et al, 2006).
- Increased water shortages, particularly in the spring and summer months, therefore increasing the water requirement for irrigation, especially in southern and south eastern Europe.
- Increased deterioration in water quality due to higher water temperatures and lower levels of runoff in some regions, particularly in summer, imposing further stress in agricultural irrigated areas.

**Table 9.1** *Effects of climate change on main water management determinants and expected social and ecological consequences*

|  | Expected intensity of negative effects | Potential consequences for agro-ecosystems and rural areas | Confidence level of the potential agricultural impact |
|---|---|---|---|
| Water resources | Changes in hydrological regime. Differences in water needs. Increased water shortage. | Variations in hydrological regime. Decreased availability of water. Risks of water quality loss. Increased risk of soil salinization. Conflicts among users. Groundwater abstraction, depletion and decrease in water quality. | High |
| Irrigation requirements | High in areas already vulnerable to water scarcity. | Increased demand for irrigation. Decreased yield of crops. | High |
| Changes in water and soil salinity and erosion | High for southern countries. | Decrease in water quality from nutrient leaching. Decreased crop yields. Land abandonment. Increased risk of desertification. Loss of rural income. | High |
| Land use | Depends on region. | Shift in optimal conditions for farming. Deterioration of soils. Loss of rural income. Loss of cultural heritage. Land abandonment. Increased risk of desertification. | High |
| Increased expenditure in emergency and remediation actions | High for regions with low adaptation capacity. | Loss of rural income. Economic imbalances. | Medium |
| Biodiversity loss | High for vulnerable regions. | Loss of natural adaptation options. Modified interaction among species. | Medium |

- Increased risk of flooding in winter due to the expected concentration of precipitation in this period, affecting significant areas of Europe. The major flood events experienced in recent years (notably 2002 and 2007) demonstrate Europe's vulnerability to floods. In addition, the projected increases in sea level will also affect flooding in low-lying coastal areas.

## The Mediterranean region

Changes in precipitation are probably the most important factor determining the likely impacts in the Mediterranean region. Despite forecasts of increased total annual precipitation in some regions, evapotranspiration is expected to increase in response to increased temperatures. Significant increased temporal and spatial variability of extreme weather is expected, increasing flooding and drought frequency leading to competition for water. Sea level rises will inundate coastal areas; rising sea levels may also lead to salination of the water supply and soil. A decrease in water availability is predicted together with an increase in water demand, leading to potential conflicts between users. Decreasing water resources in some areas may affect soil structure, while reduced soil drainage may lead to increased salinity. However, an increase in the frequency and intensity of floods is predicted in some areas where significant winter rainfall is likely. These changes are expected to reduce the diversity of Mediterranean species. In the Mediterranean region, irrigation accounts for over 60 per cent of the pressure on water resources. Box 9.1 provides an example of the potential impacts of climate change on irrigation in this region based on a number of studies.

## Spain

In Spain, the structural water deficit of many areas in the country has been aggravated during the drought episodes of the past 50 years (Iglesias and Moneo, 2005; Iglesias et al, 2008a). Past efforts to manage drought have built capacity to deal with extreme situations, but have failed to solve the conflict among users, especially with the environment (Iglesias and Moneo, 2005; Iglesias et al, 2008b). Climate change projections indicate an increased likelihood of droughts. Variability of precipitation – in time, space and intensity – can directly influence water resources availability. The combination of long-term change (e.g. warmer average temperatures and possibly lower precipitation) and greater extremes (e.g. droughts) can have decisive impacts on water demand, limiting further ecosystem services. If climate change intensifies drought impacts, Spanish water delivery systems and control may become increasingly unstable and vulnerable. Water managers may find planning more difficult. Current water management strategies based on changes in mean climate variables should be revised to account for the potential increase in anomalous events.

In Spain climate change projections indicate a decrease of precipitation in the southern regions, in some cases up to –40 per cent, by the 2050s compared to

> **BOX 9.1 POTENTIAL IMPACTS OF CLIMATE CHANGE IN IRRIGATION IN THE MEDITERRANEAN REGION**
>
> **Background:** Irrigation accounts for over 60 per cent of total water abstraction, is used on about 10 per cent of the agricultural area, and gives rise to about 90 per cent of the total value of crop production. Water resources vary greatly among basins.
>
> **Problem:** The studies focus on the evaluation of the potential impact of a change in climate on the potential crop production and irrigation demand. The aims also examine the potential increase in irrigation demand in areas already vulnerable to water use conflicts.
>
> **Methods:** Several methods including process-based agronomic models were used to estimate crop yields and crop water requirements at site and regional levels. Crop yield and irrigation demand functions were derived from the validated site results to evaluate spatial water demand and potential change in irrigation areas.
> Each of the models used in the study was validated against local data.
>
> **Scenarios:** The current baseline adopted for the socioeconomic projections was 1990 and the climatic baseline, 1951–1980. Scenarios of climate change were projected for the 2050s with several global climate models driven by a range of socioeconomic conditions.
>
> **Impacts:** Under climate change, irrigation demand is expected to increase in all southern Mediterranean regions, especially the ones with the largest current irrigation areas. The increase in irrigation demand is due to a combination of increased temperature that leads to higher evapotranspiration and decreased precipitation.
>
> **Adaptive responses:** Improvements in water delivery systems are able to supply the demand for increases in irrigation supply and the projected increase in the irrigated area in the northern half of the region, but do not achieve the same results in the south-eastern part of the region.
>
> *Source:* Bindi et al, 2000; Iglesias et al, 2000; Tubiello et al, 2000; Iglesias, 2002; Iglesias et al, 2002; Tubiello and Ewert, 2002; Moriondo et al, 2006; Salinari et al, 2006; Iglesias et al, 2007b, 2008a.

1961–1990 levels, or a small increase in precipitation in the northern regions, with changes in the annual precipitation patterns. In all cases, temperature increases of about 1.5°C are expected, and thereby increased evaporation and reduced soil moisture, resulting in more adverse regional climate conditions than presently experienced. Climate and hydrological experts are beginning to be aware of the implication of future water availability in the region (Hisdal et al, 2001; Iglesias, 2002; Iglesias et al, 2002; Lloyd-Hughes and Saunders, 2002; Iglesias et al, 2007b, 2008a).

Although projected implications of changes in the climate variables depend on the scenario, the time frame, social pressure on water resources, and the method of analysis, most studies agree that there is a likely decrease in water resource availability and an increase in water pressure, especially from agriculture (Garrote et al, 1999; Iglesias et al, 2007b, 2008a; among many others). Figure 9.1 shows that under climate change, reservoir water inflow and water resources availability decrease by −7 and −5 per cent, respectively, on average in all Spanish basins considering a range of climate change scenarios for the middle of the 21st century. These results are clearly scenario dependent and may be optimistic. By 2100 the projections may be more negative. *The Fourth Assessment Report* of the IPCC (2007) states that the reduction of precipitation in Spain may be over 20 per cent by 2100 under the scenario of high population increase and high economic growth (IPCC SRES, 2001). Under these conditions, many watersheds in the southern half of Spain (but also the right-hand side effluents in the Ebro basin) will reduce flow by 40 per cent. In contrast, irrigation demand increases in all locations under the several climate change scenarios. The results indicate increases of water demand and reductions of water supplies that surely will affect ecosystem sustainability, implying substantial future changes in water management. Water resources systems will have to adapt to the evolution of climate. If the projections become a reality, water scarcity is expected to rise in the next decades, posing additional problems for water managers and users.

# Adaptation policy context

## Defining adaptation

Adaptation is about preparing people and their assets for the impacts of climate change. It is concerned with minimizing adverse effects, or maximizing new opportunities, through taking actions that either anticipate or react to changing climatic conditions. The focus of these actions is on managing risk. Investments in risk-based actions are fundamental to reducing the environmental, social and economic costs of climate change.

The need for an adaptation policy stems from the overwhelming scientific consensus that climate change is a significant threat facing the world, its people, environment and economy. Strong mitigation measures are essential to make deep cuts in the greenhouse gas emissions that cause climate change, to avoid dangerous climate change and unprecedented environmental, social and economic disruption. However, as a consequence of present and past emissions of greenhouse gases and the inertia of the climate system, we are already committed to several decades of climate change that cannot now be avoided. Adaptation to cope with the impacts of unavoidable climate change is therefore also necessary as a complementary action to efforts to reduce emissions. In its *Fourth Assessment Report*, the IPCC (2007) recognizes that some adaptation action is occurring, but on a very limited basis, and affirms the need for extensive adaptation across nations and across sectors to address impacts and reduce vulnerability.

Various types of adaptation can be distinguished. Recent studies have highlighted the distinction between 'autonomous adaptation' and 'policy-driven

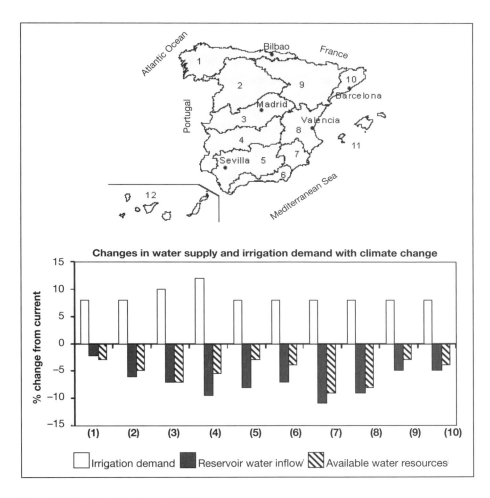

Source: Modified from Iglesias et al, 2008b

**Figure 9.1** *Changes in available water resources, reservoir inflow and irrigation water demand in the hydrological basins in Spain*

adaptation'. Autonomous adaptation describes actions taken 'naturally' by private actors, such as individuals, households and businesses in response to actual or expected climate change, without the active intervention of policy. Autonomous/endogenous adaptations are taken naturally but their extent, direction and effectiveness are a function of existing conditions, infrastructure and technologies, that are in turn a result of existing policies, not necessarily intended for adaptation. In contrast, policy driven adaptation is 'the result of a deliberate policy decision'. Policy-driven adaptation is therefore associated with public agencies, either in that they set policies to encourage and inform adaptation or they take direct action themselves, such as public investment (Stern et al, 2006). Planned policy adaptation actions focus on the vulnerability reduction of people and societies.

Adaptation strategies are put in place to deliver adaptations. An adaptation strategy is a broad plan of action that is implemented through policies and measures. Adaptation strategies are not only reactions to posed threats of climate change, but can comprise, at the same time, a large number of technical, social, economic and environmental challenges (Olensen and Bindi, 2002; Iglesias et al, 2007a, 2007c).

## Integrating climate and sustainable development

The capacity to adapt to environmental change is implicit in the concept of sustainable development. Climate change will add to the many economic and social challenges already being faced by European sustainable development, increasing the vulnerability of marginal areas and populations. Climate change is a real concern for sustainable policy development, raising major issues about the adequacy of current water and land resource management, both globally and within the EU. The unavoidable impacts of climate change put current activities, certainly at the level of individual land and water managers, at significant risk, therefore making imperative the development of both private and public adaptation strategies. These strategies must evolve taking into account the overall strategy for development in the EU.

The EU Gothenburg Sustainable Development Strategy (SDS, 2001) is the point of reference regarding the interpretation and use of the concept of sustainable development in Europe. The strategy involves a set of principles and processes for strategic planning and sustainable development, as well as a coordinated set of measures to ensure their implementation. The EU sustainable development strategy sets out a broad vision of what is sustainable (including environmental, social and economic dimensions) but does not provide an operational definition of sustainable development. It focuses on six non-sustainable trends, including global warming. The SDS is often considered an add-on to the Lisbon Strategy (Lisbon Special European Council, 2000) that focuses on the economic and social dimensions of development. The SDS adds the environmental dimension and the long-term perspective (rights to future generations). Progress on the implementation of these two strategies is achieved formally and informally (Spring reports, by using the European Environment Agency (EEA) indicators, and independent academic revisions that constitute the basis of institutional revisions such as those of the OECD).

Over a decade ago, most countries joined an international treaty – the United Nations Framework Convention on Climate Change (UNFCCC) – to begin considering what can be done to reduce global warming and to cope with whatever temperature increases are inevitable given that climate change is already happening (IPCC, 2007). Recently, a number of nations have approved an addition to the treaty, the Kyoto Protocol, which is an international and legally binding agreement to reduce greenhouse gas emissions worldwide (entered into force on 16 February 2005).

The Earth Summit in Rio (1992) ensured that the sustainable development strategy became a goal for governments around the world, by signing the Agenda for the 21st Century. This Agenda (known as Agenda 21) recognizes that broad public participation in decision making is one of the fundamental prerequisites for the achievement of sustainable development. Agenda 21 was one of the first initiatives relating to sustainable development and climate change, establishing actions and identifying actors to implement strategies on poverty alleviation, the provision of basic education and public services, environmental protection and components of sustainable development that have links with addressing climate change such as the rational use of energy and promotion of ecologically sound technologies.

The UN programme on sustainable development, Agenda 21 (1992), calls on countries to adopt national strategies for sustainable development (NSDS) that should build upon and harmonize the various sectoral economic, social and environmental policies and plans that are operating in the country. In 2002, the World Summit for Sustainable Development (WSSD) urged states to make progress in the elaboration of national strategies for sustainable development and to begin their implementation. By the time of writing, governments have continued to reiterate their commitment to develop and implement NSDS at subsequent UN Commission for Sustainable Development (CSD) sessions. The UN also provides guidelines for national reporting. At the present time, 26 countries (including Spain) and the EC have submitted SD strategies (UN, 2007).

Adaptation to climate change is an essential step towards the process of sustainable development, but the policy priorities for such adaptation in the different social sectors are often fragmented, unformulated and contradictory. The EEA provides an excellent example of a clear definition of policy priorities and adaptation strategies in relation to adaptation in the water sector (EEA, 2007). The policy priorities include: (i) reduce the vulnerability of people and societies; (ii) protect and restore the ecosystems; and (iii) close the gap between supply and demand. The adaptation strategies include: (i) sharing the loss; (ii) preventing the effect; and (iii) research and education. The EEA recognizes the value of high-quality information in order to formulate concrete strategies and the role of regulatory and institutional actions. The missing components of all current strategies are: (i) a lack of guidance related to responsibilities for implementing the strategies and actions; and (ii) a lack of a protocol for policy evaluation.

Climate change policy is a specific policy that needs to be coordinated with the EU and national sustainable development strategy processes. Climate change and sustainable development policies should be mutually enforcing, but this challenge has not been fully addressed.

## European climate change policy

Adaptation is not an alternative, but a necessary complement to mitigation; this is because the climate system responds only slowly to changes in the amounts of GHGs in the atmosphere. Climate changes over the next 40 years or so are inevitable as a consequence of present and past emissions. In recognition of this, the EC has adopted a Green Paper entitled 'Adapting to climate change in Europe – options for EU action' (EC, 2007a). This sets out options to help the adaptation process and focuses on four priority areas, including early action to avoid damage and reduce overall costs. Adaptation efforts need to be stepped up at all levels and in all sectors, and need to be coordinated across the EU.

The Green Paper first evaluates the current knowledge on impacts of climate change including explicitly the results of the Peseta study (Iglesias et al, 2007c; PESETA, 2008). Second, the Green Paper analyses the challenges to adaptation for European society and European public policy. These challenges include: taking early action and saving on future costs; timing adaptation measures; pathways to adaptation; the role of Member States, regional and local authorities; and actions at the EU level. The third is the main contribution of the Green Paper focusing on EU action and proposes priority options for a flexible four-pronged approach to adaptation (Table 9.2):

**Table 9.2** *Proposed priority options for a flexible four-pronged EU approach to adaptation*

| Pillars of the approach | Relevant policy instruments and actions |
|---|---|
| The first pillar: Early action in the EU | Existing policies:<br>• Common Agricultural Policy (CAP)<br>• European Environment and Health Action Plan (2004–2010)(EC, 2004)<br>• Water Framework Directive (WFD, 2000)<br>• Floods Directive (2007)<br>• Communication on water scarcity and droughts (EC, 2007b)<br>• EU Maritime Policy, Marine Strategy and related legislation, Common Fisheries Policy<br>• Biodiversity Communication and its EU Action Plan to 2010 and beyond (2006)<br>• Forest Action Plan (EC, 2006a)<br>• Soil Strategy (EC, 2006b)<br>• Energy and climate package (2008)<br>• Sustainable Consumption and Production Action Plan (forthcoming)<br>• Environmental Impact Assessment (EIA) Directive (1985, amended in 1997)<br>• Strategic Environmental Assessment (SEA) Directive (2001)<br>• Integrated Coastal Zone Management (ICZM) Recommendation (2002)<br>Planned policies:<br>• Industry and services Action Plan (2008) |

| | |
|---|---|
| | • Strategic Energy Technology Plan leading to the creation of a Common European Energy Policy<br>Funding programmes:<br>• Common Agricultural Policy (CAP)<br>• EU Cohesion policy<br>• European Social Fund<br>• Fisheries Structural Fund<br>• LIFE+ |
| The second pillar: Integrating adaptation into EU external actions | • EU Common Foreign and Security Policy (CFSP)<br>• Contribution of the EU to the UNFCCC effort for integrating adaptation into the national development plans though the National Adaptation Programmes of Action (NAPA)<br>• Support the 2004 EU Action Plan on Climate Change and Development (COM(2003) 85 final, 2003)<br>• Forthcoming EU strategy on Disaster Risk Reduction |
| The third pillar: Reducing uncertainty by expanding the knowledge base through integrated climate research | • EU 7th Framework Programme<br>• INSPIRE (Shared environment information system) Directive (2007)<br>• GMES (Global Monitoring for Environment and Security)<br>• Community-supported information systems (floods, forest fires, MIC monitoring and information centre for civil protection)<br>• European data centres<br>• Promote cooperation with international programmes |
| The fourth pillar: Involving European society, business and the public sector in the preparation of coordinated and comprehensive adaptation strategies | • European Climate Change Programme (EPCC)<br>• Possible establishment of a European Advisory Group for Adaptation to Climate Change<br>• Stakeholder consultation |

*Source:* Author's elaboration based on the Green Paper on adaptation (EC (2007a) and other sources)

- *The first pillar.* Early action in the EU. Early action covers policy options in the following areas: integration of adaptation when implementing and modifying existing and forthcoming legislation and policies; integration of adaptation into existing community funding programmes; and development of new policy responses.
- *The second pillar.* Integrating adaptation into EU external actions. This pillar refers to the climate change impacts and adaptation needs that would influence the relations of the EU with other countries and is based on an enhanced dialogue between the EU and developing countries, and also includes neighbouring countries and industrialized countries.
- *The third pillar.* Reducing uncertainty by expanding the knowledge base through integrated climate research. This research-based pillar focuses on the understanding of complex interactions among the climate system, environment, economic sectors and society. The process of understanding

includes cooperation and networking, support to practitioners, and improved information and communication technologies.
- *The fourth pillar.* Involving European society, business and the public sector in the preparation of coordinated and comprehensive adaptation strategies. This pillar focuses on providing guidance to society on how to make the necessary changes to adapt to climate change.

## The national climate change adaptation frameworks

Many Member States have carried out assessments of climate change impacts, including within the agriculture sector, but progress on implementing adaptation actions has been slow, due in part to the long-term nature of climate change effects or respective perceptions by policy makers and the sector alongside the complexity of the information required for decision making and in part to the number of stakeholders involved. The focus of much of the effort made to date has been on the management of flood risk, since this is the main problem in northern European countries. Immediate attention has focused on raising awareness and research activities, and these roles are often facilitated and complemented by organizations that are outside national governments, such as universities or trade and professional bodies (for example, the National Farmers' Union in the UK). National policies on adaptation in agriculture have not yet been clearly articulated.

National adaptation strategies are currently being developed. A complete review is included in the CIRCLE project (Climate Impact Research Coordination for a Larger Europe) report on the current state of National Research Programmes on Climate Change Impacts and Adaptation in Europe (15 May 2007) (CIRCLE, 2007; Medri et al, 2007) and summarized in the Adaptation report to the Directorate-General for Agriculture (Iglesias et al, 2007a).

Table 9.3 summarizes the broad range of adaptation actions that have been designed/planned at different governmental levels and in various sectors. From these efforts, both theoretical and practical knowledge has resulted in a wide range of possible options to adapt to projected climate change impacts. The review of national adaptation strategies highlights the current policy focus on reducing the risk of flooding, either from sea level rise or from increased rainfall. There are also proposals, mainly from southern Member States, to increase the capture and storage of water to ensure adequate supplies. As precipitation patterns change, their limited capacity for water storage may need to be increased to capture a greater proportion of winter rainfall than is currently the case.

Since it is not possible to review all strategies, this section of the chapter summarizes the strategies adopted by Finland – as an example of a thoroughly developed strategy – and the strategy being prepared by the UK that builds from the recognized experience of the various programmes operating in the country. Finland's Adaptation Strategy is part of the National Energy and Climate Strategy that was forwarded to the Finnish Parliament in November 2005. Its objective

**Table 9.3** *Summary of the national adaptation strategies in the EU-27 and other European countries*

| Status of the National Adaptation Strategies | Countries |
|---|---|
| Developed | Finland (published in 2005 by the Ministry of Agriculture and Forestry of Finland) Spain (PNACC is ongoing) France (National Adaptation Strategy) published in 2007 Sweden (National Adaptation Strategy) published in 2007 |
| Under preparation, to be published in the near future (EU-27) | Netherlands (most developed in the water sector) UK (Adaptation Policy Framework is already in progress, under the guidance of the Department for Environment Food and Rural Affairs, Defra) |
| Under preparation, to be published in the near future (other European countries) | Norway (currently in the process of developing adequate response strategies to the impacts of climate change, both sector by sector and as an overall strategy) |
| First steps in including climate change adaptation within the framework of their National Climate Policy in addition and complementarily to mitigation | Rest of the countries |

Note: PNACC, Plan National de Adaptación al Cambio Climatico.

is to reinforce and increase the capacity of society to adapt to climate change. Adaptation may involve minimizing the adverse impacts of climate change, or taking advantage of its benefits. While the National Energy and Climate Strategy focuses on mitigation measures to be taken in the near future, the scope of the Adaptation Strategy extends as far as 2080. The Adaptation Strategy gives a detailed account of the expected impacts of climate change and presents adaptation measures to be taken in sectors including agriculture and food production, forestry, fisheries, reindeer husbandry, game management, water resources, biodiversity, industry, energy, transport and communication, land use and planning, building, health, tourism and recreation, and insurance. Priorities identified for increasing adaptation capacities for the next 5–10 years include: (i) mainstreaming climate change impacts and adaptation into sectoral policies; (ii) targeting long-term investments; (iii) coping with extreme weather events; (iv) improving monitoring systems; (v) strengthening research and development; and (vi) international cooperation. The research programme on adaptation was initiated in 2006. The

National Strategy also identified sector-specific adaptation measures as important priorities for 2006–2015.

Action to prepare the UK for climate change has already begun. A climate change perspective is incorporated into many areas of government policy, including flood management, water resources, planning, building regulations, health, agriculture and international development. The government funds the UK Climate Impacts Programme (UKCIP, www.ukcip.org.uk) to improve the knowledge base on climate impacts and to assist stakeholders (including those in the agriculture sector) to adapt. The UK's first Adaptation Policy Framework is under development, driven by the Department for Environment Food and Rural Affairs (Defra, 2007). The recognized key priorities for adaptation for the UK over the next 30–50 years are described in the UK Climate Impacts Programme (UKCIP, 2008): water resource management; coastal and river flood defence; enhanced resilience of buildings and infrastructure; management of wildlife, forestry and agriculture; and coordinated approaches to planning.

## National climate change adaptation frameworks in Spain

There is a high degree of social and scientific awareness of the need to adapt water management to hotter and more extreme conditions. In the Mediterranean region, more adaptation measures have been adopted or are under consideration here than in the other agro-climatic zones; this is consistent with the expectation that the region will be worst affected by climate change. Although climate change is a global issue, the national adaptation plans are extremely varied from the strategic point of view, reflecting – in part – past and present efforts put into understanding the issues at stake.

In Spain, the Climate Change National Adaptation Plan, formally adopted by the ministerial cabinet on 6 November 2006, is a reference framework for the coordination of public administrations in relation to the evaluation of impacts, vulnerability and adaptation to climate change in Spain. The Plan is based on knowledge development, public participation and information dissemination. The knowledge strategy ranges from scenario development to sectoral impact evaluations. The adaptation component is not explicitly addressed. The plan establishes a complex institutional structure based in the Ministry of the Environment and coordinated by the Spanish National Office for Climate Change that coordinates the Inter-ministerial Commission, the Coordination Commission for Climate Change Policies and the National Council for Climate.

# Potential role of the current policy instruments in adaptation in the EU

## Energy and climate package

Energy is the main factor in climate change, accounting for some 80 per cent of the EU's greenhouse gas emissions. On 10 January 2007 the Commission adopted an Energy and Climate package to guide the EU towards a sustainable, competitive and secure energy policy. One of its central themes is to tackle the energy challenge by first making an effort to use energy more efficiently before looking into possible alternatives. The ambitious energy policy package proposes to pursue the objective of a sustainable, competitive and secure supply of energy.

The EU Action Plan is needed to help the EU achieve its energy goals. The European Commission has drawn up such a plan using the many contributions gathered from public consultations. The plan is comprised of several clearly defined aims that, together, will shift the EU decisively towards a more sustainable, secure and competitive low-energy economy, representing the core of a new Common European Energy Policy building on the European Strategic Energy Technology Plan (SET-PLAN) 'Towards a low carbon future' (EC, 2007c).

## Common Agricultural Policy

European agricultural policy faces some serious challenges in the coming decades – even without climate change. The most striking of these are loss of comparative advantage in relation to international growers, competition for international markets, declining rural populations, land deterioration (including salination), competition for water resources, and rising costs due to environmental protection policies. Demographic changes are altering vulnerability to water shortages and agricultural production in many areas, with potentially serious consequences at local and regional levels. Population and land use dynamics, and the overall policies for environmental protection, agriculture and water resources management are the key drivers for possible adaptation options to climate change.

The Common Agricultural Policy (CAP) plays an important role in the areas of food production, the mainstreaming of rural landscapes and the provision of environmental services. Adjustments to the CAP in the 'Health Check' of 2008 provide opportunities to examine how to integrate climate change adaptation – and mitigation – into agriculture support programmes. Consideration might be given to the extent to which the CAP can promote good farming practices that are compatible with changing climatic conditions and which contribute to protecting the environment.

The 2003 reforms of the CAP were a first step towards a framework for the sustainable development of EU agriculture. The central objective of the reforms was to promote an agricultural sector that was competitive and responsive to the market. This was founded on the principles of high standards for the environment. Decoupling brought about greater market responsiveness, whereas higher standards were achieved through cross compliance. The future direction of the CAP is clearly building on the 2003 reforms, with a continued shift from market intervention and further decoupling. Importantly, however, the CAP needs to address the challenge of climate change in order to facilitate adaptation to risks and opportunities. Here overall rules for farm support, rural development policy and crisis management will play important roles in increasing agriculture's resilience to climate change impacts.

While the CAP currently does not contain measures aimed explicitly at adaptation, there are already opportunities to support and facilitate adaptation (such as through rural development policy, see below). Since the 2003 CAP reform, which introduced the Single Payment Scheme and decoupled direct payments from production, farmers have greater flexibility to respond to climate change.

Climate change objectives have been integrated into the framework of rural development policy for the period 2007–2013 and adaptation is now recognized as one of the priorities for the EU. Member States are encouraged to incorporate climate change actions in their national strategy plans and rural development programmes. The Community Strategic Guidelines for Rural Development identify climate change as a priority for the environment and countryside (Axis 2), and recognize that agricultural and forestry practices have a role to play in adapting to the impacts of climate change. Climate change risk and adaptation is also a consideration in rural competitiveness (Axis 1), and diversification and rural life (Axis 3). The Directorate-General for Agriculture is also looking at options for the management of climate change risks and tools to aid adaptation.

As the governing policy instrument for this sector, adjustments to the CAP could provide opportunities for integrating adaptation into agricultural support measures. Consideration might be given to the extent to which the current CAP framework promotes farming practices that are compatible with changing climatic conditions and which contribute to protecting the environment. Thought also needs to be given to the longer-term structural adaptations that will be required, including changes to existing farming and land use systems, breeding to maximize yield under new conditions, and the application of new technologies such as water use efficiency techniques.

The contribution of current CAP measures towards adaptation was evaluated in a recent study (Iglesias et al, 2007a) in order to consider how existing policy instruments may be continued or extended to facilitate adaptation. The analysis also aimed to reveal where policies may present a barrier to adaptation or lead to 'maladaptation'. The strengths and weaknesses of existing CAP instruments to influence adaptation are analysed – covering both direct income support payments and rural development measures. Consideration of related legislation was also included where appropriate and the measures were grouped according to the type of adaptation option they would best support – technical, management or infrastructure. The main conclusions of the Iglesias et al (2007a) study are:

- The Rural Development Programmes have the potential to benefit further by guiding or placing an obligation on Member States to meet or consider the impacts of future climate change.
- Agri-environment schemes have the potential to support many adaptation initiatives.
- To ensure investments made through CAP bring benefits in terms of adaptation, linking funding to cross compliance should be explored.
- Mitigation to climate change is explicitly mentioned throughout the rural development regulations. This could be expanded to include adaptation.
- Adjusting the criteria for those eligible for rural development support for areas with high vulnerability to climate change may be an option to facilitate their adaptation.
- Adaptation to climate change will be needed at all spatial levels. The rural development measures can do this through careful coordination from the grass roots Leader programme all the way up to integration with river basins through the Water Framework Directive.

Supplementing current Statutory Management Requirements with new legislation that addresses climate-related impacts would create stronger incentives for Single Payment Scheme claimants to adapt. The flexibility that Member States can exercise in determining Good Agricultural and Environmental Condition (GAEC) standards allows for highly appropriate and localized management practices that assist with adaptation. The potential of GAECs would be maximized by requiring Member States to identify major environmental pressures, which may include climate impacts, and justify the inclusion or exclusion of corresponding standards. Member States should be required to make provision for training farmers on climate change issues, particularly new entrants such as young farmers. Developing the role and scope of the Farm Advisory System would be a feasible option for effective knowledge transfer. In addition to existing CAP instruments, insurance needs to be considered and encouraged to allow farmers to increase their resilience to climate change. This may provide further incentives for farmers to adapt their business and buildings in order to reduce their premiums.

## Water Framework Directive

The EU Water Framework Directive (WFD) sets out clear output targets for each of the requirements and includes a concrete timetable (Table 9.4). Some aspects of the implementation are very clear, for example, the directive recognizes the importance of leveraging a mix of policy initiatives and establishes a target for the introduction of pricing policies by 2010. The effectiveness of the pricing policies may be limited,

**Table 9.4** *EU Water Framework Directive*

| Year | Issue |
|---|---|
| 2000 | Directive entered into force (Art. 25) |
| 2003 | Transposition in national legislation (Art. 3) |
|  | Identification of River Basin Districts Authorities (Art. 23) |
| 2004 | Characterization of river basin: pressures, impacts and economic analysis (Art. 5) |
| 2006 | Establishment of monitoring network (Art. 8) |
|  | Start public consultation (at the latest) (Art. 14) |
| 2008 | Present draft river basin management plan (Art. 13) |
| 2009 | Finalize river basin management plan including programme of measures (Art. 13 and 11) |
| 2010 | Introduce pricing policies (Art. 9) |
| 2012 | Make operational programmes of measures (Art. 11) |
| 2015 | Meet environmental objectives (Art. 4) |
| 2021 | First management cycle ends (Art. 4 and 13) |
| 2027 | Second management cycle ends, final deadline for meeting objectives (Art. 4 and 13) |

*Source:* Based on the EU WFD, 2000

especially in areas with large groundwater withdrawals for irrigation. The WFD provides a consistent framework for integrated water resources management, but does not include climate change directly. The challenge will be to incorporate the measures to cope with climate change as part of its implementation, stating the first cycle for 2009 (see Table 9.4).

In many regions of Europe, inconsistent land use planning, incorrect water allocation, and inadequate water pricing automatically leads to overuse. Making water saving a priority, improving efficiency in all sectors, and applying efficient pricing policies, are already included in the WFD and are essential elements for climate change adaptation. The initial first steps for the implementation of the WFD could provide incentives to reduce water consumption and increase efficiency of use in all sectors.

## Floods Directive

The recent EU Floods Directive (2007/60/EC) on the assessment and management of flood risks entered into force 26 November 2007 (EU, 2007). This directive now requires Member States to assess if all water courses and coastlines are at risk from flooding, to map the flood extent and assets and humans at risk in these areas and to take adequate and coordinated measures to reduce this flood risk. This directive also reinforces the rights of the public to access this information and to have a say in the planning process. The directive shall be carried out in coordination with the Water Framework Directive, notably by flood risk management plans and river basin management plans being coordinated, and through coordination of the public participation procedures in the preparation of these plans. All assessments, maps and plans prepared shall be made available to the public.

## Initiative on drought and water scarcity

The communication from the Commission to the European Parliament and the Council addressing the challenge of water scarcity and drought in the European Union (EC, 2007b) is closely linked to climate change and adaptation. The communication presents an initial set of policy options at European, national and regional levels to address and mitigate the challenge posed by water scarcity and drought within the Union. The Commission remains fully committed to continuing to address the issues at international level, in particular thought the United Nations Convention to Combat Desertification and the United Nations Convention on Climate Change.

## Indicators for evaluating climate change policy

As reported in the previous section, climate change policy is likely to be fragmented and included in many European and national strategies. This framework complicates the evaluation of the policy and therefore limits the capacity of society for revising and improving the policy. The Commission developed a set of indicators to monitor the implementation of the EU sustainable development strategy. The framework for indicators designed by the Commission is based on themes (12), sub-themes (45) and area (98) directly linked to EU policy priorities. This framework intends to provide a clear and easily communicable structure for the sustainable development strategy.

The Eurostat (2005) headlines indicators for the following ten themes: economic development (gross domestic product (GDP) per capita); poverty and social exclusion (at risk-of-poverty rate after social transfers); ageing society (current and projected old age dependency ratio); public health (healthy life years at birth by gender); climate change and energy (total greenhouse gas emissions; gross inland energy consumption by fuel); production and consumption patterns (total material consumption); management of natural resources (biodiversity index; fish catches outside safe biological limits); transport (vehicle transport); good governance (level of citizens' confidence in EU institutions); and global partnership (official development assistance). Some of these indicators are already proving to be useful for evaluating climate change policy, but others – especially those related to social vulnerability – remain to be further developed.

## The Seventh Framework Programme

On 18 December 2006, the Council adopted decisions establishing the Seventh Framework Programme (FP7) of the European Community (EC) for research and technological development for the period 2007–2013, and the FP7 for nuclear research activities (Euratom) for 2007–2011. The atmospheric sciences, land use and water resources research in the Seventh Framework Programme is more locally orientated and more focused on climate change (European Union, 2008).

## Conclusion and policy implications

### Vulnerability of water resources in Spain

Water resources in Spain are increasingly unstable and vulnerable, but climate change is only one of the determinants of their vulnerability. The issue is even more pertinent where 'at risk' regions and social groups are already economically marginal or at the edge of climate tolerance. To reduce the vulnerability of water resources to climate change across Spain and the EU, robust policy options or adaptation response strategies are required. The risks are not just long term; in the short term, extreme weather events could cause major damage and loss of ecosystems, especially in marginal areas. The proportion of the rural population with limited water resources is highest in Spain and other southern regions of the EU – regions that are projected to face the greatest risks and have the fewest opportunities (from climate change). These regions are the most vulnerable. The northern regions, where water resources are less limited, rural and urban populations may be at risk of increased flooding; these regions typically have integrated the flooding-control actions reasonably into the land and water resource management plans at the national level and have the potential to invest in adaptation. If climate changes continue to intensify, many southern European regions may become increasingly unstable and vulnerable to changing climate patterns and extreme events. European society may find planning more difficult.

### Coordination between national and EU policy instruments

EU policies are the main determinant of water, land and natural resource policy in Spain. Adaptation is unlikely to be facilitated through the introduction of new and separate policies at the national level, but rather by the revision of existing local policies that undermine adaptation and the strengthening of policies that enhance it. If adaptation is to become 'mainstreamed', it will be necessary for relevant EU-wide polices, such as the CAP and the WFD, to address the issue more directly. Existing agreements also have a part to play.

Existing policy instruments can be used to stimulate and facilitate adaptation and other mechanisms must also be utilized, such as insurance, capacity building, networks and partnerships. Adjustments to the CAP and the 'Health Check' of 2008 could provide opportunities to examine how to integrate adaptation into agriculture support programmes. Consideration might be given to the extent to which the CAP can promote good farming practices that are compatible with changing climatic conditions.

Both the reformed CAP and rural development measures can assist in adapting European populations to climate change. This paper proposes that adaptation to climate change impacts in agriculture and related water management issues could be included within revised cross compliance requirements of the CAP. This will certainly modify the irrigation pattern of extensive crops. Nevertheless, the effects on largely degraded areas of fruits and vegetable production may not be accomplished. In this case, local policies may play an important role. It also proposes how options for rural development spending could include incentives for farmers and rural communities to adapt to climate change. These might include support for improved water management through the WFD, and training and capacity building through the Farm Advisory Service and Leader.

To minimize the negative impacts of climate change and to take advantage of the potential benefits, adaptation efforts will need to be introduced at all levels and may need to be coordinated across the EU. The importance and benefits of community-wide adaptation were recognized when the European Commission published its first policy document 'Adapting to climate change in Europe – options for EU action' (EC, 2007a). This Green Paper sets out options to facilitate the adaptation process and focuses on four priority areas, including early action to avoid damages and reduce overall costs. An emphasis is also placed on the need for EU coordination.

## Lessons learned from climate change policy in Spain

The vulnerability of water resources to climate change in Spain is a main component of the dissemination campaign at the national and local levels. Climate change will have both negative and positive implications and it is important that communities are given the capacity to recognize, understand and act on these. Knowledge transfer between scientists, political decision makers and the people directly affected by climate change is currently weak, and existing information is poorly used. Nevertheless, Spain is making efforts at all levels to ensure the communication and dissemination of climate change knowledge. Some difficulties challenging this knowledge transfer include: the number and range of stakeholders involved in adaptation; the inherent uncertainty in climate science and impacts projections – uncertainty can lead to confused messages and inertia if it is not communicated in the right way; and the lack of credible socioeconomic scenarios required to improve climate change impacts and provide a framework for adaptation decision making for practitioners.

Water managers in Spain have always carried out adaptive changes based on the weather and respond in the short term by altering management practices. Nevertheless, national and sectoral policies and management actions in isolation offer limited opportunities for adaptation since large changes in management may require public-funded programmes to help drive the changes.

The sectoral approach to impacts and adaptation has provided a pragmatic solution to a wide-ranging problem. However, adaptations often involve combined effort across many sectors. Water resources are sensitive to responses in other sectors, particularly agriculture, tourism and biodiversity conservation. Adaptation measures for water resources should take account of policies in other sectors. Wider influences on water resources, such as changes in non-climate driven pressures, must be considered alongside climate change. It is important to consider whether adaptations are sustainable, or rendered irrelevant by other drivers. This holistic approach should also ensure that adaptation decisions and investments are both cost-effective and proportionate to the risks or benefits that may be incurred. The main challenge of climate change policy in Spain is a clear definition of multiple priorities and responsibilities for implementing the strategic measures.

## References

Bindi, M., L. Fibbi, F. Maselli and F. Miglietta (2000) 'Modelling climate change impacts on grapevine in Tuscany', in T. E. Downing, P. A. Harrison, R. E. Butterfield and K. G. Lonsdale (eds), *Climate Change, Climatic Variability and Agriculture in Europe*, Environmental Change Unit, University of Oxford, UK, pp191–216

CIRCLE (2007) (Climate Impact Research Coordination for a Larger Europe) www.circle-era.net/ (accessed 11 November 2008)

Defra (2007) Climate Change and biodiversity in agri-environment schemes (2007) Project AC0304, www.defra.gov.uk (accessed 11 November 2008)

European Commission (EC) (1997) 'Directive 85/337/EEC on the assessment of the effects of certain public and private projects on the environment', EIA Directive (EU legislation) introduced in 1985 and was amended in 1997, European Commission, Brussels

European Commission (EC) (2002) *Integrated Coastal Zone Management (ICZM) Recommendation*, (2002/413/EC)

European Commission (EC) (2003) EU Action Plan on Climate Change and Development. Brussels, 11 March 2003. Communication from the Commission to the Council and the European Parliament 'Climate change in the context of development cooperation' COM(2003) 85 final

European Commission (EC) (2004) European Environment and Health Action Plan (2004–2010), Brussels, 9 June 2004. Communication from the Commission to the Council, the European Parliament, the European Economic and Social Committee, (SEC(2004) 729) COM(2004) 416 final

European Commission (EC) (2006a) Communication from the Commission to the Council and the European Parliament on an EU Forest Action Plan, Commission of the European Communities Brussels, 15 June 2006, COM(2006) 302 final

European Commission (EC) (2006b) Communication from the Commission to the Council, the European Parliament, the European Economic and Social Committee and the Committee of the Regions on a Thematic Strategy for Soil Protection, Commission of the European Communities Brussels, 22 September 2006, COM(2006) 231 final

European Commission (EC) (2007a) 'Adapting to climate change in Europe – options for EU action', Brussels, 29 June 2007, Green Paper from the Commission to the Council, the European Parliament, the European Social Committee and the Committee of the Regions, COM(2007) 354

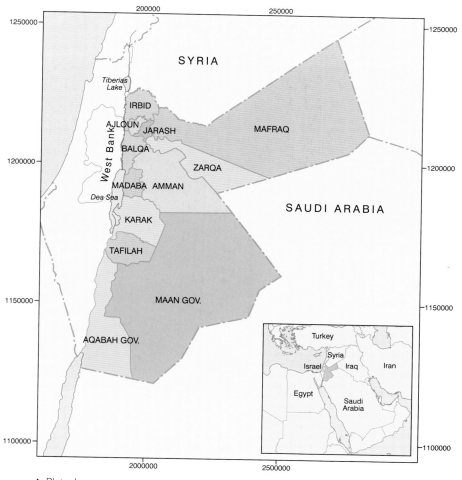

▲ Plate 1
The Hashemite Kingdom of Jordan

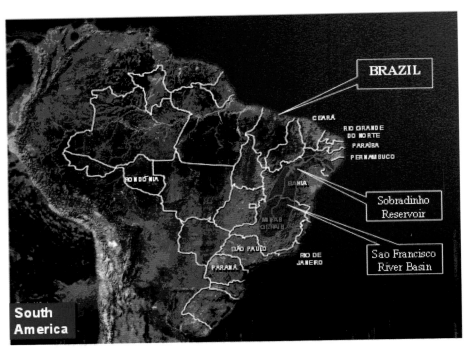

▲ Plate 2
Location of the Sao Francisco River Basin

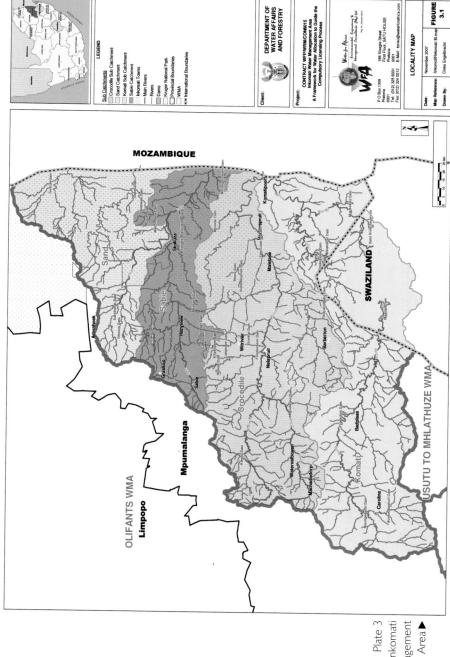

Plate 3
Map of the Inkomati Water Management Area ▶

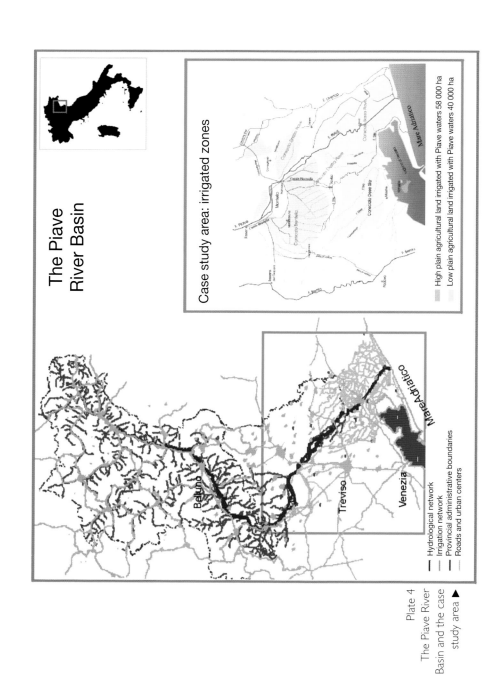

Plate 4
The Piave River Basin and the case study area ▲

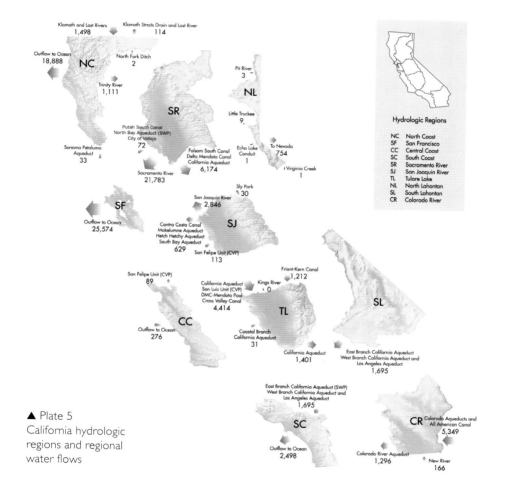

▲ Plate 5
California hydrologic regions and regional water flows

▲ Plate 6
California rivers and water facilities

European Commission (EC) (2007b) 'Water scarcity and drought (2007)', Communication from the Commission to the European Parliament and the Council addressing the challenge of water scarcity and drought in the European Union, 28 July 2007, COM(2007) 414 final

European Commission (EC) (2007c) 'A European Strategic Energy Technology Plan (SET-PLAN) Towards a low carbon future', Communication from the Commission to the Council, the European Parliament, the European Economic and Social Committee and the Committee of the Regions, Brussels, 22 November 2007, COM(2007) 723 final

European Commission (EC) (2007d) INSPIRE (Shared environment information system), Directive 2007/2/EC of the European Parliament and of the Council of 14 March 2007 establishing an Infrastructure for Spatial Information in the European Community (INSPIRE)

European Commission (EC) (2008) 'Energy and climate package', integrated proposal for Climate Action from the European Commission, launched 23 January 2008, http://ec.europa.eu/energy/energy_policy/index_en.htm (accessed March 2008)

European Environment Agency (EEA) (2007) Climate change and water adaptation issues. EEA Technical Report No. 2/2007 EEA, Copenhagen

European Union (EU) (2007) 'Floods Directive', EU Directive 2007/60/EC on the assessment and management of flood risks, *Official Journal of the European Union*, 6 November; entered into force 26 November 2007

European Union (EU) (2008) The Seventh Framework Programme, research policy http://ec.europa.eu/research/fp7/index_en.cfm (accessed 11 November 2008)

Eurostat (2005) Indicators for monitoring the EU Sustainable Development Strategy, http://ec.europa.eu/eurostat/sustainabledevelopment (accessed March 2008)

Fankhauser, S. and R. S. J. Tol (2005) 'On climate change and economic growth', *Resource and Energy Economics*, vol 27, pp1–17

Garrote, L., I. Rodríguez and F. Estrada (1999) 'Una evaluación de la capacidad de regulación de las cuencas de la España peninsular', VI Jornadas Españolas de Presas. Libro de Actas, Madrid [An evaluation of the regulation capacity of continental Spain. Sixth Spanish Conference of Dams] pp645–656

Global Monitoring for Environment and Security (GMES) www.gmes.info/ (accessed 11 November 2008)

Hisdal, H., K. Stahl, L. M. Tallaksen and S. Demuth (2001) 'Have streamflow droughts in Europe become more severe or frequent?' *International Journal of Climatology*, vol 21, no 3, pp317–333

Iglesias, A. (2002) 'Climate changes in the Mediterranean: Physical aspects and effects on agriculture', in H. J. Bolle (ed.), *Mediterranean Climate*, Springer, New York

Iglesias, A. and M. Moneo (eds) (2005) 'Drought preparedness and mitigation in the Mediterranean: Analysis of the organizations and institutions', *Options Méditerranéennes*, CIHEAM, Centre International de Hautes Etudes Agronomiques Méditerranéennes, Paris

Iglesias, A., C. Rosenzweig and D. Pereira (2000) 'Agricultural impacts of climate in Spain: Developing tools for a spatial analysis', *Global Environmental Change*, vol 10, pp69–80

Iglesias, A., M. N. Ward, M. Menendez and C. Rosenzweig (2002) 'Water availability for agriculture under climate change: Understanding adaptation strategies in the Mediterranean', in C. Giupponi and M. Shechter (eds), *Climate Change and the Mediterranean: Socioeconomic Perspectives of Impacts, Vulnerability and Adaptation*, Edward Elgar, Cheltenham, pp75–93

Iglesias, A., K. Avis, M. Benzie, P. Fisher, M. Harley, N. Hodgson, L. Horrocks, M. Moneo and J. Webb (2007a) 'Adaptation to climate change in the agricultural sector', AGRI-2006-G4-05. Report to European Commission Directorate-General for Agriculture and Rural Development. ED05334. Issue Number 1. December

Iglesias, A., L. Garote, F. Flores and M. Moneo (2007b) 'Challenges to manage the risk of water scarcity and climate change in the Mediterranean', *Water Resources Management*, vol 21, no 5, pp227–288

Iglesias, A., L. Garrote, M. Moneo and S. Quiroga (2007c) 'Projection of economic impacts of climate change in sectors of Europe based on bottom-up analysis (PESETA)', Lot 4: Agriculture, Physical Impacts on Agriculture, Universidad Politécnica de Madrid

Iglesias, A., A. Cancelliere, F. Cubillo, L. Garrote and D. A. Wilhite (2008a) *Coping with Drought Risk in Agriculture and Water Supply Systems: Drought Management and Policy Development in the Mediterranean*, Springer, The Netherlands

Iglesias, A., L. Garrote and F. Martin-Carrasco (2008b) 'Drought risk management in Mediterranean river basins', *Integrated Environmental Assessment and Management* (in press)

IPCC (2007) *Climate Change 2007: Fourth Assessment Report of the Intergovernmental Panel on Climate Change*, Cambridge University Press, Cambridge

IPCC SRES (2001) *Special Report on Emission Scenarios*, Intergovernmental Panel on Climate Change, http://sres.ciesin.org (accessed 11 November 2008)

Levina, E. and H. Adapmas (2006) 'Domestic policy frameworks for adaptation to climate change in the water sector', OM/ENC/EPOC/IEA/SLT(2006)2, OECD, Paris

The Lisbon Special European Council (2000) 'Towards a Europe of innovation and knowledge', http://europa.eu/scadplus/leg/en/cha/c10241.htm (accessed 11 November 2008)

Lloyd-Hughes, B. and M. A. Saunders (2002) 'Seasonal prediction of European spring precipitation from ENSO and local sea surface temperatures', *International Journal of Climatology*, vol 22, pp1–14

Long, S., E. A. Ainsworth, A. D. B. Leakey, J. Nösberger and D. R. Ort (2006) 'Food for thought: Lower-than-expected crop yield stimulation with rising $CO_2$ concentrations', *Science*, vol 312, pp1918–1921

Medri, S., S. Castellari and M. Konig (eds) (2007) *Report on the Current State of National Research Programmes on Climate Change Impacts and Adaptation in Europe*, CIRCLE (Climate Impact Research for a Larger Europe), Vienna

Mendelsohn, R., A. Dinar, A. Basist, P. Kurukulasuriya, M. I. Ajwad, F. Kogan and C. Williams (2004) *Cross-sectional Analyses of Climate Change Impacts*, World Bank Policy Research Working Paper 3350, Washington, DC

Moriondo, M., M. Bindi, C. Giannakopoulos and J. Corte Real (2006) 'Impact of changes in climate extremes on winter and summer agricultural crops at the Mediterranean scale', *Climater Research*, vol 31, pp85–95

Nicholls, R.J. and R. S. J. Tol (2006) 'Impacts and responses to sea-level rise: A global analysis of the SRES scenarios over the 21st Century', *Philosophical Transactions of the Royal Society A – Mathematical, Physical and Engineering Sciences*, vol 361, no 1841, pp1073–1095

Nordhaus, W. D. (2006) 'Geography and macroeconomics: New data and new findings', *Proceedings of the National Academy of Science*, www.pnas.org/cgi/doi/10.1073/pnas.0509842103 (accessed 11 November 2008)

Olensen, J. O. and M. Bindi (2002) 'Consequences of climate change for European agricultural productivity, land use and policy', *European Journal of Agronomy*, vol 16, pp239–262

Projection of Economic impacts of climate change in Sectors of the European Union based on boTtom-up Analysis (PESETA) (2008) http://peseta.jrc.es (accessed 11 November 2008)

Sachs, J. (2001) 'Tropical underdevelopment', Working Paper 8119, National Bureau of Economic Research, Cambridge

Salinari, F., S. Giosue, F. N. Tubiello, A. Rettori, V. Rossi, F. Spanna, C. Rosenzweig and M. L. Gullino (2006) 'Downy mildew epidemics on grapevine under climate change', *Global Change Biology*, vol 12, pp1–9

SDS (2001) EU Gothenburg Sustainable Development Strategy. 'A sustainable Europe for a better world: A European strategy for sustainable development', European Commission, Brussels, http://ec.europa.eu/environment/eussd (accessed 11 November 2008)

Stern, N., S. Peters, V. Bakhshi, A. Bowen, C. Cameron, S. Catovsky, D. Crane, S. Cruickshank, S. Dietz, N. Edmonson, S. L. Garbett, L. Hamid, G. Hoffman, D. Ingram, B. Jones, N. Patmore, H. Radcliffe, R. Sathiyarajah, M. Stock, C. Taylor, T. Vernon, H. Wanjie and D. Zenghelis (2006) *Stern Review: The Economics of Climate Change*, HM Treasury, London

Strategic Environmental Assessment (SEA) Directive (2001) European Directive 2001/42/EC (the SEA Directive) on the assessment of the effects of certain plans and programmes on the environment, European Commission, Brussels

Tol, R. S. L. (2006) 'The Stern review of the economics of climate change: A comment', Economic and Social Research Institute Hamburg, Vrije and Carnegie Mellon Universities, 2 November, 2006, www.fnu.zmaw.de/fileadmin/fnu-files/reports/stern review.pdf (accessed 11 November 2008)

Tubiello, F. N. and F. Ewert (2002) 'Simulating the effects of elevated $CO_2$ on crops: Approaches and applications for climate change', *European Journal of Agronomy*, vol 18, no 1–2, pp57–74

Tubiello, F. N., M. Donatelli, C. Rosenzweig and C. O. Stockle (2000) 'Effects of climate change and elevated $CO_2$ on cropping systems: Model predictions at two Italian locations', *European Journal of Agronomy*, vol 13, pp179–189

UK Climate Impacts Programme (UKCIP) (2008) www.ukcip.org.uk/ (accessed 11 November 2008)

United Nations (UN) (2007) Status of national reporting for CSD-16/17 December 2007. The third cycle of the UN Commission on Sustainable Development (CSD), CSD-16/17

UN Commissions for Sustainable Development (2007) National Reports for CSD-16/17, www.un.org/esa/sustdev/natlinfo/natlinfo.htm (accessed March 2008)

United Nations Framework Convention on Climate Change (UNFCCC) (1992) FCCC/INFORMAL/84. GE.05-62220 (E) 200705

WFD (2000) Water Framework Directive, Directive 2000/60/EC of the European Parliament and of the Council establishing a framework for the Community action in the field of water policy (EU Water Framework Directive, WFD), European Commission, Brussels

*Part II*

# Issues in Water Resource Strategy
---

# 10

# Water Conflicts: Issues in International Water, Water Allocation and Water Pricing with Focus on Jordan

*Munther J. Haddadin*

## Strategic importance of water

The various needs for water by all living beings are well known; the social needs are manifested in water inputs for hygiene, for the maintenance of public health, and for job creation. Environmentally it is a crucial input for a clean environment in urban and rural settings. And economically it is so vital that there is practically no economic activity that can run without water input. As a matter of fact, life as we know it exists here and nowhere else in the universe because water, in its liquid form, exists here on our planet and nowhere else that we know of.

The strategic importance of water stems from its role outlined above. Clean blue water meets the needs of society for domestic, municipal and most industrial purposes. Waterfalls, natural or man-made, generate energy for various social and economic purposes. Water is also crucial in the process of food production through rain-fed and irrigated agriculture. Food is the major source of supply for energy in the human body, and solar energy is indispensable for plant growth and fruit production.

Economic and social development cannot proceed and be managed without water. Such development improves the standard of living of people and prompts a greater demand for water as living styles improve. Municipal and industrial water needs are usually a function of the economic category of the country in question, as are the food needs. As such, the water need of any country for municipal, industrial and agricultural purposes is a function of its economic standing. Provided

**Table 10.1** *Water needed in various uses (cubic metre per capita) by income categories of Middle East countries*

| Income category | High income | Upper middle income | Lower middle income | Low income |
|---|---|---|---|---|
| Agricultural | −940 | −1250 | −1500 | −1780 |
| Municipal | −100 | −85 | −75 | −55 |
| Industrial | −260 | −165 | −125 | −65 |
| Total | −1300 | −1500 | −1700 | −1900 |

*Note:* A negative sign denotes need while a positive sign denotes supply.
*Source:* Haddadin (2006) and World Bank Country Groups based on GNP for the year 2002 (World Bank, 2002)

environmental conditions allow the production of the food components needed by a given country, the water needs per capita to meet the said needs for the four income categories of countries of the world are shown in Table 10.1.

The benefits drawn from water use in agriculture are worth looking at. In resource terms, agriculture is the only user of soil water (green water) stored in the soil after rainfall. Soil water is an important component in the renewable water resources of any country. In Jordan, its blue water equivalent exceeds the indigenous blue water. Agriculture can use treated wastewater, a fact that adds to the useable renewable water resources. In demographic terms, agriculture could be a natural incentive for balancing the spatial distribution of the population. Without properly functioning agriculture, the population will crowd in cities with associated economic, social and environmental burdens on society.[1] In terms of equity, rain-fed agriculture assures a better distribution of benefits than irrigated agriculture, unless land reform is implemented in the latter case.

Water use in agriculture has immense positive impacts on rural societies. It creates jobs at cheaper rates per job than industry. In Jordan, for example, the average capital cost per job in industry during the development efforts of the 1980s was about US$26,000 equivalent compared to about US$3000 in irrigated agriculture. A unit flow of water (one million cubic metres per year) is capable of settling about 150 families whose living earnings will be rooted in agriculture and supporting services. The experience in Jordan has shown that, when irrigation development is supported by development of the social infrastructure, the integrated development results are very impressive (Haddadin, 2006). Per capita income improved to match the country average; life expectancy at birth improved to match the country average; education services benefited males and, even more so, females. The role of women in economic and social development became more pronounced. The morbidity rates and mortality rates among children were drastically reduced. The cultural habits witnessed some transformation toward more modern societal communities.

Water use in industry has its economic and social benefits as does its use in services. The export and import substitution potentials have positive economic and financial impacts. All told, the economic, social, environmental and public health importance of water impart a political and strategic position upwards along the scale of priorities for any given country. If water falls short of its industrial and agricultural needs, a country can rely on 'shadow water' – a term coined by the author (WWF II, 2003; Haddadin, 2007) – that is water replaced in the water-short country by water used in exporting countries to produce agricultural and industrial goods to the water-short country.[2]

Shadow water exemplifies the importance of choosing trading partners. The world has witnessed several cases where a trade embargo has been used as a political weapon by certain countries against others. Shadow water dependence makes it even more important that the trading partner is as reliable and politically friendly as possible.

## Issues in international water conflicts

The author would like to emphasize his position vis-à-vis the outcry of water wars, a position he has maintained ever since the outcry was launched in Washington, DC, in 1986 during a water conference organized by the Center for Strategic and International Studies (CSIS) (Starr and Stoll, 1988). The conference focused on Middle East international rivers and the moderator initiated that outcry and identified the Middle East as the stage of impending water wars, despite advice to the contrary offered by the author.

From a natural point of view, water is used to extinguish fires, never to ignite them. Events following the water wars outcry proved it to be wrong. The wars that succeeded that outcry in the Middle East were: the Iraq invasion of Kuwait in 1990, the eviction of Iraq from Kuwait by an alliance led by America in 1991, the Palestinian–Israeli war of 2002 and, thereafter, the invasion of Iraq by an American led alliance in 2003 and the Israeli war against Lebanon-based Hizbullah in 2006 that destroyed much of Lebanon's infrastructure. None of these wars was triggered or caused by a water conflict. It is interesting to note that the outcry of water wars identified no other theatre for it but the Middle East, where oil reserves are abundant. Oil, unlike water, is flammable and can cause war, which it has done in several of the instances of war in the Middle East since 1986 listed above. The author would also like to emphasize that despite his search for cases of water wars between states he could find none.

Wolf and Yoffe of Oregon State University (Yoffe, 2002) identified 264 international water courses in their work in 2001. The number is likely to increase as new states emerge from old ones. There are yet untold cases of transboundary groundwater aquifers shared or to be shared among neighbouring states, and these aquifers are both renewable and non-renewable. It will not be surprising if we find out, when time-consuming studies are carried out, that no adjacent states are free from a common transboundary aquifer. The Middle East and North Africa countries are examples of this proposition.

Water alone does not trigger wars, which are usually the result of disputes over territories and other complicating factors related to national security. Despite the strategic value of water and its importance to national security, it does not alone trigger wars. Water gains secured by war are usually the outcome of a zero sum game where the victor may acquire waters belonging to the loser in that war, but that is usually linked to territorial gains. Since a state of war cannot last forever, neither ending it nor building peace between neighbours can be achieved without a restoration of rights, primarily giving up the territories that were acquired by war, and the waters that go with such territories. The neighbours can, through peace negotiations, settle water disputes, but the water gains of war are short lived and are not worth the cost.

But water conflicts are anticipated between neighbours who share the same watercourse or a water body. These are, not surprisingly, frequent and could lead to tensions in state relations. Water conflicts can be triggered and sustained for a variety of reasons, as discussed below.

## Conflicts over water sharing

The primary cause of dispute among neighbours is the dispute over water sharing. Upstream riparian parties on an international watercourse favour a water regime that gives them a free hand to do as they please with the water that crosses their territory. This, of course, is sometimes to the detriment of the downstream riparian parties. More often than not, upstream and downstream riparian parties do not agree on the principles by which they can share the waters of a watercourse. Legal regimes have been formulated to attend to such principles and have been pronounced by special international conferences and international law associations and bodies. The most recent is the United Nations Convention on the Non-Navigational Uses of International Water Courses passed by the General Assembly on 21 May 1997. The number of member countries that have ratified the convention is as yet modest compared to the 'yes' votes in the General Assembly adopting the convention. Conflict of this nature exists today in the Middle East between Turkey on the one hand and Syria and Iraq on the other. The Tigris and the Euphrates originate in Turkish territories and flow through Syria (Euphrates) or on the Syrian borders (Tigris) before they enter Iraq, meet at Qarnah and then form the Shatt El Arab waterway. Turkey advocates her rights to exploit the waters of the two rivers in her territories to develop Southeast Anatolia through the Güneydoğu Anadolu Projesi (GAP) project, and no agreement has been reached to share the waters of the two rivers with the exception of an interim accord between Syria and Turkey in 1989 whereby Turkey would release 500m$^3$ per second to cross the Syrian borders for use by both Syria and Iraq. The latter two states agreed in April 1990 to share that flow at proportions of 58 per cent for Iraq and 42 per cent for Syria.

## Conflicts over compliance

Even in cases where there is agreement on water sharing, conflicts arise because of non-compliance issues by a riparian party. Circumstances could develop such that a riparian can take advantage of the inability of a neighbour to stop unauthorized water use and may violate the terms and conditions of a treaty between them. Such a case existed between Jordan and Syria over the Yarmouk treaty of 1953, and does exist today over their treaty of 1987 concerning the same river. Syria, the upstream riparian party, has been withdrawing more water than the bilateral treaty assigns to it.

## Conflicts over territory

A watercourse or an aquifer may exist in a territory disputed between two neighbours. The water sharing or exploitation becomes, among other issues, a conflict factor. Such cases exist in the Middle East today. One concerns the Orontes river that emerges from springs in Lebanon, flows through Syria to the Alexandretta Province, now inside Turkey, before it discharges into the Mediterranean. Syria has a claim on the Alexandretta Province (Turkey's Hatay Province today) that was annexed to Turkey in 1939 by France, the mandatory power over Syria between 1921 and 1946. Since Syria claims that province as part of its territory, Turkey, says Syria, does not have legitimate access to the river and therefore is not a riparian on it. Syria's denial of riparian status for Turkey on the Orontes triggers a negative reaction by Turkey, which spills over to Turkey's attitude towards Syria on the Tigris and the Euphrates where Syria is a downstream riparian with respect to Turkey, which is the upstream riparian on both rivers.

Another example is seen in the Palestinian Territories occupied by Israel since 1967. Israel claims Palestinian territories are as yet undefined and may yet claim all or most of the West Bank. The underlying aquifers are the subject of disputes between Israel and the Palestinian Authority, as is the Jordan River on which the Palestinians claim legitimate rights by virtue of their territories adjoining the river in the Jordan Valley. Israel, on the other hand, occupies these territories and claims that they are disputed, not occupied, territories.

## Conflicts over water quality

The potential for quality conflict is high on the Euphrates River. Turkey's expansion of irrigation in the Harran Plains of the GAP region produces agricultural drainage water that flows underground and discharges into the Euphrates basin, raising its salinity. At Qaim, where the river crosses into Iraq, the salinity is reported to be high in August, at roughly twice its historic level. This is likely to be exacerbated when Syria and Turkey reach their projected expansion of irrigated areas in the Euphrates basin, causing appreciable harm to Iraq, the downstream riparian party.

In the Jordan basin, a case of contention over Lebanon's use of part of its share in the Hasbani tributary brought tension to the region in 2002 prior to the American invasion of Iraq. That contention, aside from the factors of enmity between Lebanon and Israel, was due to water quality issues. Lebanon's share, as defined in 1955 by an American sponsored plan, was meant for the irrigation of Lebanese lands located in the Hasbani tributary sub-basin. Lebanon, in 2002, decided to serve a small number of villages with municipal water from the Wazzani Springs that emerge very close to the borders with Israel and discharge into the Hasbani. The Wazzani Springs have no arable lands in Lebanon to irrigate and therefore, Israel contended, Lebanon is entitled to no share in the springs, whose water should be left to flow to the Israeli territories where the Hasbani would join the other tributaries and form the Upper Jordan. The water quality of the Wazzani Springs is superior to the quality of water in the Hasbani. and therein lies the source of the conflict between the two countries that was eventually resolved with American mediation.

A quality conflict exists on the Yarmouk between Syria and Jordan, but that is, at this time, subsidiary to the water sharing conflict. Syrian users upstream take the better quality water and the Yarmouk quality downstream is degraded by the inflow into the river of water of higher salinity from hot springs on the Jordanian side and on the northern Yarmouk bank occupied by Israel. This conflict is yet to be resolved.

The water annex of the Jordan–Israel peace treaty alludes to the rehabilitation of water quality in the Lower Jordan River. No action has been taken to implement that clause. NGOs have become involved and are currently pressing for the rehabilitation of the Jordan River, even if it requires drying up the agricultural areas that depend on its water for irrigation.

## Repercussions of international water conflicts

International water conflicts have several repercussions in addition to the tension they create among the parties to the conflict. These tensions could overflow and affect diplomatic relations among states of the region and their international alliances. Some of the primary repercussions are discussed below.

1   *On-going water operations and use.* Such a case exists in Jordan where infrastructure was built using borrowed funds from Germany to extend the main irrigation canal in the Jordan Valley to its limit and add some 6000 hectares of arable land to the irrigated area. Overuse by Syria of the surface and groundwater of the Yarmouk basin taxed Jordan's share in that river, a share Jordan uses to irrigate the arid Jordan Valley. The operation of that costly irrigation infrastructure has not been possible. Additionally, irrigation water

to the northern areas of the Jordan Valley has been diminished because of the Syrian practice. This has resulted in suppressed agricultural yields and greater hardship for farmers.

2  *Allocations made to municipal water.* The erosion of a country's share in international water can affect water allocations, including the quantity allocated to municipal uses. This had been the case in Jordan between 1986 and 1995 as a result of Syrian overuse of the Yarmouk waters. The Peace Treaty between Jordan and Israel mitigated that impact, when more water became available to Jordan from Israel.

3  *Future water plans.* Plans made to make use of the international water shares would be amended when such shares are eroded. Usually more expensive schemes will have to be worked out to compensate for the eroded share. Again, such was the case in Jordan when it relied on the Yarmouk shares to fill the Wehda dam to allocate municipal water to the northern provinces of the country. Today, more expensive projects have had to be planned to supply those provinces with municipal water, because the water that the dam was designed to regulate has mostly been withdrawn by upstream users in Syria.

4  *Water quality* disputes could affect public health if the water in question is allocated to municipal uses. This was the case in 1998 when the Yarmouk water, containing nutrients from Syria, saw a rapid growth of algae that originated in Israel and came with the Jordan share from Lake Tiberias. The proliferation of algae in July and August of that year created an excessive load on the water treatment plant to the extent that its finished water, pumped to the capital Amman, was not fit to drink. The event led to the resignation of the Minister of Water and Irrigation and, a week later, of the entire cabinet (*Al Rai*, 1998).

5  *Social, economic and environmental gains.* The benefits outlined above that accrue from irrigated agriculture would be lost as a result of an erosion of water shares.

## Issues in water allocation

In water-strained countries the number of people per unit flow is above the comfortable average. The sector users of municipal, industrial, service businesses, environmental, agricultural and other areas face reduced supplies. Priorities have to be set to allow all the users to share the available, useable water resources.

Agriculture has no competitors for the use of soil water supporting rain-fed agriculture. As such, some 45 per cent of the renewable water in Jordan (soil water) is automatically allocated to agriculture for lack of an alternative use. There are some competitors in the reuse of treated wastewater (grey water), for example industrial and environmental uses are such competitors. The allocation of grey water does not usually entail tough competition. The main area of competition is in the allocation of fresh drinkable water and this increases as the water strain increases.

Under conditions of high water stress, that is, when there are a high number of persons per unit flow of water, competition emerges not only among sector users, but also among the users in the same sector as well. Some of the issues to be accommodated within water allocation are summarized below.

1. *Equity*. This issue is of primary importance because it touches on security and the sense of citizenship and equality as stipulated in the constitutions of most countries of the world. It is important at the time when allocations are made and is particularly important in distributing the flows allocated to user sectors.
2. *Priorities*. In allocation, municipal water usually has first priority, followed by industrial water and then agriculture. However, the priority given to these sectors entails different meanings and these are usually spelled out in the water policy of the subject country. In Jordan, the first priority given to municipal water entails the supply of a certain minimum compatible with the standard of living of the population. It does not entail the allocation to municipal water of the entire flow needed by the population. The same applies to industrial allocations. When it comes to agriculture, allocations made to irrigation take into account both social and economic factors. Perennial crops should not be exposed to high water stress that would threaten their sustainability, and seasonal crops should be allocated a flow that is capable of producing a crop and keeping the farmers financially viable.
3. *Re-allocation*. In water-strained countries pressure builds on officials managing the water sector to re-allocate water, usually in favour of municipal users (including services). Cases emerge when the beneficiaries of reallocation are certain industries. Jordan witnessed a need to reallocate water planned for irrigation expansion to municipal uses in Amman (Haddadin, 2006). It also witnessed a case where agricultural water, already used in irrigation, was diverted to help maintain the production of the Arab Potash Company in Safi to meet export obligations. Similar to the case of the decision taken in 1978 to divert agricultural water to Amman, another situation arose in 1997 whereby water from the Mujib, then planned for expanding irrigation in the southern Ghors south of the Dead Sea, had to be shared with industry (Potash production expansion) and also with Amman.
4. *Augmenting water resources*. The high water strain and the decisions to reallocate water planned for irrigation expansion prompted the adoption in the water policy of the treatment and reuse of municipal wastewater in irrigation. This measure compensates agricultural users for the diversion of water originally planned for their use. The water quality, however, is not the same nor is the cropping flexibility offered by the diverted water.

5 *Risks of reallocation.* In all of the above measures triggered by water strain, there has been conflict between the water authorities and the party at whose expense the measure was taken, namely, the farmers of the Jordan Valley. The replacement of the diverted water with treated municipal wastewater impacted the marketing potential of the entire Jordan Valley produce. The export markets shied away from Jordan Valley produce, which is out of season and therefore commands good prices. Farm income was depressed and the complaints and protests of farmers prompted the authorities to quieten them by rescheduling debts, a measure that negatively impacted the financial standing of the government-owned Agricultural Credit Corporation. The reallocation of planned freshwater to municipal uses left stretches of land in the southern Jordan Valley by the Dead Sea and others south of the Dead Sea uncultivable because of the lack of irrigation water. In the case of the southern Jordan Valley, the irrigation infrastructure has been in place since 1988, but no return from irrigation has accrued, a real economic and financial loss and an embarrassing situation given that funding to build the infrastructure was secured from the German Capital Aid acting through Kreditanstalt für Wiederaufbau.

6 *Authority to allocate and reallocate.* The authority of allocation and reallocation is vested in the Boards of the Water Authority and the Jordan Valley Authority. Their decisions, however, are not without political repercussions and they therefore seek to have their decisions ratified by the Council of Ministers. The beneficiaries themselves are represented by two members on each board, a small minority indeed. But their voices are heard and they can influence decisions. Processes directly involving the beneficiaries have started in the Jordan Valley where irrigation users have been given a role in certain areas to allocate and distribute irrigation water (Salman et al, 2008).

## Issues in water pricing

The last in the set of issues that trigger conflict in the water sector are the issues of water pricing. In Jordan the authority to set water prices is vested in the Council of Ministers on the recommendation of the board of the respective authority (the Water Authority or the Jordan Valley Authority). Several issues come into play as the water tariff is examined or re-examined by the respective board of directors as summarized below.

1 *Cultural factors.* The majority of Middle Eastern countries started knocking at the gate of modernity only in the mid-20th century. Water services were taken up by municipalities whose coverage was limited to a few urban areas. The rest of the population centres had to rely on central government for water supplies. The prevailing outlook towards water among the Arab and Muslim societies is that it is a gift from God, and that people are partners in water,

fire and pastures. This last cultural notion is borrowed from a Hadeeth, or a saying of the Prophet Mohammad. As such, people were used to hauling water from springs free of charge and they would carry the burden of its transport to their homes. Indeed, there is yet another saying by the Prophet that says that 'water is not sold unless an effort is made to obtain it'. This saying allows charges for water supply to recover the cost: capital and operational. However, water charges, at its source, are not allowed provided the amounts are taken by the beneficiary to satisfy only his needs.

2 *Public awareness.* Credible decisions by government are those taken in line with the public understanding of the issues addressed by those decisions. It is important that the public be made aware of the factors that prompt the government to take the subject decisions. Failing that, conflict arises between the public and government and protests are organized against unpopular decisions. In Arab Islamic countries it is important to clarify to people the compatibility between the religious *fatwas* or rulings and the decisions concerning water. Water pricing is one issue to be clear about. In this respect it is important for the public to realize that water is a gift from God only at its source and that what is charged for it is paid to recover the cost of the service.

3 *Economic factors.* It is usually an acceptable average to have the water charges approximate 2 per cent of gross national product (GNP), including wastewater cost recovery. Percentages above this average, if charged to consumers, will impact on their family budget, forcing deficits or reducing the quality of service of the rest of the basket of expenditure. Depending on how pressed the budget is, people will react. But, in general, any price increase is met with resentment among consumers. In a country like Jordan the government contributes whenever the cost exceeds that percentage. The increased cost of water and wastewater service in Jordan is not attributed to natural developments, but rather to an unnatural rate of population growth. When it comes to a tariff for irrigation water, the matter becomes even more political. Irrigated agriculture helps put food on the table, food that has otherwise to be imported, incurring foreign exchange payments. Agriculture is a major employer, especially when the downstream activities from it are taken into account. It has a reasonable value added, and it helps to maintain a balance in the distribution of population between rural and urban areas. A major factor to take into consideration when the irrigation water tariff is decided is the status of agricultural marketing and farm gate income as compared to the cost of production and the indebtedness of farmers. Moreover, seasonal factors affect the decision, such as whether or not adverse weather conditions affected the produce quantity or quality. Again, any increase in the water tariff is met with resentment on the part of farmers.

4 *Income disparities.* As an expression of equity, the government in Jordan assigns a municipal water tariff on rural and the disadvantaged areas lower than the

tariff for the capital city of Amman. Within the same water tariff structure, an element of cross-subsidy is built in. Higher consumers of water are presumed to be the more well-off segment of the population and are charged higher values per cubic metre of water consumed. Less well-off people are charged less and the disadvantaged, as indicated by their quarterly water consumption, are charged the least. In effect, the rich subsidize the poor and the government, at this time, subsidizes all groups. However, the tariff is set so as to recover the full cost of operation and maintenance, with a small surplus for replacement. In terms of irrigation water, the water tariff is a function of use: higher level consumers, presumably the owners of orchards and water-consuming trees, pay roughly double what the farmers of seasonal vegetables are made to pay per cubic metre.

5  *Water quality.* A source of dispute between the farmers and the Jordan Valley Authority (JVA) is the fact that the authority charges the same tariff for all qualities of water in the Jordan Valley. It is known that water quality in the Middle Jordan Valley, served in the dry months by the King Talal dam on the Zarqa River, is almost marginal because of the high percentage of treated wastewater flowing into the reservoir behind the dam. It is also believed by the farmers that better quality water is being diverted to Amman – in the farmers' words – to 'take showers and wash cars and buildings'. This aspect of water pricing caused protests in 1993 and farmers went to court to settle their disputes with the JVA.

6  *Population and demographic factors.* Jordan has been host to several waves of refugees and displaced persons caused by the major conflict in the Middle East. The resolution by the World Community (League of Nations) in 1919 to assign mandate powers to Britain and entrust her with the task of implementing the Balfour Declaration put the region on a collision course with those attempts. Israel was nonetheless proclaimed in 1948 and the first major wave of Palestinian refugees formed a substantial influx into Jordan. Voluntary transfer of Palestinians from the West Bank to Jordan took place between 1950 and 1988. A forced migration wave of Palestinians was made during the war of 1967 and another from the Gulf States in 1990. Jordan was host to all these additional population movements. The towns and urban centres in Jordan grew at an accelerated speed, and refugee camps were set up in various parts of the country and are still there today. The cost of water and wastewater services escalated as local water sources were outstripped by demand and remote water sources had to be tapped. The capital cost, as well as the operational cost of the water projects, escalated drastically to the extent that the cost of water service in the country approximated 6.5 per cent of GNP around the turn of the century. This is a very high percentage and the government treasury, despite the requirements of economic structural adjustments, is contributing about half that percentage. This

50:50 split between the government and consumers is about the same proportion as the capital expenditure and operations expenditures of the water service.

7  *Water distribution*. Under conditions of severe water strain, water has to be rationed. Such has been the case for the municipal water service even in the capital city of Amman since the early 1970s. The capital city today is served once a week with municipal water and consumers have to store their weekly needs in storage tanks on the premises. Water rationing creates a social strain and financial burden as some people, denied their regular service for reasons of maintenance or system breakdown, have to purchase water from vendors at many times the network water cost. The strains suffered by the water distributors are immense and the complaints section of the service provider can barely cope with the number of complaints.

## Issues involving foreign trade

In the above discussion, reference was made to the importance of 'shadow water' and how it exemplifies the importance of choosing trading partners. In fact shadow water is a primary component of the population–water resources equation of any country, but is even more vital in water-strained countries. The import and export of commodities entails the consumption of water. In Jordan, for example, shadow water contained in imported food commodities accounted for an average of 77 per cent of the need for agricultural water between 1994 and 2002 (Haddadin, 2008).

Under conditions of scarcity, several politicians and professionals argued for more water to be diverted from the Jordan Valley to municipal uses in the cities. They cited the export of Jordan Valley fruits and vegetables as a means of exporting water. They were countered by other professional practitioners who showed the social and environmental gains of irrigated agriculture and the comparative economic advantages of the Jordan Valley out-of-season produce in regional markets.

The heavy weight that shadow water occupies in balancing Jordan's population–resources equation deserves particular attention to secure the continuity of flow of imported goods from their sources. Trade embargoes, either by individual countries or by the world community, have been imposed in the past against certain countries for political reasons.[3] Long-term trade agreements will serve a strategic objective of stabilizing a 'water source' and the price thereof. It is for these reasons that choosing reliable trading partners is so important.

Political conflicts between trading partners can affect the flow of shadow water and would thus negatively impact the balance of the population–water resources equation.

## A conflict prevention and management centre

Conflicts between states may be better avoided through dialogue in a multinational forum in the form of a Conflict Prevention and Management Centre. In such a forum, affected parties can start a dialogue with the help of intermediaries. The centre might list conflicts over water resources as a start, including the conflicts over shadow water. If conflict prevention is not possible then its resolution might be sought through the forum. Failing either, the centre might assist in managing the conflict within certain bounds until resolution is achieved.

## Conclusions

Since water is invaluable to all forms of human activities and the survival of living creatures, it should be viewed as a strategic commodity and a social good. It should also be managed in an integrated way in which resources, users, economic status, social well-being, human resource capacities and overall economic and social output are weighed in the process of water allocation and reallocation, water pricing, legislation and institutional arrangements.

The author believes that water is a social good at its source and an economic dimension is added to it once investments are undertaken to make it available for users. In this respect users should pay the cost of the service and a margin of profit for the investor. Factors that urge governments to contribute in the form of subsidies are basically strategic social, economic and political factors.

International water shares should be protected by all possible peaceful means and the environmental integrity of water resources should be maintained. International water conflicts are best resolved through negotiation. The creation of a Conflict Prevention and Management Centre in the Middle East may be a good starting point towards debating and resolving water conflicts.

## Notes

1. We indeed see in many countries the rural-to-urban migration that is a result of poorly functioning agricultural sectors.
2. The term 'virtual water' was coined in 1996 to refer to the water 'contained' in imported food commodities; no known location is specified when that term is used. The new term, 'shadow water', refers to a location specific water saving brought about by food and industrial imports.
3. The US embargo against Cuba in the 1960s, and the UN sanctions against Iraq in 1990 are examples.

# References

*Al Rai* (1998) daily newspaper, 10 August and 20 August, Amman, Jordan

Haddadin, M. J. (2006) *Water Resources in Jordan: Emerging Policies for Development, the Environment and Conflict Resolution*, RFF Publishers, Washington, DC

Haddadin, M. J. (2007) 'Shadow water: Quantification and significance for water strained countries', *WaterPolicy*, vol 9, pp439–456

Haddadin, M. J. (2008) 'Water challenges in Jordan and a strategy to face them', lecture at the Al Asriyya Forum, Amman, Jordan, 19 February (in Arabic), printed in *Al Urdun* weekly newschapter 20 February 2008

Salman, A., E. Al Karablieh, H. J. Regner, H. P. Wolff and M. J. Haddadin (2008) 'Participatory irrigation water management in the Jordan Valley', *Water Policy*, vol 10, pp305–322

Starr, J. R. and D. C. Stoll (1988) *U.S. Foreign Policy and Water Resources in the Middle East*, CSIS Panel Reports, Washington, DC

World Bank (2002) www.worldbank.org/data/countryclass/classgroups.htm

WWF III (2003) World Water Forum III convened in Kyoto, Japan, March 2003

Yoffe, S. (2002) 'Indicators for future water dispute', PhD dissertation, Department of Geosciences, Oregon State University, Corvallis, OR

# 11

# Good and Bad Forms of Participation in Water Management: Some Lessons from Brazil

*Jerson Kelman*

On his first hunger strike in protest against the Sao Francisco river inter-basin transfer project, Dom Luiz Cappio – a bishop of the Catholic Church who lives in a small city on the banks of the Sao Francisco River – was successful in getting an agreement with the Federal Government. When completed the inter-basin scheme would benefit 10 million people outside the Sao Francisco basin that, on average, have a very low income, mostly due to climatic uncertainties. Dom Luiz halted his hunger strike under the Federal Government's commitment to promote a national debate about the project and restore the river basin, essentially through investments on sewage collection and treatment, as well as on the protection of the riverbanks against erosion.

The government kept its word but Dom Luiz changed his mind and went into a second hunger strike. This time it seemed that he was prepared to die, if necessary, in order to halt the construction works. This hunger strike lasted almost one month. It was followed closely by the media and aroused intense debate and deep emotion in millions of Brazilians. Many thousands went through short duration fasts to indicate their solidarity with Dom Luiz.

A bishop is not a naive person. He knows that a democratic government cannot give in to blackmail. So, what was his intention? To achieve sanctification through self-sacrifice? Perhaps, but more likely, he counted on beating the government once again. After all, from the government's point of view, it would be unthinkable to allow him to become a martyr. If, in fact, the government had given up, Dom Luiz would become an important religious leader. This is less than being a saint. But there is no need to die.

However, the government stayed firm. Minister Patrus Ananias, a devout Catholic responsible for social security, classified the bishop's behaviour as blackmail by suicide. He asked what would happen if another bishop started a hunger strike for the project? When Dom Luiz's health reached a critical condition, he accepted the advice from his physician and ended the hunger strike.

Dom Luiz was not interested in a win–win alternative. After the first hunger strike, he was received by Lula da Silva, the Brazilian president, in a meeting where technicians tried to explain that the infrastructure being built would not harm the river basin because less than 3 per cent of the mean flow of the river would be diverted. From a positive perspective, this small quantity of water would benefit enormously the population living in the receiving area. Dom Luiz kept silent during the entire presentation. At the end, he said that he was not interested in technical or economical explanations. He knew in his heart that the project was evil. He refused to discuss the issue within a rational framework. His point of view was supported by faith, not by reason.

The current Brazilian democracy was installed more than 20 years ago, but it still carries the scars of the previous autocratic regime, ruled by the military, for a similar period of approximately 20 years. These scars are engraved in the current constitution, approved in 1988, when the record of the violations of human rights by the previous regime were still fresh in the memory of the population. The Brazilian constitution strongly supports individual rights, frequently at the expense of collective rights. This explains why Dom Luiz's standpoint has become such an important issue in the country.

In this chapter a number of cases related to water resources allocation and river use dispute will be described. The focus will be on the description of an environment where the search for utopist, unanimous decisions often obliterates the functioning of the democratic decision making process. This is an issue common to many countries that lack strong democratic traditions and institutions. For these countries, extreme care should be adopted when enunciating a problem to be solved. One should resist the temptation of assuming that stakeholders will act in an objective, rational manner. It would be helpful to make the problem solvable by some mathematical friendly decision making process, but that could produce a 'solution' that would lack political feasibility.

## The Sao Francisco River Inter-basin Diversion Project

Table 11.1 shows the main features of the Sao Francisco River. Despite its impressive drainage area (larger than France), it is entirely located in Brazil, covering 5 of the 27 states (Figure 11.1). A brief description of the Inter-basin Transfer Project (from now on referred to as the Project) is presented in Box 11.1. Its purpose is to convey excessive water of the donor basin in wet years, that otherwise would flow into the Atlantic Ocean, to be stored in the existing reservoirs of the recipient region (Figure 11.2). The Project will work in dual mode. The high mode will be activated in wet years, when the main reservoir of the Sao Francisco (Sobradinho, with a storage capacity of 34 billion cubic metres) will be spilling or close to spill. In this case, pumping will be at its highest capacity of $127m^3/s$ and the opportunity cost of the electricity used in the pumps will be close to zero. Otherwise, the low mode will be activated, with a pumping rate of only $26m^3/s$. Roughly, this is equivalent to $80m^3$ per capita per year. Simulation studies have shown that the probability of operating in low mode is close to 60 per cent.

---

**BOX 11.1 DESCRIPTION OF THE SAO FRANCISCO RIVER INTER-BASIN DIVERSION PROJECT**

The Rio Sao Francisco River rises in the state of Mina Gerais in the Serra da Conastra in Brazil at an elevation of approximately 1600m and flows for 2700km north and east. The river system is considered to be key to the future economic development of the semi-arid areas of Brazil. The flows of the drainage provide the hydropower to fuel the industry and water supply for fruit and vegetable production. Non-riparian semi-arid states to the north east have long coveted the waters of this river system, as these states have periodically suffered long and severe droughts that have decimated the economy, caused innumerable deaths and persistent migration of rural people to urban areas and proposals for major trans-basin diversions to the north and east of the drainage have been put forth for over 75 years. The Sao Francisco Inter-Basin Diversion Project has been prepared to fulfil the aspirations of these areas.

The proposed scheme will have two major points of diversion. The first will divert from the river just below the existing Sobardinho Dam, at a point known as Cabrobo and will divert an average flow of 99 cumecs from the river through a series of three major pumping stations, 15 regulatory reservoirs, 229km of canals, 23km of tunnels and 3km of aqueducts.

The second diversion will be from the existing Itaparica Dam and reservoir located further downstream and will divert 29 cumecs to a system of canals, pipelines and a reservoir. This sytem will include six pumping stations, 297km of canals, 84km of pipelines, 8.2km of tunnels and 25km of aqueducts. The project will also involve en-route construction of two hydroelectric plants of 52MW capacity each. The project is estimated to cost over a billion dollars. The total annual diversion will be of the order of $1.5km^3$/year.

*Source:* Compilation of International Experiences in Inter-basin Water Transfer, published by the International Commission of Irrigation and Drainage – ICID, in September 2003

---

As became clear in the opening paragraphs, the Project has caused heated discussions. On one side are those who view any water exportation as the bleeding of a dying river. They think that diverting water is analogous to forcing an unhealthy

**Table 11.1** *The Sao Francisco River Basin*

|  | Metric | Imperial |
|---|---|---|
| Drainage area | 630,000km$^2$ | 240,000 sq.mi. |
| Mean flow | 2600m$^3$/s | 68M ac.ft/year |
| Minimum flow | 600m$^3$/s | 16M ac.ft/year |
| Regulated flow | 2100m$^3$/s | 53M ac.ft/year |
| Projected mean diverson | 65m$^3$/s | 1.7M ac.ft/year |

person, under intensive care, to donate blood. On the other side there are those who prefer the analogy of the Sao Francisco River being a healthy person donating blood in order to save the life of a moribund region located outside the river basin.

It is regrettable that both sides appeal to these dramatic and emotional images because it limits the discussion to irrelevant topics. But before revealing what the relevant topics are, it is necessary to give a brief description of the recipient region.

The water availability of the recipient region, formed by the states of Ceará, Rio Grande do Norte, Paraiba and Pernambuco (Figure 11.1, Plate 2), considering the regulated outflow of the existing reservoirs, would be sufficient to meet, for several years in the future, the basic needs of the population (roughly 40m$^3$ per year is all one person needs to drink, bathe, clean and cook). Dom Luiz thinks that all the government should do is to help people have these basic needs attended. This may mean the construction of pipelines to connect remote villages to the local reservoirs or the building of individual tanks, one per household, capable of storing rain that falls on the roofs.

However, water is not used exclusively to satisfy human consumption. It is also used as an input for agricultural or industrial production. Taking all the water uses into account, it something in the order of 1500m$^3$ per year and per capita are required for a technologically unsophisticated community to achieve a reasonable income level and, by association, a reasonable quality of life.

Because the water availability of the recipient states is around this threshold level, there are two possible policies for the region: export people or import water. The first alternative has been implicitly applied for decades, as a significant proportion of the population of the Brazilian southeast, including President Lula as a child, migrated from the dry northeast. This alternative is defended by those who propose public investments where water is easily available, as it is in the Sao Francisco River valley. These investments would create an immigration flux from the dry area, and the problem would be solved. The second alternative is more political than economical. It tries to avoid the suffering of moving millions from places that have been inhabited for centuries and where successive generations have built an infrastructure to survive in the semi-arid conditions.

# Good and Bad Forms of Participation in Water Management 193

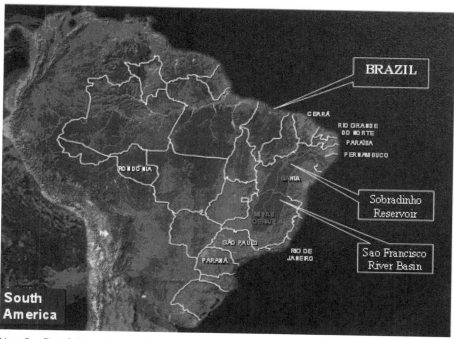

Note: See Plate 2 for a colour version.

**Figure 11.1** *Location of the Sao Francisco River Basin*

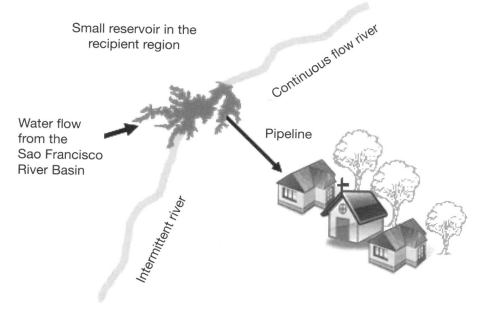

**Figure 11.2** *Schematic description of the water supply problem at the recipient region*

The recipient region has very limited groundwater and non-perennial rivers. The obvious solution has been to store water, which the region has done. However, water managers face a major problem in reconciling two conflicting objectives. Objective one is to ensure that the cities of the northeast have water during the droughts that regularly ravage the region. To achieve this objective, water managers have to hold water in the reservoirs for years. Objective two is to maximize the number of jobs and economic production from available water. The conflict arises because pursuance of objective one means that, with very high temperatures and low humidity, immense quantities of water are lost through evaporation, resulting in neither jobs nor economic production. The reconciliation of this conflict requires that managers have some other mechanism for meeting basic needs in times of drought, so that evaporative losses can be reduced and existing water resources can be used more productively. In other words, managers need to have the possibility of using water from the Sao Francisco in case a drought lasting several years occurs, such as the one that happened in the 19th century, when close to one million people died.

Most of the irrigated land in the recipient region yields low economic value crops, such as beans. Considering that a hectare planted with mango trees is much more profitable than if planted with beans, why would someone choose to grow beans rather than mangos?

The main reason is the lack of water security. In this situation and applying the minimax criterion, it makes sense to decide in favour of beans. The explanation for this is that if there is a water failure, the damage for a bean grower is limited to one year; on the other hand, lack of water for someone who has planted mango trees may imply a much greater loss: the tree may die and a new one would take several years to yield the first fruits. For the same reason, few employment intense industries decide to open factories in the region, despite the low labour cost. The Brazilian northeast has been the location of this 'vicious cycle': people are poor because there are few investments to raise high-value crops; there are few investments because there is no firm water supply; there is no firm water supply because people are poor and cannot pay the firm water supply cost.

There is a strong correlation between poverty and lack of water security, although world statistics based on mean values may fail to capture this fact. Indeed, depending on the size of the region/country and on the internal hydrological diversity, mean values may mean little. Brazil, for example, which covers roughly half of South America, has a per capita availability of 36,000m$^3$/year, which is much higher than the threshold level. Nevertheless, scarcity of water is the major problem in the semi-arid Brazilian northeast because: (i) the variance of the annual river flows is very high; (ii) the rainy season is short, typically three months; and (iii) most of the rivers are intermittent due to the low water retention capacity of the shallow soils (the Sao Francisco is an exception and for this reason it is called the Brazilian Nile). In this environment, life would be impossible without the existence of hundreds of reservoirs that were built during the last decades.

Most of the prosperous regions of the world do not suffer water scarcity. Exceptions, like the American west, have benefited from heavy investment in water infrastructure. However, water security is a necessary but not sufficient condition for achieving progress. An example is the population of the donor region, living in the Sao Francisco River basin. Although there are a few prosperous zones in the basin, due to the production of fruits for exportation, most of the population is still very poor.

Worse, they believe, insufflated by Dom Luiz and others, that if the Project is constructed they will become poorer. In a strange way, they could be right. The financial resources devoted to the Project, all other factors remaining constant, would probably exhaust the government's capability to invest in other projects that could benefit the population of the donor river basin. In other words, the dispute for water between the donor and the recipient regions is unjustified, but for money it is real.

The government understood the nature of the problem and decided to invest in the Sao Francisco River Basin Revitalization, hoping to calm down the opposition, lead by Dom Luiz. This was undertaken at the expense, obviously, of other parts of the country. But Dom Luiz and others ignored the government's initiative, putting at risk the possible win–win outcome. They preferred to centre the discussion on a moral issue: the pretend right of the population of the donor basin to decide freely how to allocate the water.

This is a false premise. First, because there is more water than would be required to satisfy, in the foreseeable future, all the consumption needs in the basin and for exportation to the recipient region. This has been demonstrated in the Sao Francisco River Basin Plan, developed by the Brazilian water regulatory agency – ANA. The second reason why it is false is that the population of the river basin does not own the river. It is a natural asset of the country and should be used for the benefit of all Brazilians, those living inside and outside the basin.

The suspicion, highly inflated by the media, that the Project could 'kill' the Sao Francisco River, led to a diversion of the government's focus from some relevant topics. First, the Project did not include the implementation of a capillary network of shorter and smaller channels and pipelines, both in the recipient and the donor regions, in order to convey the water from local reservoirs to wherever people live and work (Figure 11.2). In fact these hydraulic structures should be built before initiating the major construction works that will allow exportation of water. Unfortunately Dom Luiz and his followers failed to understand that these smaller hydraulic structures, although necessary, would not be sufficient to solve the drought problem. Local reservoirs need to receive water from the Sao Francisco River, in addition to the intermittent flow of local rivers.

Second, water users in the recipient region did not commit themselves to pay for the use of the diverted water before the beginning of the construction works. This means that it is very likely that they might be allowed to get benefits from the Project without assuming any responsibility. In this unfortunate possibility, the maintenance cost of the Project would depend on the government budget. In developing countries this is not a good option because it is easier to convince politicians to build a new structure than to maintain an existing one.

The government should have set the political, legal, institutional and financial arrangement for the operation and maintenance of the Project before hiring any contractors. In reality it only achieved to set the first one: the political arrangement. Governors of the recipient states signed a political pact recognizing their interest and joint responsibility in the Project, but the water users, public and private, have not been obliged, as they should, to commit themselves to pay at least the operation and maintenance (O&M) costs of the Project through firm contracts. As it is, there is no proper ownership and the whole infrastructure could become a white elephant.

Third, although a water rights system has been successfully implemented by the ANA in some river basins in the semi-arid Brazilian northeast, it is still necessary to extend this experience to all basins of the region and adopt enforceable rationing procedures (Kelman and Kelman, 2002). The key is to have a system that recognizes that there are many 'waters' – at the extremes low reliability and high reliability (and similarly for quality). Entitlement and tariff systems must differentiate between these different products.

The lesson from this case study is that policy makers should focus attention on the real problem to be solved and refrain from paying excessive attention to issues raised by people who in fact do not want the status quo to be changed.

## Water supply to the Metropolitan Region of Sao Paulo

The hydraulic connection between the Piracicaba River basin and the Metropolitan Region of Sao Paulo (MRSP), located outside the basin, occurs through a series of reservoirs, tunnels and channels, forming the so called 'Cantareira System'. The mean natural inflow to the system is $40m^3/s$ and the maximum authorized flow out of the system, through the conveyance structures, is $33m^3/s$. The authorization for the diversion was granted by the Federal Government in 1974, at a time when there was no dispute for the use of water. The authorization was valid for a period of 30 years. In recent years the political leaders of the donor basin resented the effects of water shortages and claimed that the authorization should be revised in order to decrease the water quantity at the authorization renewal date in 2004. Their purpose was to remove the bottleneck to the development of the valley itself. But, contrariwise to the political and religious leaders of the Sao Francisco valley, they never denied the possibility of an authorization. Interestingly enough, the percentage of diverted flow in the Piracicaba case was 83 per cent of the mean flow, as compared to less than 3 per cent proposed in the Sao Francisco case!

Unlike the Sao Francisco case, in the Piracicaba case it was possible to keep the discussion within a rational framework. The leadership of the donor basin knew that an interruption to the flow to MRSP would create a chaotic situation that would harm everyone.

Both the donor basin and the recipient region are intensively populated and highly industrialized. The MRSP has a population of 18 million people, 39 counties, contributes a significant share of Brazilian GDP and demands a supply flow of 65m$^3$/s. Many important big cities are located in the donor basin and their interests are defended by a capable political leadership that acts through the River Basin Committee.

In order to decide on allocations to the MRSP and downstream to the Piracicaba basin, the constitution of an 'allocation authority' was proposed. However, it was soon realized that this would only transform a problem that could be solved once and for all into a recurrent problem to be solved every month. Worse, the political battle would be around who would be entitled to implement the authority, rather than on how to evaluate the costs and gains of each possible allocation.

Instead of that, ANA and the Sao Paulo State Government chose a 'mathematical solution' that was very simple to understand. It is based on a pro rata allocation of the inflow proportional to the basic water needs respectively in the donor and in the recipient regions. Volumes of water that are eventually not used are counted, for later use, as if there were stored in a 'water bank'. Occasional overflows are also counted and subtracted from the 'savings' of each region, proportional to the volume each region decided to keep in storage. This naive proposition was the object of intense debate, particularly in the Piracicaba Basin committee. In the end, it was approved by all.

The lesson from this case study is that stakeholder participation in the decision making process is a necessity. Nevertheless, this participation should go beyond the mere selection of representatives that will make the actual decisions. Stakeholders need to understand the rationale of the water allocation criteria and agree that it respects common sense. The best water allocation rule is not necessarily the optimal one, from the economical point of view. A simple rule, but one that is highly understandable and accountable, may be the best choice. This is the case of the Piracicaba–MRSP hydraulic connection.[1]

## Sewage treatment

Considered as a country with abundant freshwater, Brazil has been using its rivers in a disorganized manner. The previous case study of the MRSP is a good example. The local water and sewage company must get bulk water from a river located some 100km away, as the rivers in the metropolitan area are, to a large extent, too polluted to be used for water supply.

The Brazilian Water Act of 1997 allows the implementation of water charges, both for diverting bulk water from the rivers or for polluting them, so as to prevent abuses. The Act was enacted as a reaction to the business as usual scenario, in which rivers would continue to be degraded, penalizing current and future generations on their water capital.

The Water Act adopts the 'polluter pays' principle, an idea successfully implemented in Europe: whoever pollutes more pays more. The River Basin Committee decides how much and who to charge for discharging polluted effluents. The committee is a sort of 'water parliament', with representatives of federal, state and local governments, civil society and the productive sector. To this last segment belong the companies that use the river, such as water and sewage companies, irrigation districts, hydroelectric plants, navigation companies and some industries located along the river.

The funds raised are invested in programmes aimed at the improvement of the rivers' conditions, according to the priorities set by the committee. In order to boost the committee's activities, the Brazilian water regulatory agency, ANA, launched, during its first year of existence (2001), a revolutionary programme based on 'output based aid'.

The River Basin Pollution Abatement Program, PRODES, focuses on cleaning up river basins. It does not subsidize engineering work or equipment, but pays for the final result, which is treated sewage. The programme consists of providing economic incentives for the construction of new sewerage treatment plants, aiming at the environmental recovery of the country's most polluted river basins.

Paying for treated sewerage is an innovative response to decades of ineffective subsidies, allocated to water and sewerage companies in Brazil and other developing countries. A considerable part of these subsidies were used up to build 'white elephants', that is, huge ineffective infrastructure works. PRODES also depends on taxpayers' money. However, it has improved the quality of public expenditures through reliance on a simple concept: it is more effective to pay for an actual result than for a promise of result.

Within the PRODES programme the sewerage treatment is paid for throughout the first five years of the sewerage treatment plant operation. The disbursement, however, is subject to an adequately provided service. If the service provision does not meet the required standards, the allocated funds, which have been deposited in a development bank, return to the National Treasury. The required standards are set in terms of sewage quantity and on the quality of the treatment. This arrangement reduces the risks for both sides. It ensures the service provider that there is no *non-compliance* risk due to government budget cuts as the committed funds were set aside in a development bank. The government, on the other hand, does not run the risk of having to pay for inadequately implemented services.

The most interesting result of PRODES is not what went right, but what went wrong. At the beginning of the programme, many municipal authorities responsible for sanitation approached the ANA, at the negotiation stage, with excessively ambitious projects, in terms of quantity and quality. When they realized the difficulty of fulfilling their promises, either because the sewage collection system did not work as satisfactorily as originally thought or because the pollution removal process was not as efficient as foreseen, they would return to ANA seeking a renegotiation. They came to the obvious conclusion that it is preferable to receive less than to receive nothing. The population, on the other hand, stopped paying for a service that was not being rendered.

The lesson from this case study is that a policy of subsidies for sanitation is not necessarily bad, as it benefits the whole community, rather than individual citizens, contrariwise to the water supply case. However, these subsidies should be used to pay for results, rather than promises.

## Hydropower

Wealthy countries have already utilized most of their hydropower potential (on average 70 per cent), and developed their economies in the process. Brazil and other developing countries, on the other hand, still have a long way to go. Brazil has developed 24 per cent and Africa 3 per cent of the potential for this low-cost, renewable source of energy (Figure 11.3). Many developing countries could emulate Brazil's successful hydropower programme, which accounts for more than 80 per cent of the country's electricity production. Natural and technological conditions are ideal for major hydropower programmes in many developing countries, including the mountainous countries, which are among the poorest in the world.

Hydropower technology is mature, proven and applicable wherever there is falling water. Brazil has shown that the much-publicized social and environmental problems associated with hydropower can be addressed. Local people can and do benefit as Brazilian laws mandate that 3 per cent of revenues from hydropower are fed back to local communities. Also, the environmental footprint can be

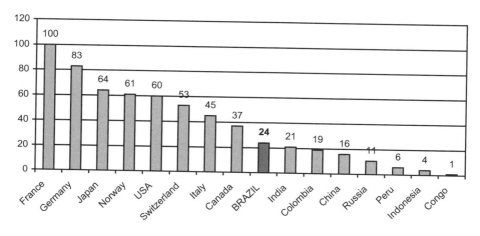

**Figure 11.3** *Developed hydropower as a percentage of potential hydropower*

dramatically reduced: the next generation of hydropower plants in the Amazon will, per unit of energy generated, inundate about 1 per cent of the area inundated by the previous generation of technologies.

Given all these favourable conditions, one would think Brazil would stay clean, in terms of energy production, for many years to come. Unfortunately the future may be the other way round, thanks, mainly, to the efforts of some dam-hating NGOs (for short, DAHNGOs), both national and international. The DAHNGOs are doing their best to impede any new hydropower development, often in open conflict with other environmental NGOs that are truly concerned with sustainable development. The DAHNGOs work is facilitated by the very liberal Brazilian system, in which a single, unaccountable prosecutor can hold up any project for years, on any pretext, irrespective of the standpoint of his peers. A DAHNGO's power becomes immense when it manages to gain the heart and mind of just one prosecutor. And there are thousands of them! The result in the energy sector is that Brazil is not using its abundant, cheap, climate-friendly hydropower, and instead is using more and more fossil fuels and now more nuclear energy, with its considerably higher costs.

In all probability, there is not one single hydropower plant in the world with sufficient merit to receive the seal of approval of the DAHNGOs. They are few in number but capable of making a lot of noise. In Brazil, they act to increase the complexity of the already cumbersome socio-environmental licensing process, including judicial disputes. They have successfully created a major obstacle for the timely and predictable expansion of generation capacity. This, in turn, is a serious threat to economic growth, the elimination of poverty and control of the emission of gases that enhance the greenhouse effect.

The DAHNGOs' efforts are facilitated by the fact that much of Brazil's unexplored hydropower potential is in the Amazon, which is an environmentally sensitive region. It is understandable that people around the world are concerned at the prospective of constructing dams there. They fear that the rainforest could be destroyed, although only 0.25 per cent of the Amazon land has been inundated or will be inundated in the next ten years by hydropower reservoirs.

DAHNGOs take advantage of this fear. For example, the International Rivers Network (IRN) proclaims on its website that they are 'working with a coalition of civil society organizations based in the region to stop the construction of these projects and promote viable alternatives to meet Brazil's energy needs'. It is reassuring that IRN accepts that a developing country has energy needs. This eliminates the simple and wrong alternative of freezing the per capita consumption at a level so low that the country would be condemned never to develop (at the time of writing, per capita consumption in Brazil is 15 per cent of that in the US). What then would be the viable alternatives?

Let us start with the most desirable of all: solar. In a tropical country, with plenty of sunshine all year round, this seems to be an interesting alternative. Indeed, most forms of energy derive from solar. For example, hydropower depends on rainfall provided by the hydrological cycle, which, in turn, depends on solar energy for 'pumping up water' through evapotranspiration. Also, solar energy provides the photosynthesis for producing sugar-cane-based ethanol.

Incidentally, blending of ethanol with gasoline may be the simplest and most effective act to control the emission of greenhouse gases. In Brazil, ethanol accounts for 32 per cent of all energy used in automobiles, either through being mixed with gasoline (20 per cent ethanol) or in flex-fuel vehicles. The boost of ethanol consumption in the developed countries could help to mitigate poverty in several tropical countries. Brazilian ethanol production alone could easily increase sixfold, to roughly 100 billion litres per year, without decreasing food production or cutting one single tree from the Amazon forest (presently the area planted with sugar cane is less than 3 per cent the area dedicated to low density cattle raising). This would be sufficient to substitute 5 per cent of the worldwide forecasted consumption demand for gasoline in 2025.

Solar energy can also be used directly to heat water for industrial and household use. This saves electricity and natural or petroleum derived gas and, obviously, is a good practice. But, when it comes to producing electricity directly from solar energy, unfortunately the technology has not yet delivered a process in which the cost could be competitive. Presently, its unit cost is roughly tenfold that of hydropower.

In second place in the preference of environmentalists is wind power. It is the world's fastest-growing energy industry with an average annual growth rate of 29 per cent over the decade 1996–2005 (Florence, 2006). Unfortunately wind power availability is intermittent, as is the wind. Unlike water, which can be stored in reservoirs, there is no possibility of storing wind. Therefore, wind power can only be used as complementary to some other energy source, to be turned on whenever necessary. Also, it is very expensive; not as costly as solar, but still the unit cost is roughly twice that of hydropower. It is a reasonable choice for countries that have to decide between nuclear or wind power, such as Germany or Spain. But for a developing country that still has hydropower to develop, it would mean abdicating from gaining competitiveness in the global economy, which would be a strategically unacceptable decision.

Nevertheless, some countries are subsidizing wind power in order to keep pace with the advances of technology. This is the case in Brazil where a programme to construct 1100MW of wind generated power, subsidized by consumers, was launched by the government. Although this is a significant amount of power, it will account for only 1 per cent of the country's installed capacity. This is not surprising: wind power is always a small percentage of total power, even in countries that invested heavily in it. Germany, for example, the country with the highest installed wind power capacity, gets only 6 per cent of its electricity from this source. All things considered, although wind power can be used, it cannot be considered an alternative solution to hydropower.

The next alternative is one that is very competitive with hydropower: bio-electricity. It consists in burning the remains of a seasonal harvest to produce electricity. It is neutral in terms of greenhouse gas emissions because the same quantity of carbon released to the atmosphere during the burning stage was previously trapped into vegetal tissue at the plant growing stage.

Because burning sugar-cane bagasse is an efficient way of producing bio-electricity, it may play a major role in developing countries, in case developed countries decide to mix ethanol into gasoline. In Brazil, this is already happening. In the next three years, some 5000MW of bio-electricity plants are expected to be installed on the agricultural frontier, located in the states of Goias and Mato Grosso do Sul. These new plants, associated with sugar and ethanol production, are economically efficient and can compete without subsidies. Again, this is an impressive quantity of energy, but it is equivalent to just one year of the country's demand growth. All should be done, and is being done, to expand bio-electricity. However, this alternative by itself is not sufficient to entirely meet the new load.

Much less desirable are the alternatives based on fossil fuels. The least harmful to the environment and least costly would be natural gas. But very few countries can presently produce large quantities of natural gas. Brazil is not one of them, mainly because the most important offshore fields have only recently been discovered and a number of years will be needed before they enter production. But some neighbouring countries, namely Bolivia and Argentina, have been known for a long time to be rich in natural gas.

Only a few years ago the energy integration of the southern part of South America seemed to be a win–win situation. Accordingly a 2200MW transmission line was constructed to connect Brazil and Argentina, with the main purpose of transporting electricity that would be produced in Argentina by natural gas fired plants and consumed in Brazil. But the energy flux could be reversed when Brazilian reservoirs were spilling. Unfortunately the majority of investments in the energy sector in Argentina ceased when energy prices were frozen by the government in a move to control a major economical crisis. Now this transmission line is used only sporadically. For example, in 2007 it served for four months to transport energy from Brazil to Argentina, which was experiencing a particularly severe winter.

The energy integration between Bolivia and Brazil is more successful but much more could be achieved. Presently some 30 million cubic metres of natural gas flows daily from Bolivia to Brazil through a long pipeline. It is used by industry, vehicles (some of them can run on natural gas, gasoline and ethanol) and gas fired thermal plants. This last use, for electricity production, only occurs when the reservoirs of the hydro plants are low. Presently, if all the gas fired plants were producing, there would be either a natural gas or an electricity shortage. Perhaps both shortages would occur simultaneously. As an emergency solution, Brazil decided to build two gasification plants capable of processing 20 million cubic metres per day of liquefied natural gas, which will be transported from producing countries by ships. This energy supply deficiency would not exist if in the last ten years hydropower entrepreneurs, both public and private, had not experienced many rejections of their licence requests to build new plants, either from the socio-environmental administrative branch or from the judiciary.

Alternative fossil fuel energy sources other than natural gas can be oil and coal. Definitely no environmentally concerned person or institution would think that this is a reasonable option. Oil and coal are the villains of the greenhouse effect.

All things considered, and contrary to the standpoint of the DAHNGOs, Brazil needs to develop its hydropower potential. The challenge is to do it wisely in order to avoid the errors committed in the past, mainly in the 1970s, when some hydropower plants that today would not be allowed, were permitted to go ahead by the ruling military dictatorship. This is what is happening – the area submerged per unit of power developed at the recently approved 4000MW Rio Madeira project is 3 per cent of that of the infamous 250MW Balbina project. Serious consideration has to be given to the socio-environmental constraints, which in practice means having the ability to differentiate the good from the bad dam sites, and adopting a holistic point of view. To do that, it is necessary to evaluate the trade-offs between the local (in general negative) and the global (in general positive) effects associated with the proposition of a new hydropower plant.

This task has become more difficult in recent years because of the fundamentalist wave, headed by people and institutions like Dom Luiz and the DAHNGOs, against the construction of any new hydraulic structure. They have followers (fortunately not many) working in the licensing agencies (state and Federal), in the judiciary and in the Public Prosecutor's Office (incidentally, the power granted by the constitution of 1988 to this institution is virtually without precedence in other countries). For them, there is no good dam site and their focus is on what bad things can occur at the local scale if the proposition of a new infrastructure gets the go ahead. They never ask what would be the consequences on the global scale if it gets the red light. They focus entirely on the sins of commission, and avoid entirely the sins of omission.

As a result of this myopic ideological environmentalism, close to 70 per cent of all energy to be produced in Brazil by new plants in the next 15 years will burn oil or coal. Before the fundamentalist wave, the share of oil and coal in the Brazilian electricity matrix was limited to 20 per cent.

This terrible result was achieved through the conception of an environmental licensing process that is performed case by case, without an overall view of the system. Both the executive and judiciary tend to decide about the social–environmental feasibility of a new plant based mainly on local considerations, which is a criterion that benefits the thermal option. The explanation for this is that hydro plants occupy, in general, large areas, they are site specific and displace local people, even in remote areas of the Amazon; thermal plants, on the other hand, occupy relatively small areas and can be built in carefully selected sites so as not to disturb local people.

The implicit objective function applied by the decision makers is to minimize local disturbance. Very little attention is given to the fact that if a large hydro plant cannot be built because of local socio-environmental concerns, it is very likely that it will be replaced by several small thermal plants that, although disturbing very little the local socio-environment, will collectively severely disturb the global environment though the emission of greenhouse gases. Or even worse, no plants will be built and the electricity will not be produced, causing an economic crisis. In any case, either the electricity will be more expensive because of the use of oil rather than water, or it will be unavailable. This would harm the country's competitiveness, decrease the number of jobs and increase poverty.

It is difficult to believe that even a DAHNGO would align itself to such evil purposes. But, perhaps unintentionally, some do. For example, in December 2007, the Brazilian electricity regulatory agency, ANEEL, organized an auction to decide which company would get the concession to build the Santo Antonio hydro plant of 3300MW. It is located on the Madeira River (a tributary of the Amazon River that runs from Bolivia to Brazil), a few kilometres upstream of the capital city of the Rondonia State (Figure 11.1, Plate 2). The concession is for a period of 35 years. The bids were in terms of the unit price of energy to be sold for 30 years, through firm contracts, to a set of distribution companies. The winner would be the one to offer the lowest bid.

The auction was the last step of a lengthy and laborious process of public hearings held by the Federal Socio-Environmental Licensing Agency (Ibama) and of disputes in the judiciary. The DAHNGOs, including IRN, tried their best to impede the issuing of the environmental licence. However, they had a difficult task for four reasons.

1 Santo Antonio is a run-of-the-river plant. This means that its 'inundated area–installed capacity' ratio is much smaller than that of the old hydro plants built in the 1970s. For example, its ratio is 3 per cent of Balbina's, which is a 'bad' hydro plant located on one of the Amazon tributaries that would not be built nowadays.
2 The Law ensures fair compensation to the people to be resettled. Although they constitute a minority, their rights must be respected. But the rights of the majority must also be respected. Almost 200 million Brazilians who will receive electricity from Santo Antonio, transported by high voltage transmission lines, belong to this majority.
3 Despite the efforts of international DAHNGOs in transforming this matter into an international dispute, the backwater of the Santo Antonio reservoir does not reach Bolivia.

4   Santo Antonio's energy will replace at least 25 per cent of the energy presently produced in the Amazon region by oil burning thermal plants. Annually they cause the emission of the equivalent of 5 million tons of carbon dioxide and cost around US$2 billion. Because the 5 million consumers that live in the Amazon region cannot afford such a large bill, this oil cost is shared by all 50 million consumers scattered throughout Brazil.

For all these reasons, all legal obstacles were removed and the auction could proceed. But the DAHNGOs respect the democratic process only when it serves their own interests. Otherwise they follow the twisted stakeholder participation theory, which 'asserts that any group that has an interest in, or could arguably be affected by the outcome of a public policy debate, has the right to pressure the decision makers until they accede to the activists' demands' (Driessen, 2003). In the auction of the Santo Antonio power plant the activists exercised this 'right'.

In the early morning of the auction day, around 6.00 am, a group of some 150 activists invaded the headquarters of ANEEL. Most of them belonged to a Brazilian DAHNGO called MAB, which is closely associated to the IRN. They demanded the cancellation of the auction. The police were called and acted firmly. The invaders were expelled, fortunately without serious injury. The auction was realized and the winning lowest bid of US$48 per MWh (rate: 1 US$ = 1.8 R$) was US$30 per MWh below the mean unit cost of thermal plants. For the consumers, this means an annual saving of almost US$700 million.

The lesson to be learned from this case study is that a democratic government should respect the rights of the people targeted for resettlement. However, the directly affected people and the NGOs have no right to veto the construction work and condemn a developing nation to remain as such in the foreseeable future.

## Conclusion and policy implications

Elected governments – much abused by the so-called progressives – are the only institutions capable of reconciling the full range of interests in complex decision making processes. The progressive project, particularly in countries that only recently became democratic, is to improve the performance of state institutions. Those groups who systematically undermine the state and who self-proclaim themselves to 'represent the people' have to be identified as the enemies of democracy.

Some stakeholders just want to preserve the status quo and are not interested in win–win outcomes. As a procrastination technique, they often call for the application of the 'precautionary principle'. However, for government officials the sins of omission should be as undesirable as the sins of commission. For this reason, the 'precautionary principle' should be used with caution.

Stakeholders need to understand the rationale of each decision. Therefore, the best water allocation rule is not necessarily the optimal one, from the economical point of view. A simple rule, but one that is easily understandable and accountable, may be the best choice.

A policy of subsidies for sanitation is acceptable, as it benefits the whole community, rather than individual citizens, contrary to the case of water supply. In poor communities, subsidies for water supply are also acceptable, provided the beneficiaries pay at least the O&M costs. Otherwise there will be no ownership.

A democratic government should respect the rights of the people to be resettled, in case of the construction of a hydropower infrastructure. However, the right to veto is not included among these rights.

The socio-environmental cost resulting from the implementation of a hydropower plant is not so high. However, the cost resulting from an inefficient socio-environmental licensing process is extremely high, resulting in the substitution of hydropower by thermal power. The first alternative is inexpensive, environmentally friendly, sustainable and it uses water. The second one is just the opposite, and it uses oil or coal. What is needed is a socio-environmental licensing process capable of evaluating not only the consequences of implementing a proposed project, but also the consequences of not implementing it.

## Acknowledgements

The comments and suggestions of two close friends that read the first manuscript – Ben Braga and John Briscoe – are greatly appreciated.

## Note

1  The record of water allocation and use of the two regions can easily be accessed on the following website: www.ana.gov.br/bibliotecavirtual/pesquisa.asp?criterio=cantareira&categoria=0&pesquisar=Pesquisar&NovaPagina=1

## References

Driessen, P. (2003) *Eco-Imperialism*, Free Enterprise Press, Madison, WI
Florence J. (2006) 'Global wind power expands in 2006', *Publishing on the Internet*, www.earth-policy.org/Indicators/Wind/2006.htm (accessed 6 January 2007)
Kelman J. and R. Kelman (2002) 'Water allocation for economic production in a semi-arid region', *Water Resources Development*, vol 18, no 3, pp391–407

*12*

# Issues of Balancing International, Environmental and Equity Needs in a Situation of Water Scarcity

*Barbara Schreiner*

South Africa is a water scarce country with a large proportion of its water in international river basins. Internally, within the country, competing and increasing demands are imposing strains on limited water resources, with significant negative impacts on the ecological functioning of rivers and wetlands, and a failure to meet international agreements. The Department of Water Affairs and Forestry (DWAF) and related institutions are faced with complex decisions in applying policy regarding the allocation of water as the real trade-offs between the competing demands are highlighted.

As a result of apartheid, a large proportion of the black population is poor, and has little or no access to water for productive purposes. In accordance with national policy, the government is currently in the process of reallocating water in order to meet international requirements, to meet the requirements of the aquatic ecosystems and to achieve redress in relation to race and gender. Implementation of the policy indicates that there is significant potential for cooperation in the reallocation of scarce resources. It also indicates considerable scope for a redistribution of resources based on improved water use efficiency by existing users. However, implementation poses its own challenges. The complexity of procedures can negatively affect the ability to implement good policy; the possibility of legal challenges to the process still exists, and limited human and financial resources may test the ability of the government to deliver on expectations.

This chapter looks at the challenge facing South Africa in relation to the fair distribution of scarce water resources between ecological requirements, international requirements and the people of South Africa, and the strategic issues that must be confronted in the process. The Inkomati Water Management Area in the north eastern part of the country is used as a case study to examine the potential for reallocation of water and the process that has been followed to date. This is followed by an examination of some of the lessons and policy implications of the process.

## Water availability in South Africa

South Africa is a country of erratic, low rainfall, punctuated by droughts and floods. Overall the country receives about 500mm of rainfall per annum, but even this is unevenly spread across the country from a relatively wet eastern side of the country to an extremely dry western edge where annual rainfall is less than 200mm per annum. With high temperatures and sparse vegetation, the potential evaporation in the western part of the country is extremely high, contributing to a shortage of water. Groundwater is relatively limited, but is a key water source in particular areas. Per capita water availability is low, in the region of 1100m$^3$ per annum.

Most of South Africa's major river basins are shared with neighbouring states. In all of these shared river basins, some form of agreement and intergovernmental structure is in place to ensure equitable sharing of the water or of the benefits derived from the water. These arrangements are guided by the Revised Southern African Development Community (SADC) Protocol on Shared Watercourses of 2000, which sets out the processes, procedures and institutions for management of shared watercourses in the SADC area.

To complicate the picture, much of South Africa's urban and industrial development has been driven by the presence of minerals rather than the presence of water and some of the major urban and industrial areas are in areas poorly served by rivers or groundwater. This has resulted in a complex network of inter-basin transfers, for example to provide water to Gauteng (the industrial heartland of the country) and to provide water for power generation in the north eastern part of the country.

Water scarcity is further exacerbated by increasing degradation of water resources through, inter alia, water pollution, habitat destruction and the presence of invasive alien species. Mining, poorly maintained sewage systems, industrial effluent and agricultural chemicals all contribute to water pollution.

### Racially skewed access to water

Within the scenario of general water scarcity, there are differing degrees of scarcity experienced by different sectors of the population. Under the apartheid government, black South Africans suffered from water deprivation in two respects.

The first was that access to water for productive purposes was tied to access to land. As the apartheid government deprived the black population of their access to land, through, for example, the 1913 Land Act, and drove them into limited 'homeland' areas, they equally deprived black people of their access to raw water for productive purposes. As a result, despite the white population making up less than 10 per cent of the total population, 95 per cent of agricultural water use was in the hands of these same whites. This has led to conflicting demands in over-allocated and closed catchments as black communities wish to obtain water for productive purposes.

At the same time, while the white sector of the population received water supply and sanitation services of the highest quality, black residential areas were provided with poor, if any, water services. At the time of liberation in 1994, it was estimated that around 12 million people did not have access to safe drinking water and around 21 million did not have access to adequate sanitation (DWAF, 1994).

The lack of access to water for productive purposes by black South Africans is a mirror of their general exclusion from the mainstream economy. Despite being a middle income country, in the region of 40 per cent of the population are poor – the vast majority of them being black.

## Current access to water

While the policy and legislation are clear on the need for and the processes for reallocating water to achieve better water management, and to achieve racial and gender redress, implementation has proved to be more difficult, and it is not yet clear whether the approach adopted will have served to reduce the conflict over access to water or merely postponed it.

In the 12 years since the advent of democracy in 1994, a great deal has changed in relation to the provision of water services. Over 12 million people have been provided with safe drinking water through a massive government infrastructure programme. Nonetheless, as a result of population growth over the years since 1994 and smaller household sizes, around 4 million people have access to water services but not of an adequate standard, and 2.9 million do not have access to improved water supplies at all (DWAF, 2007b). While there are concerns regarding the sustainability of some of the infrastructure provision of the past 12 years, that is not the focus of this chapter.

On the water resources side of the picture, which is the focus of this chapter, less has changed in access to water for black South Africans, and what has changed to date has been more as a result of the land reform programme than as a result of a deliberate water reform programme.

Currently over 60 per cent of the raw water used in the country is used by irrigated agriculture, largely white and male. Other key users are power generation (coal fired, not hydropower), mining, large industry and the municipal sector.

While the National Water Act provides the tools for the reallocation of water through a process called compulsory licensing, water reform and reallocation is only now beginning to be implemented in three different catchments, the Jan Dissels in the Western Cape, the Mhlatuze in northern KwaZulu Natal and the Inkomati Water Management Area (WMA) in Mpumalanga Province. This is happening nearly ten years after the promulgation of the legislation that enabled it. This chapter will use the Inkomati WMA experience as a case study for the consideration of the water allocation reform programme.

## Development and economic growth drivers

Despite being a middle income country, South Africa has one of the highest Gini co-efficients in the world (0.69) – a disparity in wealth that is indicated in disparities in access to water for productive purposes as well. South Africa is 55th in the world in terms of gross primary productivity (GPP) and yet it is 121 on the list in terms of Human Development Indicators.

The South African government has set a target of 4.5 per cent economic growth between 2005 and 2009 and 6 per cent between 2010 and 2014. The economic development programme of government is set out in the Accelerated and Shared Growth Initiative of South Africa (ASGISA), which states, inter alia, that 'we need to ensure that the fruits of growth are shared in such a way that poverty comes as close as possible to being eliminated, and that the severe inequalities that still plague our country are further reduced' (ASGISA, 2007).

South Africa has nine provinces, each of which has developed a Provincial Growth and Development Strategy that sets out the growth plans and trajectory for the province. At local government level, Integrated Development Plans are designed to capture the economic growth and development intentions of the municipality. All these plans should be nested and aligned. Unfortunately, such planning is relatively new and many of the plans are somewhat lacking in detail and specificity. Many of the plans, particularly at local government level, are developed by consultants on behalf of the municipality, in order to comply with legislation, and are seldom referred to again once they have been finalized.

Job creation, economic growth and poverty eradication are key to the future of South Africa, but in the context of water scarcity, crucial decisions must be made as to the optimal use of water in support of economic growth, poverty eradication and social development – water can only be provided to meet some demands, and not all. A strategic view of the appropriate development trajectory of any particular area is necessary to examine the trade-offs of such decisions and to decide on the optimal use of scarce water resources. Such decisions are often a challenge in the light of the somewhat vague plans at local, provincial and even national level. Such decisions are a particular challenge when the majority of catchments are in deficit and options for future infrastructure development to increase water availability are becoming increasingly scarce and expensive.

The drive for shared economic growth runs in parallel with the government's drive for ensuring broad based black economic empowerment (BBBEE) and racial redress after the years of racial discrimination under apartheid. BBBEE is a deliberate policy of government to ensure that black South Africans gain access to the economy after centuries of exclusion. This includes ensuring that black South Africans can get access to water for productive purposes.

It also runs parallel to a drive for gender equality that has posed particular challenges in the redistribution of natural resources such as land and water.

# Policy and legislative framework

The reallocation of water is taking place in a policy and legislative framework shaped, overall, by the need to redress the inequities created under apartheid (both race and gender) and to ensure the management of scarce resources in the best interests of the people of South Africa, while also recognizing the water needs of neighbouring states. Meeting such a complex array of needs in a water scarce country is potentially conflictual, and the Department of Water Affairs and Forestry has been developing approaches to sharing and reallocating water that are designed to minimize potential conflict, including through intensive public participation in the reallocation processes. The National Water Act provides the tools for sharing water between competing demands according to a hierarchy of needs that places water for basic human needs and for aquatic ecosystems first, followed by water for international purposes and, only then, water for economic purposes.

## The National Water Act

Water resources management in South Africa is governed under the National Water Act (Act 36 of 1998), which is premised on the principles of equity, efficiency and sustainability.

### Equity
The principle of equity is there to ensure the equitable allocation of water for productive purposes in the face of the historically inequitable distribution of water under apartheid. The Act is, therefore, fundamentally redistributive in nature and provides the tools for the reallocation of water to ensure such equity in access to water. The primary tool is that of compulsory licensing. Under compulsory licensing, the Department of Water Affairs and Forestry can call for all existing and potential users in a given geographical area to apply for new licences to use

water. The process of compulsory licensing allows for the reallocation of water in order to:

> *Achieve a fair allocation of water from a water resource which is under water stress when it is necessary to review prevailing water use to achieve equity in allocations*
> 
> - *To promote beneficial use of water in the public interest*
> - *To facilitate efficient management of the water resource, or*
> - *To protect water resource quality.* (RSA, 1998, p60)

The process of compulsory licensing has begun, as mentioned above, in three catchments in South Africa, under the rubric WAR – Water Allocation Reform.

Another aspect of equity is that of international equity. As mentioned briefly above, many of South Africa's key river basins are shared with neighbouring states, such as the Orange-Senqu, the Limpopo and the Inkomati. The National Water Act is very clear on the need to provide water to meet international requirements and the Act sets this as a priority use second only to the provision of water for basic human needs and to meet ecological requirements. The Act states that international water resources will be managed in a manner that optimizes the benefits for all the parties in a spirit of mutual cooperation and that agreed allocations for downstream countries will be respected.

## *Environmental sustainability*

Environmental sustainability is built into the Act through the concept of the Reserve. The Reserve is that amount of water of an appropriate quality to meet the ecological requirements of the water resource to maintain ecological functioning, and to ensure sufficient water at 25 litres per person per day for people without access to safe domestic water. The Act confers on the Reserve the status of a right – the only raw water use given this status in South Africa.

In many catchments in the country water is already over-allocated, and the requirements of the Reserve are not being met. Compulsory licensing, referred to above, is in the process of being used in order to reallocate water to meet the requirements of the Reserve as well as to reallocate water for equity purposes and for international requirements.

## *Efficiency*

In a water scarce country it is important that water resources are used efficiently – currently not the state of affairs in South Africa. Compulsory licensing will not only allow for reallocation of water but will also enable the state to drive water use efficiency more effectively in particular catchments. The Inkomati WMA case study will be used to show how this can be done.

# International agreements

As mentioned earlier, many of South Africa's major river basins are shared with neighbouring states. Management of these basins has a direct impact on domestic water issues.

## Revised SADC Protocol on Shared Watercourses

South Africa is one of the member states of the Southern African Development Community (SADC), as are all the neighbouring states with whom South Africa shares river basins. South African joined SADC in 1995 after the ending of apartheid. Prior to this, SADC was a structure of other governments in Southern Africa opposed to the apartheid policies of South Africa and to the military and other intervention by South Africa in neighbouring states. It is, therefore, relatively recently that South Africa has entered into formal cooperation with neighbouring states at the SADC level. Despite the isolation of South Africa under apartheid, there is little history of conflict over the management of shared river basins. Indeed, negotiations for sharing water resources took place even while South Africa and neighbouring states were locked in political conflict (Tekateka and Malzbender, forthcoming).

SADC is one of the few areas in the world with a fully ratified international protocol or convention on the sharing of international rivers: the Revised SADC Protocol on Shared Watercourses (2000), to which South Africa is a signatory. This protocol sets out the framework for the management of shared river basins within SADC. The original SADC Protocol on Shared Watercourses was already in place in 1997 when the UN Convention on the Law of the Non-Navigational Uses of International Watercourses (1997) was drawn up. This document resulted in the revision, in 2000, of the SADC protocol, with many of the clauses of the UN Convention being incorporated into the Revised SADC Protocol.

The Revised SADC Protocol aims to foster closer cooperation between the SADC states for the judicious, sustainable and coordinated management, protection and utilization of shared watercourses, and to advance the SADC agenda of regional integration and poverty alleviation.

Amongst other things, the Revised Protocol deals with the sustainable, equitable and reasonable use of shared watercourses, and the need for the environmentally sound development and management of shared watercourses. It sets out certain general principles, which include that in utilizing a shared watercourse within its own territory, each state must take all appropriate measures to prevent the causing of significant harm to other watercourse states; a balance must be maintained between resource development and environmental protection; and that information and data on hydrology, water quality, hydrogeology, etc. should be shared.

## Interim IncoMaputo Agreement

The Inkomati river basin is a shared one, between South Africa, Mozambique and Swaziland. South Africa is not only the upstream country, but is also more developed and therefore using more of the water. Over the years, the equitable sharing of the water has been a potential source of conflict, particularly as demands grew. Surprisingly, despite very poor political relations between South Africa and Mozambique after Mozambican independence in 1975 (driven by ideological differences and the funding, by South Africa, of Renamo, the armed opposition to the Mozambican Frelimo government), the two countries continued to meet and to discuss the sharing of the waters of the Inkomati river basin. Various meetings, agreements and negotiations took place over a period of 30 years. These culminated in the signing by South Africa, Mozambique and Swaziland of the Interim IncoMaputo Agreement. This agreement was signed in August 2002 during the World Summit for Sustainable Development, in Johannesburg. It aims to promote cooperation among the three countries to ensure the protection and sustainable utilization of the water resources of the Incomati and Maputo watercourses (Article 2). It supersedes the Piggs Peak agreement of 1991 and is valid until 2010 or until a new, comprehensive water agreement is signed.

The agreement uses the definition of watercourse of the UN Convention and the revised SADC Protocol, as well as the general principles of the SADC Protocol. These include the principles of sustainable utilization, equitable and reasonable utilization and participation, the prevention principle and the cooperation principle.

Article 4 of the agreement sets out the responsibilities of the three countries, which include that the countries must individually or, where appropriate, jointly, develop and adopt technical, legal, administrative and other reasonable measures in order to, among other measures:

- coordinate management plans and planned measures;
- monitor and mitigate the effects of floods and droughts;
- provide warning of possible floods and implement agreed upon urgent measures during flood situations;
- exchange information on the water resources' quality and quantity, and the uses of water;
- implement capacity building programmes.

The last point is interesting in that this is one of the few, if not the only, international water sharing agreement that specifically refers to the need for capacity building.

Article 7 refers to 'Sustainable Utilization' and sets out that each country is entitled to optimal and sustainable utilization of the relevant water resources, as long as the interests of the other countries are taken into account and as long as there is adequate protection of the watercourses for present and future generations.

This article also refers to the need to promote and implement water conservation and improved water use efficiency.

Article 9 states that any abstraction of waters from the Incomati or Maputo Watercourses must be in conformity with the flow regimes set out in Annex I of the agreement.

## Case study: Water allocation reform in the Inkomati Water Management Area

The policy and legislative framework is thus in place in South Africa, and in SADC, to ensure equitable, sustainable and efficient use of water and to balance environmental, international and equity needs for water, in the face of increasing water scarcity and competing demands. However, the translation of the legislative mandate into implementation is where the real test lies and where the requirement for strategic decision making raises its head. South Africa has recently moved into the implementation of compulsory licensing in order to achieve redress in relation to access to water, to meet ecological and international requirements, and to bring over-allocated catchments back into balance. As mentioned, one area in which the process towards compulsory licensing has been initiated is the Inkomati Water Management Area and this section examines the process so far in this water management area in order to examine and illustrate the implementation challenges. The process is by no means complete, so the ultimate success of the process cannot yet be evaluated, but there are some useful lessons to be learned from the process so far, including lessons regarding the development of appropriate strategic approaches towards meeting the intention of the policy.

The Inkomati Water Management Area covers around 29,000km² in Mpumalanga Province in the north eastern part of South Africa. It falls into an area that has seen considerable political turmoil and conflict over the past 40 years in particular, through apartheid repression and resistance in South Africa and civil war in Mozambique (aggravated by South Africa's support of Renamo). Political stability in both countries has only been achieved relatively recently.

The introduction of DDT in 1945 as a means to control malaria (Packard, 2001) meant that white commercial farmers could move into the area, prized for its good soils and access to water. The black farmers of the area were forcibly removed from these areas into the KaNgwane 'homeland' along the lower Komati.

Since then, this area has seen major agricultural development and a huge growth in commercial afforestation with exotic species (mainly pine and eucalypt). This expansion has dramatically increased abstraction from rivers while pollution and habitat destruction have also increased. The rivers of this area are therefore under considerable strain. At the same time, Mozambique, which is downstream of all of the rivers of the Inkomati Water Management Area, is seeing good economic growth and is in need of water for both economic and social development.

Thus, while the Inkomati Water Management Area falls into one of the wetter parts of South Africa, only one of the four sub-catchments (the Sabie River Catchment) has water still available for allocation. The other three sub-catchments (the Komati, Sand and Crocodile river catchments) are over-allocated to the extent that the Reserve is not being met fully. Nor are international requirements fully met as per the current agreements.

The catchment has specific requirements in relation to ecological protection. Nearly 40 per cent of the Kruger National Park game reserve falls into this water management area, raising the level of ecological protection required in the lower reaches of the rivers. The Kruger National Park is an internationally recognized conservation area, now part of a transboundary park with Mozambique. Both in terms of protecting tourism revenue and in terms of ecological protection, it is important to maintain good environmental flows through the park. The Incomati estuary in Mozambique is also ecologically and economically significant, being an important habitat for breeding colonies of aquatic birds, while also providing water and ecological services to local populations. It is, in addition, an important area for shrimp, fin fish and shellfish breeding. It has extensive mangrove forest that requires protection.

Approximately 1.6 million people live in the water management area, with the majority living in rural areas. As with the rest of South Africa, access to water in the catchment is racially skewed with very little water in the hands of black users. Irrigated agriculture and commercial forestry are two of the most important economic activities in the WMA. Tourism, recreation, manufacturing and a small amount of coal mining are also important contributors to the economy. Approximately 8 per cent of the water is transferred out of the WMA for power generation.

The water management area contributes only 2.3 per cent of GDP despite having 3.7 per cent of the national population. This discrepancy arises due to a lack of major industries or manufacturing in the area, and a lack of beneficiation of raw materials. Some of the water resources of the Inkomati WMA are exported to the neighbouring Olifants WMA for power generation. The catchment is also shared with Mozambique and Swaziland and water must be made available for these countries.

The Inkomati Water Management Area has four sub-catchments, the Crocodile, Komati, Sabie and Sand river catchments. This chapter will focus on water allocation in the Crocodile and Sand river catchments in order to illustrate the potential and complexities of balancing equity, environmental and international requirements.

## Crocodile river catchment

The challenges in the Crocodile river catchment are particularly acute with a number of conflicting demands on the scarce resource. The Crocodile river forms

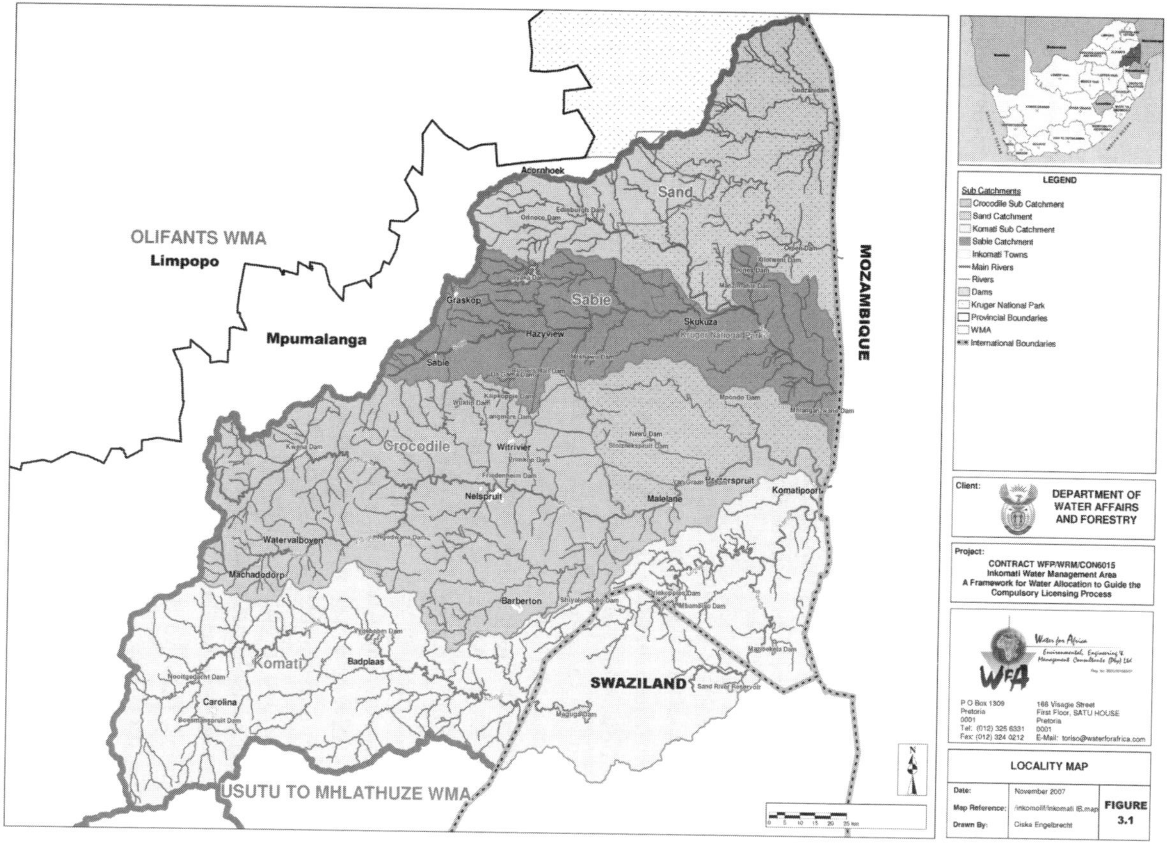

**Map 12.1** *Map of the Inkomati Water Management Area*

Note: See Plate 3 for a colour version.

the southern boundary of the Kruger National Park, with a resultant high level of ecological protection required due to the conservation status of this area. The river then flows across the border into Mozambique. Sufficient water needs to be made available to meet the Reserve and international requirements. At the same time, water needs to be made available to black users and women to achieve the government's objective of racial and gender redress. These allocations must be made in the context of a catchment that is already severely over-allocated. Historically there have been problems relating to ensuring that sufficient water is allowed across the border, the quantity of water assigned for the aquatic ecosystem, and the quantity of water available to small black farmers. The Department of Water Affairs and Forestry, as the responsible authority, is expected to mediate these conflicting demands, but does not always do so successfully.

Currently the water requirements for Mozambique are determined by the Piggs Peak agreement as a minimum of 0.9m$^3$/s (63 million m$^3$/a).[1] However, South Africa and Mozambique subsequently signed the Interim IncoMaputo Agreement and new requirements will be negotiated under this agreement. While there is no negotiated quantity yet, it can be estimated on the basis of Mozambique's downstream requirements. South Africa's current operating rule in relation to cross-border flows takes 45 per cent of the allocation from the Crocodile River and the rest from the Komati system. Using this approach, it is likely that the total minimum flows that will be required from the Crocodile catchment at the border is 50.5 million m$^3$/a and the annual average flow is 130.5 million m$^3$/a – a considerable increase from the previous agreement.

Since the current 0.9m$^3$/s is not always met (a matter that has resulted in some tensions between South Africa and Mozambique in the past[2]) (Vaz and Van der Zaag, 2003), ensuring water for the international requirements at the new level will require some curtailment of water use on the South African side of the border.

Finding sufficient water to meet the Reserve requirements will also require curtailment of current water use. Protection of aquatic ecosystems and the environment is one of the elements of the Interim IncoMaputo Agreement as well as being enshrined in South African and Mozambican law (Vaz undated).[3]

To achieve this reallocation of water, a process of compulsory licensing has been initiated in the catchment. Compulsory licensing is a lengthy process, with a number of steps. Prior to the actual call for licences in which all current and would-be water users must apply for a licence, a Water Allocation Plan must be developed that is based on the Catchment Management Strategy.

The Catchment Management Strategy and the Water Allocation Plan need to be able to achieve the multiple goals of meeting international requirements, providing sufficient water for ecological purposes and providing water for equity purposes. To achieve this, all role players and stakeholders, including all levels of government, the various economic sectors and poor communities must be involved in the process, and, to the greatest extent possible, agree on the reallocation approach and targets.

The Inkomati Water Management Area is, so far, the only WMA in South Africa with a functioning Catchment Management Agency (CMA) – an agency established by government to manage the water resources of a particular area as part of a policy of decentralization and democratization of water management. In the other water management areas the Department of Water Affairs and Forestry is still responsible for management of water resources. However, the Inkomati CMA is not yet fully staffed or fully functional, and, as a result, both the DWAF and the CMA are performing functions in the WMA. This has the potential for conflict due to unclear roles and overlapping functions. For example, the CMA is in the process of preparing a Catchment Management Strategy for the WMA, which will set out the vision and strategies for managing the WMA. This is developed on the basis of wide consultation with stakeholders.

In parallel, DWAF, which has not yet delegated most of the water resource management functions to the CMA, began the process of developing a water allocation framework (WAF). This document is seen as the precursor to a Water Allocation Plan (WAP) that will be developed once the catchment management strategy is complete. Strictly speaking, and as set out in the National Water Act, the Catchment Management Strategy should be developed prior to the development of a WAP in order to set the strategic direction for the WAP, but the availability of donor funding (within a limited time frame) to drive the development of the water allocation framework, has resulted in the two processes running simultaneously. In order to address this, once the Catchment Management Strategy has been completed, it will be used as the basis of a consultative process to develop the WAF further into a more specific and detailed WAP that will set out clearly what water is to be allocated to which sectors, and what curtailments of water use will be put in place to achieve reallocation.

Also running in the WMA is a process of validation and verification of existing water use of both surface and groundwater. The intention of this process is to assess whether the current water use has (i) been correctly registered with the Department; and (ii) is lawful. The completion of this process will enable a better understanding of current lawful use in the area, versus the quantity of available water and, once again, in an ideal world, this process should have been completed prior to the development of a WAF. Unfortunately, due to the technical complexity of this process, it is taking a long time and intensive resources to complete this process and it was decided to run the development of the WAF in parallel with this process. It is possible that the results of the verification and validation may be open to legal challenge by water users who may wish to contest what has been determined as their existing lawful use, and this too may hinder the finalization of this process.

## Developing the WAF

The development of the WAF began with an assessment of the current water use status in the WMA, which identified, inter alia, the high levels of over-allocation in most of the area, and the challenges in terms of meeting ecological and international requirements. Eighty-three per cent of water in the WMA is used by agriculture (DWAF, 2007a) and the bulk of this is used by white farmers.

This was followed by a three pronged approach in support of water reallocation in the water management area, one leading to the development of a draft WAF, one focusing on promoting rainwater harvesting in poor rural communities, and one focusing on empowering disadvantaged stakeholders to be able to apply for authorization to use water, and to be able to use any water allocated effectively. Although the National Water Act puts forward compulsory licensing as the pre-eminent tool for reallocation of water, the department and the project team adopted a more strategic approach, looking for parallel and complementary strategies that could provide water for productive purposes to the poor in particular, while compulsory licensing ran its course. Despite the catchment being over-allocated, it was felt that the quantity of water required to support a rainwater harvesting project was more than justified by the social impact that it would bring to poor households in particular.

The rainwater harvesting project aimed at building demonstration tanks in rural villages to begin the promotion of rainwater harvesting in this area as a way to alleviate poverty in poor rural households. Water for household food gardens falls under Schedule 1 of the National Water Act and therefore does not require authorization for use. Households are entitled to take water for such purposes without any reference to the department or any other authority. This arm of the project was intended to ensure that poor households were aware of the possibilities of rainwater harvesting in support of food gardening and of the potential financial support from the department to develop rainwater tanks. One driver behind this element of the project was the concern that water allocation reform only deals with authorized water use, and the users of very small amounts of water, like household food gardeners, do not fall within this category. It was considered important to ensure that the very poorest households benefited from the process of water allocation reform, and not only those already sufficiently well resourced to be using larger amounts of water. This chapter will not be dealing with this arm of the project.

The second leg of the project was an educational programme aimed at disadvantaged communities to communicate to them the water allocation reform programme, to explain to them their rights and options within the programme, and to capacitate them to apply for water for productive purposes under compulsory licensing. One aspect of this part of the project also entailed the identification of potential projects for productive water use, mainly in the irrigation sector, for poor black communities.

The third leg of the programme was the actual development of a water allocation framework. The approach adopted in the development of the WAF was to begin with the development, in consultation with stakeholders, of a set of principles specific to the water management area. These built on national level principles already developed by the department in consultation with national level stakeholders. It was felt by the project team that, in order to achieve buy-in to the process from stakeholders, the project should begin with the development of key principles that would guide the implementation of the water allocation reform programme in the WMA. Through consultation with key stakeholders, a set of principles was developed to guide the development of the WAF and to guide compulsory licensing in due course.

Ten principles were developed that cover the full WMA. These are attached in Annex A. Amongst other things, these principles imply that stakeholders have agreed to:

- give priority allocations to the Reserve and to meet international requirements;
- a target for redistribution of water to black users (40 per cent of water use in the WMA in the short term and 50 per cent in the longer term) and to women (50 per cent of allocated water)
- give priority to ensuring that all water users and uses are water use efficient before any supply side management options are considered;
- water use licences including time frames to allow users to progressively reduce their use in line with their new allocations, aligned with the gradual uptake of water by new users.

Further to these, certain principles were developed that relate to specific catchments. In the Crocodile catchment these included:

- a target of 25 per cent of water to be allocated to black users in the short term, to be achieved through land reform;
- that any water use reduction to existing water entitlements will be done only to meet the priority needs of the Reserve and the international requirements and not for reallocation to other water users in the area;
- that the implementation of the Reserve will be undertaken progressively but that the current state of the aquatic environment must not be further degraded.

The catchment specific principles for the Sand River Catchment included:

- a target of a minimum of 50 per cent of irrigation and forestry water use in the Sabie and Sand river catchments to be allocated to black individuals and/or institutions addressing black needs;
- the determination of the ecological Reserve requirements for the Sabie and Sand River systems will recognize the importance of these river systems to the Kruger National Park.

## Limits to consultation

These principles were discussed with stakeholders in four workshops, one held in each sub-catchment in the WMA. Representatives at these workshops included farmer groups (water user associations), mines, small farmers and some local government representation. Unfortunately, provincial government and local government were not adequately represented and it has proved difficult to get the necessary buy-in and support to the process from these key government actors, no doubt partly due to the high workloads of government officials, and partly due to a lack of understanding of the significance of the project for economic and social development in their areas. At the workshops, a draft set of principles developed by the project team were presented to the stakeholders for discussion. After the consultation the project team revised the principles on the basis of the issues raised in the workshops. The revised principles will be further consulted with stakeholders in the next phase of the project, that is when the call for compulsory licensing is actually issued.

There was valuable input received from stakeholders on the principles, and substantial revisions were made as a result. There was also general agreement on most of the principles from all of the stakeholders present. There were, however, three key areas of concern for stakeholders. The first was that the international requirements were too large and that the water would be wasted by allowing it to flow downstream to Mozambique – they wished to see the international allocation substantially reduced. The second was concern regarding the allocation of water to women, with concerns ranging from the sense that 50 per cent was too high, to concerns about whether water should be allocated to women at all. It is, perhaps, worth noting that the presence of women in the workshops was extremely low despite a substantial proportion of the small black farmers of the area being female. The third concern that was voiced strongly was whether the department has the capacity to implement compulsory licensing and to issue, in a timely manner, the number of licences that will be required in the process. This arises from experience in the water management area and elsewhere where it has taken several years for a water use licence to be issued by the department.

Despite these concerns, the general consensus was support for the principles. The principle that the project team expected to be most contentious – the allocation of water to black users – was not contested at all. This is perhaps because this water will mainly be transferred through the process of land reform, which has already been running for several years, allowing people to adjust to and accept the idea.

Despite the apparent consensus, there are two concerns regarding the future response of stakeholders to compulsory licensing. The first is that the stakeholders present in the workshops were a relatively small group of people – probably 100 in all. While many of them were representatives of groups (e.g. groups of farmers and water user associations), it is unlikely that the consultation reached the majority of water users in the catchment. It is possible that, in the actual process of compulsory licensing, water users who were not consulted directly may reject the principles and therefore the process built on those principles.

The second concern relates to the possibility of legal challenge. The National Water Act is written in such a way as to enable the department to curtail existing water use without having to pay compensation, unless severe economic prejudice is caused. This is based on an interpretation of water use as a use right, not a property right. As a use right, the constitution allows a change to that right without compensation, in the interests of redress, in a manner that would not be possible should it be considered to be a property right. In the late 1990s, during and after the passing of the National Water Act, there was unhappiness expressed by farmer groups with this notion of uncompensated curtailing of what they considered a property right. It was understood by senior officials in the department, that the farmers had been advised by their lawyers not to attack the constitutionality of the Act per se, but rather to wait for implementation of the compulsory licensing clause and to contest the constitutionality of the Act in implementation. It is, therefore, very possible, that despite the apparent support for the principles that are designed to underpin compulsory licensing, there will be a legal challenge to the implementation of compulsory licensing, including a legal challenge to the constitutionality of such actions.

## Key challenges and issues

There are some interesting challenges that arise from the process of WAR in the Inkomati WMA that require reflection.

Reallocation in the Inkomati WMA has only focused on meeting the requirements of equity, the Reserve and international requirements, and to achieve efficiency. It has not attempted major inter-sectoral reallocation – this will be done, and is done, through water trading. This is mainly taking place between farmers (usually from lower value to higher value crops, or from less productive to more productive farmer) or from irrigation agriculture to mining. Currently trading is purely on a willing-seller, willing-buyer basis, although the process is mediated through the department in the sense that the department must cancel the seller's licence and issue a new licence to the buyer – on condition that the buyer's proposed water use is in accordance with the requirements of the Act. Representatives of the mining sector, in the consultative workshops, proposed a more proactive approach from government, one that would see government actively intervening to move water from the irrigation sector to the mining sector due to the greater returns from water in the mining sector than in irrigation. Such intervention from government is currently under consideration in the form of water banking, but is not yet factored into the process of water allocation reform.

The process that has been adopted to reallocate water is one with significant administrative burdens. Not only will compulsory licensing require the consideration of large numbers of licence applications, it has also been accompanied by other expensive and administrative processes, such as validation and verification.

Validation and verification are the two steps required to determine whether a user's water use has (i) been registered correctly with the department; and (ii) is legal. The process is time-consuming and resource intensive (both financial and human), and in many cases is inconclusive, since it is difficult to prove what the existing lawful use is in volumetric terms. The administrative burden and long time frames associated with compulsory licensing have lead the department to begin to consider whether there are, perhaps, alternative ways to achieve the same result. State regulated water banking and trading is one approach that is currently under consideration to assist water allocation reform through a less administratively cumbersome approach. The challenge, however, is the need for strong regulatory and institutional capacity to make water banking work effectively.

A significant challenge that has arisen in the process of water allocation reform in the Inkomati Water Management Area is the issue of unintended negative consequences – in this case, possible loss of jobs. In order to free up water for the Reserve and for international requirements, water must be taken away from existing users. In many areas, this reduction can be compensated for by increased efficiency, and the areas under cultivation will not necessarily be affected. In some areas, however, the efficiency gains will not be sufficient to meet the curtailment requirements, and water uses will have to be cut back further. In this case, it is clear that there is a strong possibility of job losses. The counter-intuitive possibility, which warrants further investigation, is that the former also results in job losses. Current evidence points to a lower labour requirement associated with more efficient irrigation systems. Thus, pushing farmers towards more efficient irrigation systems could result not only in water savings, but also in job losses. The promotion of water conservation and demand management may, therefore, result in significant negative impacts for the poorest households in the water management area, thus defeating one of the primary objectives of water allocation reform nationally, namely the provision of water for redress and equity. This is an area that requires further investigation and analysis.

## Conclusion and policy implications

The water allocation reform process being implemented in South Africa is innovative and unprecedented. As with all new approaches it has some teething problems, but potentially some more significant challenges. It also, as an innovative and path breaking programme, has potential lessons for other countries wishing to balance environmental, international and equity issues in water allocation.

First, South Africa is fortunate in that a strong policy and legislative framework supports the reallocation of water and the balancing of environmental, international and equity needs, the latter both internally and between states. This policy and legislative framework is indicative of and supported by strong political support for the reallocation process, particularly in relation to the international and equity elements. Redress of apartheid injustices is a key element of South African government policy. There is also a strong commitment from the South African government to support economic and social development in neighbouring states, partly due to a historical debt to the countries that supported the liberation movements, and partly due to a pragmatic recognition that it is in the economic interests of South Africa to see the economies of neighbouring states grow. It remains to be seen what the political response will be to the idea of providing water for ecological requirements at the expense of jobs in the area. In the face of the immense need for job creation to reduce poverty, it is likely that a compromise position will be adopted where ecological requirements are sacrificed in the short term in the hope of retaining jobs and finding a way to reduce water use over time to meet the Reserve without compromising jobs. The ideal solution would be to provide alternative jobs, but with a shortage of water this can only be done if water is moved into sectors that are more labour-intensive.

As with all state processes, the policy and legislative framework only provides the strategic direction and the foundation. Many of the most significant challenges occur in implementation. And it is in the implementation that some of the successes and some of the challenges of this process lie.

The process that has been adopted by the department is technically sound but administratively heavy. The validation and verification process has not yet been completed because of the technical complexity of approach and its resource intensive nature. A comprehensive reserve determination for the Crocodile catchment has not yet been completed, resulting in the current work being based on a relatively low-confidence assessment of ecological flow requirements. The comprehensive reserve determination considers both the social and economic implications of various levels of protection of the aquatic ecosystem and is developing scenarios for the various options in order for decision makers and stakeholders to understand the implications of certain choices in an informed manner (Harrison Pienaar, pers. comm.). This work has not yet been completed due to the resource intensive and complex nature of the work. It is well accepted that there are many complexities involved in understanding and managing a catchment, but processes must be adopted that are commensurate with the capacity of the institutions to deliver. This is not always easy to determine in advance, but it is important to ensure that processes and approaches are appropriate to local conditions, particularly in relation to capacity to deliver. The principles of simplicity and implementability must underpin the approaches taken. It is possible, otherwise, for the perfect to become the enemy of the good.

The shortage of skilled and qualified staff has posed a problem in the Inkomati, and in other catchments in South Africa. This has resulted in the process of water allocation reform in the Inkomati WMA relying significantly on consultants. This means, unfortunately, that some of the critical intellectual property and lesson learning relating to this process is resident in the consultant teams, rather than in the department or the Inkomati CMA. This raises concerns regarding the replicability of the process over the rest of the country and the cost that may have to be incurred to bring other teams of consultants up to speed with experience to date. The high turnover of staff within the department also poses a challenge in terms of institutional memory in relation to water allocation reform.

Shortage of staff in other government departments and agencies, and lack of understanding on their part of the importance of this programme to the development programme of the area has undermined the cooperative government relationships necessary to make water allocation reform work effectively. Despite the rhetoric of integrated water resources management, there is a lack of integration still between water resource planning and economic and development planning. There are several underlying causes, including the poor planning happening at local and provincial level. The department has put in place several processes to improve planning between government departments as they relate to water, such as provincial water summits to align development plans with water plans, and much progress has been made. However, water is not yet a sufficiently high priority for all government spheres and officials.

A lack of resources and capacity within the department and the Inkomati CMA raises another challenge – that of policing the new allocations. Experience has shown, both in this area and elsewhere, the need to police water allocations carefully, particularly in times of drought. Local organizations, such as water user associations, can be, and have been, extremely useful in ensuring that water users do not over abstract. The role of such institutions is particularly important when capacity in the department and Inkomati CMA is limited. However, it is important that clear roles are defined for the various institutions in this regard in order to ensure that duplication does not take place and that gaps are not left between the roles of the three key players.

Despite the involvement of stakeholders in the process, and their apparent support for the approach and the principles driving compulsory licensing, the response of stakeholders to the actual curtailment of their water is a key risk facing this project. Despite the best consultation and attempts to obtain buy-in from stakeholders, the possibility still exists of a legal challenge to the constitutionality of the approach to water allocation reform. Legal challenge could tie up the process for many years in the courts, partially, if not totally paralysing the process of water allocation reform. The challenge to the department is to develop a mitigatory strategy and a 'plan B' on how to address such a situation.

Linked to this is the need to continue to investigate how best to achieve water allocation reform and whether there are other, more effective approaches that can be adopted, such as cap and trade approaches or facilitated water banking. The department, working with experienced consultants, is now looking at alternative, faster and more efficient ways of achieving the same end.

It is interesting that the gender issue was one of the most contentious. Clearly, the need and drive for racial redress in the country has sufficient momentum and authority for it not to be questioned even by those who may not fully endorse it. The gender redress drive, however, is clearly weaker, despite the existence in South Africa of a National Gender Commission and an Office on the Status of Women. The land reform programme is not particularly engendered and there is still a lot of work to be done to translate gender policy into real change on the ground, and even more work to be done to get that change accepted within a still strongly patriarchal society.

It is also clear that, despite government policy on prioritizing the allocation of water for international agreements, there is a different response from those who might see their own water use curtailed in order to provide water to Mozambique. This highlights the tension between local, parochial interests and the interests of government at the national level. It highlights too, the need for government intervention in order to ensure that the greater good, as embodied in policy, is implemented on the ground.

The Inkomati case study shows that there is the possibility, and the desire, both internationally (between South Africa and Mozambique) and nationally, to move towards cooperation in the face of water scarcity, rather than conflict, but the process is a vulnerable one, subject to a range of possible challenges. Using a consultative process is important in reducing the areas of potential conflict but is unlikely to remove them altogether.

# Annex A: Principles for the Reallocation of Water in the Inkomati Water Management Area

## Principle 1

The amount of water available to allocate will be determined after accounting for the Reserve,[4] the international water requirements as stipulated in the Interim IncoMaputo Agreement, and the water requirements for power generation outside the catchment. This determination will use the latest available information.

## Principle 2

The potential impacts on the quality of the resource (this includes the quality of all aspects of the water resource including water quality, the integrity of riparian and instream habitats and aquatic organisms), will be considered when granting a licence under the compulsory licensing process.

## Principle 3

Water allocations will be made after considering the allocation priorities outlined in the Nation Water Resource Strategy (NWRS), the strategic importance of the use, water use efficiency criteria and the potential financial impacts on existing lawful water use.

## Principle 4

In reconciling the water available for allocation and the water requirements of the different sectors, an integrated resource planning (IRP) approach which involves considering a range of options including water trading, improving water use efficiency, reuse and recycling of water and investment in supply augmentation, must be used. However, priority will be given to ensure all water users and uses are water use efficient before any supply side management options can be considered.

## Principle 5

An appropriate assurance of supply for each water use sector will be determined after consultation with stakeholders, and after considering the options for managing abstractions on a year-by-year basis, as well as the impacts of assurance of supply on the users and the economy of the region.

## Principle 6

A minimum target of 40 per cent[5] of water use by irrigation, forestry or industry (by volume) to black users[6] is envisaged for the whole Inkomati WMA. Where this target cannot be met immediately, sufficient water will be reserved for future use by black users in order to meet these targets as soon as possible.

A supplementary target of a minimum of 50 per cent of the water to be authorized to women will also apply. In the longer term reallocation processes must ensure 50 per cent of water is authorized to black users.

## Principle 7

Notwithstanding Principle 6, water allocations will only be made to viable enterprises, wishing actually to take up that water. Allocations will be consistent with the capacity of the land, and allocations will only be made where the users have lawful access to the land.

## Principle 8

Preference will be given to applicants complying with the BBBEE codes of practice, DWAF's BBBEE guidelines, or users in equity share schemes.

## Principle 9

Water allocations and licences will include timeframes to allow users to progressively reduce their use in line with their new allocations, aligned with the gradual uptake of water by new users.

## Principle 10

In developing the water allocation framework, the process will take into account the possible negative and positive impacts of the water allocation reform on the economy of the WMA.

## Notes

1   All three countries benefited in some way from the Piggs Peak agreement. Swaziland not only got the water agreements that it wanted but was responsible for enabling the negotiations between the other two countries. South Africa won support for the Komati Development Plan and Mozambique, in supporting the Komati Development Plan, got agreements from South Africa about the Sabie River and the minimum flow at Ressano Garcia.
2   On 6 July 1982, as a result of a severe drought, the Incomati River at the border between South Africa and Mozambique was reduced from an average $6m^3/s$ to $40l/s$. Two months later it dried up completely for the first time in 30 years of recording water levels. After the signing of the Piggs Peak agreement in 1991 there were further problems with meeting cross-border flows, resulting in a promise by the then Minister of Water Affairs in 1997 to ensure that the flows were met in future. Although this promise was met in 1998, there have been problems in meeting the flow subsequently.
3   Article 6 of the Revised SADC Protocol on Shared Watercourses deals with the 'Protection of the Environment' and states that the three countries shall, individually and, where appropriate, jointly, protect and preserve the aquatic environment and ecosystems of the Incomati and Maputo watercourses, taking into account generally accepted international rules and standards.
4   The Reserve includes the water quantity and quality needed to maintain aquatic ecosystems in a particular.
5   Current estimates indicate that between 24 per cent and 34 per cent will be in black hands after the land reform process.
6   Black users will be defined as per the BBBEE Act.

## References

ASGISA (2007) 'Accelerated and shared growth initiative for South Africa (AsgiSA)', South African Government, www.info.gov.za/asgisa (accessed 26 February 2008)

DWAF (1994) 'Water – an indivisible national asset', Water Supply and Sanitation Policy White Paper, Department of Water Affairs and Forestry, Cape Town, South Africa

DWAF (1999) *Resource Directed Measures for Protection of Water Resources*, Volume 3: *River Ecosystems*, Version 1.0. Department of Water Affairs and Forestry, Pretoria, South Africa

DWAF (2001) 'Methods for assessing water quality in ecological Reserve determinations for rivers', Version 2, Draft 13. Department of Water Affairs and Forestry, Pretoria, South Africa

DWAF (2004) *National Water Resource Strategy*, 1st edn, Department of Water Affairs and Forestry, Pretoria, South Africa

DWAF (2007a) *Inkomati WMA Catchment Assessment Report*, Department of Water Affairs and Forestry, Pretoria, South Africa, January

DWAF (2007b) *Annual Report 2006/7*, Department of Water Affairs and Forestry, Pretoria, South Africa

Incomati Basin (Mozambique, Swaziland, South Africa) Negotiating a water sharing agreement: Main characteristics of the basin, www.unesco.org/water/wwap/pccp/zaragoza/basins/incomati/incomati.pdf (accessed 28 April 2008)

Packard, R. (2001) '"Malaria blocks development" revisited: The role of disease in the history of agricultural development in the Eastern and Northern Transvaal Lowveld, 1890–1960', *Journal of Southern African Studies*, vol 27, no 3, pp591–612

Reed, D. and M. de Wit (eds) (2003) *Towards a Just South Africa: The Political Economy of Natural Resource Wealth*, WWF Macroeconomics Program Office, Washington, DC

Republic of Mozambique, Kingdom of Swaziland and Republic of South Africa (2002) 'Tripartite interim agreement between the Republic of Mozambique and the Republic of South Africa and the Kingdom of Swaziland for co-operation on the protection and sustainable utilisation of the water resources of the Incomati and Maputo watercourses', www.jointmaputobasin.org/index.php?option=com_docman&task=doc_details&gid=4&Itemid=4 (accessed 21 April 2008)

RSA (1996) White Paper on a National Water Policy for South Africa, www.dwaf.gov.za (accessed 17 April 2008)

RSA (1998) National Water Act No 36 of 1998, Republic of South Africa www.dwaf.gov.za (accessed 28 April 2008)

SADC (2000) *Protocol on Shared Watercourses*, Southern African Development Community, Gaborone

Tekateka, R. and D. Malzbender (forthcoming) 'Transboundary water management issues under the NWA and regional collaboration, policies and conventions', in R. Hassan and B. Schreiner (eds), *Water Policy in South Africa*, Resources for the Future, Washington, DC

Vaz, A. C. and P. van der Zaag (2003) *Sharing the Incomati Waters Cooperation and Competition in the Balance*, UNESCO IHP WWAP, IHP-VI Technical Documents in Hydrology, PC+CP Series no 14, UNESCO, Paris

*Part III*

# Interaction between Policy and Strategy

# 13

# Modelling Negotiated Decision Making under Uncertainty: An Application to the Piave River Basin, Italy

*Carlo Carraro and Alessandra Sgobbi*

A quick look at the media highlights the fact that the world is experiencing increasing concern over the availability of water resources. Recent publications by international organizations have confirmed how water scarcity levels are on the rise, both in terms of quantity and quality, and several studies have identified water scarcity as a serious and growing problem, particularly in arid and semi-arid areas (see, for instance, Gleick, 2000 and Raskin et al, 1998). Yet, despite the recognition that water is fundamental for life and a major component of current strategies for actions in many areas of national and international policy development,[1] water management is still problematic.

In recent years, one of the responses to water scarcity has been to promote collective negotiated decision making procedures, both at the national and international level. Through negotiations, proposals are put forward by the interested parties, who have both common and conflicting interests (Churchman, 1995). The idea is that negotiated decisions can lead to management choices that are better adapted to local conditions, and can result in easier implementation, less litigation and improved stability of agreements. In real life, there are many examples of international negotiations over global and regional natural resources, such as the atmosphere, the seas, biodiversity, fish stocks and others (for some examples, see for instance Breslin et al, 1990, 1992). Similarly, there are many examples of international cooperative efforts to manage shared water resources. One of the oldest attempts is the Baltic Sea agreement, linking environmental quality issues with nations' development policies. Similar attempts can be found in the Aral Sea and in the Caspian Sea, as summarized in Conca and Dabelko (2002).

It has been shown that negotiated policy making can indeed represent a constructive way forward, yet a model of formal negotiation theory that can deal with both the characteristics of water resources and the processes of strategic negotiation is still lacking. In fact, relatively little is understood about the interactions between the structure and the outcomes of the negotiating process, and most of the analyses of real negotiations adopt an empirical, ad hoc, approach, without exploring the underpinning theories, or attempting to develop a unified, formal theory of negotiation (see, e.g., Yoffe et al, 2004). As a result, we are still far from understanding, in a broad enough range of situations, which factors may affect agreements, and where and how bargaining processes can be shaped to obtain a more desirable state of affairs with respect to a shared resource. There is a rising demand for investigation into negotiation theories and techniques, as well as in applied simulation models, which can help address these questions, and that can be used by decision makers as 'negotiation-support-tool'.

The main objective of this chapter is therefore to explore the usefulness of formal negotiation models – based on game theory – to help better manage water resources. In particular, we will use a game theoretic set-up to simulate a bargaining process among water users, who have to decide how to share a river's water, with the ultimate aim of identifying which factors are likely to affect their negotiation strategies, and the resulting agreement – if one is achieved.

Game theory is a mathematical tool that provides a formal language for describing the strategic interactions among individuals. In strategic games, players – in our case, water users – will choose strategies that will maximize their return, given the strategies the other players choose. The essential feature of game theory is that it provides a formal modelling approach to social situations in which decision makers interact with other agents. It is thus an economic instrument well suited to the study of bargaining situations: using a rigorous mathematical framework, it can provide a relatively simple and intuitive representation of the forces driving the negotiation process, with an explicit recognition of the fact that choices made independently by each individual will be influenced by other individuals' actions and strategies – that is, an individual's choice will largely depend on what others will do. It is this interdependence of players' strategies and action that is indeed at the core of negotiation processes.

In this context, a 'bargaining problem' can be defined as a situation in which (i) individuals (players) have the possibility of concluding a mutually beneficial agreement; (ii) there is a conflict of interests about which agreement to conclude; and (iii) no agreement may be imposed on any individual without his approval (Muthoo, 1999), or agreements may be difficult to implement and enforce with the opposition of the regulated subjects. Such a problem can be characterized by the set of utility functions[2] that, for each and every feasible agreement, allow the estimation of the payoffs received by each player taking part in the bargaining, including the payoffs derived from the disagreement point – that is, the payoffs from the allocation resulting when no agreement among the parties is reached. This approach focuses on individual incentives, and it rests on the assumption that parties have predetermined positions as opposed to flexible interests (Fisher et al, 1991). In fact, most negotiations follow this paradigm (see, for instance, Churchman, 1995).

The model applied here has three main distinguishing features. First, it builds upon existing non-cooperative, multilateral, multiple issues bargaining models (e.g. Rausser and Simon, 1992; Adams et al, 1996; Thoyer et al, 2001; Simon et al, 2006), and applies a similar tool to a different reality – both in terms of players and policy space, as well as the underlying geography and economy of the area. The exercise offers a good test case for this modelling approach and the robustness of its results – which will ultimately determine its usefulness. Second, we introduce uncertainty into the analysis, that is our model explores the influence of stochastic variations in water resources on players' strategies, thus allowing the identification of a more robust space of feasible policies for water management. And finally, the involvement of local actors in determining management policies is indirectly addressed, through the construction of players' preferences and payoffs with their direct input.

The remainder of this chapter is organized as follows. We begin by outlining the underlying framework in the next section, whereas the subsequent section describes the application of the model to the Piave River Basin water allocation problem. The four parts of this section first set the context of its application, discuss the estimation of the relevant parameters, present the results of the simulation exercises and finally the results of policy relevant exercises. The final section concludes this chapter by summarizing the results and drawing generalized policy lessons on the usefulness of the proposed approach.

## The underlying bargaining framework

The model presented in this chapter builds upon the multilateral, multiple issue negotiations model developed by Rausser and Simon (1992) – in itself an extension of the two-person, one issue Rubinstein–Ståhl model (Ståhl, 1972; Rubinstein, 1982) – and extended to include uncertainty over water supply in Carraro and Sgobbi (2008). In this framework, a finite number of players have to select a policy for sharing water resources from some collection of possible alternatives. If the players fail to reach an agreement by an exogenously specified deadline, a disagreement policy is imposed. The disagreement policy is known to all players: it could be an allocation that is enforced by a managing authority; it could be the loss of the possibility to enjoy even part of the negotiated variable; or it could be the continuation of the status quo, which is often characterized as inefficient. The constitution of the game as a finite horizon negotiation is justifiable empirically

– as consultations over which policies to implement cannot continue forever, but policy makers have the power (if not the interests) to override stakeholders' positions and impose a policy if negotiators fail to agree. In finite horizon strategic negotiation models, it is unavoidable that '11th hour' effects play an important role in determining the equilibrium solution. In fact, last minute agreements are often reported in negotiation. In our model, unanimity is required for an agreement to be reached. Although this may seem excessively restrictive – in some cases, such as government formation, simple or qualified majority rules may be more realistic – unanimity is justifiable empirically when no cooperation is the status quo, when there is no possibility of binding agreements, or enforcement of an agreement is problematic – all cases in which the agreement must be self-enforcing and voluntary. Unanimity may also be appropriate when a compromise among different perspectives is sought, with the objective of identifying a family of agreements that could be politically acceptable to all negotiators.[3]

Finally, in our formulation of the game, (part of) the players' utility is not known with certainty, as it depends on stochastic realizations of a negotiated variable. Players' strategies will then depend on their expectations about future water availability.

The game is played as follows. At each round of the game, provided no agreement has yet been reached, a player is selected to make a proposal. The order in which players make a proposal is determined by an access probability, which can be interpreted as the players' ability to influence the process. The importance of players' access probability is intuitive, and ways to approximate it in simulating a real bargaining process will be discussed in the application of the tool. The proposer will then propose a policy package – which specifies the resources' shares for each of the negotiating parties, including himself. Next, all the remaining players respond to the offer in the order specified by the vector of access probabilities. If all players accept the proposal, the game ends. If there is at least one player that rejects the offer, the next period of the game starts. In the next period of the game, the next player in the sequence specified by the access probabilities proposes a policy package, which the remaining players can in turn either accept or reject. The game continues in this fashion until either all players agree to a proposed policy package, or the deadline is reached, at which point the disagreement policy is implemented.

As shown by Rausser and Simon (1992) and the applications of their model, the solution to such a negotiation game can be characterized intuitively. Assume that no agreement is reached until the last round of the negotiation: in this situation, players will only accept an allocation yielding them at least as much utility as the utility they would derive from the disagreement policy. Similarly, at bargaining rounds prior to the deadline, a player will accept a proposal if and only if it yields at least as much utility as his expected continuation payoff. This is defined as the utility he expects to derive from rejecting the offer, and moving on to the next round of negotiation. In an offer round, players will first of all identify those allocations that are feasible, and then select, among the feasible allocations, those maximizing their utility. When identifying feasible proposals, players will not only take into account the physical constraints on water availability, but they will also consider only those proposals that are likely to be accepted by the other players. The characteristics of the disagreement outcome ensure a solution: by assumption, the disagreement policy yields players a lower utility than any agreement. The player who is selected to propose an allocation in the last round, therefore, can get his preferred share. Players in the previous rounds anticipate this outcome, and the game is theoretically solved in the first round of negotiation.

## The Piave River Basin

In order to demonstrate how the use of computer simulations based on game theoretic assumptions may help water managers in a context where multiple players have conflicting interests over water resources, but where unanimity is needed for a policy to be determined, we will apply our model to the water allocation problem of the Piave River Basin. Computer simulation tools may indeed offer valuable support for investigators, who may not be able to observe the position of negotiators with a high degree of precision: simulations thus allow investigators to explore the negotiation process for a wide variety of preferences and policies, and identify robust and stylized facts that may help them in their strategies.

The main objective of our exercise is to compare different constitutional features of the negotiation process, to determine the factors that grant more bargaining power to different players. We are also interested in comparing different water allocation rules in terms of their implications for individual and society's welfare, with a particular emphasis on the performance of different allocation rules in the face of uncertain water supply. Indeed, uncertainty is a key feature of water management, and it will become even more so in the face of climate change, with its likely impacts on both the water cycle and water demand.

## The 'battle of the Piave'

The Piave River Basin (PRB) is among the five most important rivers in the north of Italy (see Figure 13.1, Plate 4). Traditionally, water management was primarily aimed at favouring irrigated agriculture and hydroelectric power production. However, the increase of other, non-consumptive, uses of water – such as recreation and tourism in the Dolomite valleys – and the rise in environmental awareness, coupled with variation in the water flow, have led to increasing conflicts. Tensions over water management become fierce in the summer season, when the combination of dry months and peaks in demand often lead to local water scarcity situations.

The highly political and strategic nature of the problem has led to what is called 'the battle of the Piave', with the problem of exploitation of water resources at the centre of the debate, especially in relation to the requirements of the tourism industry in the Dolomite valleys, and the needs for agricultural water uptake downstream (Baruffi et al, 2002). We can thus identify three major areas of conflict with respect to water management in the Piave River Basin:

1 *Hydroelectric versus environment.* Even though hydroelectric power generation does not consume water per se, its storage and release patterns have a significant impact on water availability for other uses – water temporarily stored for power generation is not available downstream. Furthermore, part of the water used for hydroelectric power generation is diverted away from the Piave River to the neighbouring Livenza River.
2 *Agriculture versus environment.* Lakes and reservoirs in the mountainous part of the river are managed to guarantee enough water for irrigation. As a consequence, much of the river flow in the middle part of the basin is reduced significantly for a large part of the year: many of the river inlets, and the main riverbed itself, are completely dry for long stretches.
3 *Tourism development versus consumptive water uses.* With the current level of water abstraction permits, only the release of water stored in reservoirs can guarantee meeting the demand for water downstream, with important negative impacts on the tourism industry of the upstream areas, where water reservoirs are located.

It is now widely accepted that the current exploitation regime for the Piave River Basin is not sustainable, as it has significant negative impacts on the river balances and ecological functioning, with consequent risks for the safety of local communities and economic activities (Franzin et al, 2000). Dalla Valle and Saccardo estimate an average water deficit of 3.4 million m$^3$ and, in dry years, this deficit can reach 75.5 million m$^3$, as happened in 1996 (Dalla Valle and Saccardo, 1996). In 1994, for instance, the average abstractions amounted to about 106m$^3$/s and 82.6m$^3$/s in the summer and winter periods respectively – compared to a virtual water availability of 89m$^3$/s and 115m$^3$/s (Dalla Valle and Saccardo, 1996).

The current situation with respect to water users and management plans in the PRB represents a good test case for the proposed non-cooperative, multilateral bargaining model: the planning authority (the River Basin Authority of Alto Adriatico[4]) intends to take into account the interests and needs of all major stakeholders in planning for water use, yet its initial attempts have encountered their opposition. Exploring the key issues of conflicts in allocating water in the PRB within the proposed framework may highlight management strategies that are good compromises among different users, thus helping reduce conflicts over the resource and promoting its sustainable development.

The model is necessarily a simplified representation of the existing problems with managing water resources in the Piave River Basin, but the substantial contribution of local actors in identifying the key negotiation variables and players'

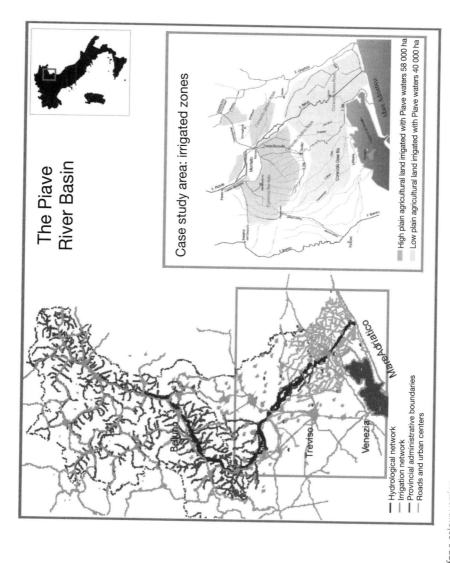

**Figure 13.1** *The Piave River Basin and the case study area*

Note: See Plate 4 for a colour version.

utility functions make it nonetheless useful for starting to explore the problem more formally and suggesting some policy entry point for conflict management. The process of transposing reality into a simplified model, and determining players' utilities and preferences, is intrinsically subjective: to avoid as much as possible introducing biases in the estimates, a suite of tools has been used in this research. The framework approach adopted is inspired by the NetSyMoD[5] framework (Network Analysis, Creative System Modelling and Decision Support). The problem itself, the information, the choice set and the judgement are defined with the contribution of different actors, who may be various experts in the disciplines relevant for the solution of a certain problem, or they may be the actors and decision makers that are formally or informally involved in the participatory process of decision making, for instance during the definition of a local development plan.

## Model application

The management problems of the Piave River Basin are complex, involving a multitude of actors at different levels, and issues of different urgency. After reviewing the literature on the subject area, a series of interviews were held with the main actors – both at the individual and at the institutional level.[6]

The existing conflicts of interests among the Province of Belluno on the one hand, and the agricultural water users in the Province of Treviso on the other, was thus singled out as the most important aspect of controversy of the Piave River. The downstream agricultural users are represented by the two major Land Reclamation and Irrigation Boards (LRB), the LRB of Destra Piave and the LRB of Pedemontano Brentella di Pederobba. These are the institutions mandated with managing irrigation infrastructure and water distribution. The role played by ENEL – the electricity production company – also appeared as important in determining water availability in the area. Finally, a municipality representative of all the municipalities in the lower part of the Piave River was also included, defending environmental interests of the river downstream of Nervesa, the closing section of the basin.

Players negotiate over two main issues, namely the abstraction permits for consumptive uses of water in the summer (dry) months and in the winter (wet) months. Water allocations in winter and summer are considered separately as the needs are very different in the two periods of the years, and so is water availability. For both the winter and summer periods we use the average flow ($m^3/s$), and players negotiate over the share of the total resource that they can abstract for their own use.

To mimic the existing conflict, the Province of Belluno is modelled as negotiating to prevent water release from the reservoirs, as it is in its interests to maintain them relatively full, given their landscape and tourist uses. We only consider ENEL's water diversions at a specific point in the electricity producing system, namely the Fadalto power plant system, directly linked to one of the major water reservoirs in the mountainous part of the river. The rationale for this choice is that virtually all the water diverted at the Fadalto system is not returned to the Piave River, but transferred to the Livenza River. Water use for power production at this point of the system can thus be effectively considered consumptive use.

Given the wealth of studies and political processes aimed at finding a compromise solution to the 'battle of the Piave', our players can be assumed to have common knowledge of their opponents' preferences. Because of this, players will not be able to behave strategically in the sense of trying to deceive their opponents.

Finally, it should be pointed out that the RBA is not included explicitly as a player in the game. The authority is interested in exploring the conflicts and potential solutions to tailor its management plan accordingly, thus identifying policies and allocation patterns that may represent a good compromise among competing water users and environmental protection. For our purposes, the RBA is assumed to establish the rules of the negotiation, as well as the default policy to be implemented in case of no agreement. This formulation is in line with our main objective – that is, test a framework that could prove a useful instrument for the management authority in identifying a set of water sharing agreements that could be politically acceptable to all water users.

In the formulation of our model, players' utility is a declining function of the distance between the (proposed or implemented) policy package and their ideal winter and summer allocations, weighted by the relative importance that each player assigns to satisfying his requirements in the summer versus the winter. The closer the proposed or agreed allocation to a player's ideal point, the higher his utility will be.[7]

In order to improve on the applicability of game theory to real life negotiation problems, we estimate players' ideal points, inferring them directly from players' stated preferences.[8] It clearly emerges that players have conflicting preferences: the two LRBs prefer more water in the summer as opposed to the winter season, as agriculture requires more irrigation during this period. On the other hand, ENEL would rather release water from the reservoirs in the winter for hydroelectric production, when demand for energy is higher. The remaining two players have more or less symmetrical preferences – even though they require different amounts of water. We infer players' minimum acceptable allocation, which may be determined by technical reasons (as in the case of agriculture or hydroelectric power generation), economic purposes (as in the case of the Province of Belluno and tourism in the reservoirs) or environmental considerations (as in the case of the minimum water flow, determined by the RBA on the basis of Law 183/1989 and the Ministerial Decree 28/07/2004).

We directly elicit the relative importance that players attribute to the two negotiated variables by asking them to rank the two dimensions of the agreement – and water price – in order of importance, signalling also the intensity of their preferences by the distance between any of the two variables (Roy and Figueira,

2002). The weights attributed to the two variables by the five players are a reflection of the underlying demand for water: for tourist and environmental purposes, water demand is assumed constant throughout the year; for irrigation purposes, water is crucial in the summer; for hydroelectric power generation, on the other hand, demand is higher in the winter months, when the price for electricity is higher, and so water in the winter is more important.

Finally, we derive players' access probability, which represents the political influence of players, from both legislation and players' observed behaviour. The political weights play a critical role in determining the equilibrium outcome. In the literature, players' access probability is approximated by a random variable with a well specified probability distribution, or its value is assumed ad hoc (as in Adams et al, 1996; Thoyer et al, 2001; Simon et al 2006). In this latter case it is easier to see how access probabilities can be considered as a proxy for players' relative power, but the choice of the values may be hard to justify empirically.[9] In this paper, we characterize players' political power on the basis of two elements: on the one hand, the influence of each actor will be partly determined by the policy priorities set by the dispositions of Italian legislation on matters related to water management and local development (Law 36/1994). As we shall also see in our policies exercises below, the priorities currently set in the Piave River Basin in case of water scarcity are an important source of (in)efficiency when there is uncertainty over water availability, and allocation is based on water flow rather than share of water flow. But players' influence will also depend on their role within the social network of institutions operating in the area – with more active and central actors being more able to influence the opinion and behaviour of others. This latter component is estimated using Social Network Analysis techniques.[10]

Finally, we assume players have the same aversion to risk and, in the case of uncertain water availability, we explore the impact of players' beliefs about the availability of water in the summer months on their negotiation strategies.[11]

Players maximize their utility, subject to a series of constraints. In particular, they cannot propose to allocate all the water available to themselves. When evaluating an opponent's proposal, a player will not accept an allocation scheme that yields him a utility lower than a minimum acceptable compromise, to reflect the fact that water users may not be able to undertake their normal production activities with a water allocation below a certain threshold level. In addition to these individual constraints, players must abide by the physical constrains of water availability, determined by the hydrological balance of the Piave River. The total water available for consumption in each period is determined by the natural flow of the river and the water that is released by the upstream water reservoirs, minus the amount of water that must be left in the river to guarantee its continuing ecological functioning (the minimum water flow, MWF). Thus, to estimate the water constraint, we consider three factors: the average (virtual) water flows at the closing sections of the river basin, as reported in ADB (1998); the theoretical maximum releases from the reservoirs, as derived from the release curves of the

**Table 13.1** *Summary of players' utility function parameters*

|  | Ideal winter allocation (m³/s) | Ideal summer allocation (m³/s) | Relative importance of winter water | Relative importance of summer water | Access probability |
|---|---|---|---|---|---|
| LRB of Pedemontano Brentella (CBPB) | 28 | 44.4 | 0.2 | 0.8 | 0.30 |
| LRB of Destra Piave (CBDX) | 16.5 | 34.45 | 0.3 | 0.7 | 0.30 |
| ENEL | 42.7 | 42.7 | 0.7 | 0.3 | 0.17 |
| Province of Belluno (PRBL) | 46.8 | 64.3 | 0.6 | 0.4 | 0.15 |
| Riverside communities (COMRIV) | 20 | 20 | 0.5 | 0.5 | 0.08 |

main reservoirs (Carlini and Sulis, 2000); finally, to the above, we subtract the MWF set by legislation – which varies between the summer and winter months (ADB, 1998).

In the stochastic version of the model, we introduce uncertainty in summer water availability, thus implicitly assuming water available in winter can be predicted with reasonable accuracy. Although this simplification may not seem realistic, it can be justified empirically, since water consumption patterns are relatively stable in the winter months. Thus, the choice is to focus on summer variations, as the summer period is clearly more critical, and weather conditions make accurate predictions more difficult.[12]

## The negotiation game: Main results

The results of the baseline simulation exercise show that, if the last round is reached, when selected to be a proposer, each player would propose his ideal allocation, which would be accepted by the others. In fact, in the final round of the negotiation, players would attain their highest payoff, should they be selected to be proposers.

The value of this numerical simulation is not in the quantitative results but, rather, in the qualitative intuition that can be derived, it is therefore instructive to look at the equilibrium quantities more in detail. In the equilibrium allocation agreement, it is clear that the PRBL ends up with a much lower allocation with respect to its ideal than the other players (about −37 per cent), while the remaining players experience more or less the same proportional reductions. The exception is CBPB, which ends up with approximately its ideal allocation.

But what are the underlying forces at play that drive these results? There are several exercises that can be performed to explore these issues and similar questions. We will summarize here the main results of interest from a policy perspective, while the full details can be found in Carraro and Sgobbi (2007).

The results of these simulation exercises conform to both intuition and previous results in the literature, in particular:

- *Result 1.* With certain water supply, players' access probability – which can be interpreted as their political effectiveness – is an important source of bargaining power. Shifting access probabilities confers an advantage to the player with higher access, yet, as we reach the agreement, this advantage decreases, while the relative disadvantage experienced by other players levels off. The relation between players' access probability and the equilibrium agreement is, however, not straightforward: increasing the strength of one player does not only improve his position, but may also improve the relative position of other players. What is the reason determining this build-up effect that progressively worsens the situation of this player, while leaving substantially unchanged the outcome for the player whom we have weakened by construction? Unlike in the experiments with the Rausser and Simon model (for instance, in Simon et al, 2006), we cannot explain this outcome by the similarity of players' ideal points, since we are dealing with a situation of pure conflict. The driving force behind this effect is likely to be players' participation constraint and the complex interactions among different sources of bargaining power. This result brings us to the hypothesis that access probability is not the only source of bargaining power for players.
- *Result 2.* Access probability is not the only source of bargaining power. The size of players' acceptance region – that is, the set of negotiated variables that they are ready to accept, before rejecting any offer – does influence to a significant extent the equilibrium agreement. Players who have a smaller acceptance set are able to extract a larger share of the pie than otherwise: by construction, the other players are forced to concede more, as they all prefer any agreement to no agreement, and are thus keen to find a compromise. This result is akin to the observation that players' default strength – that is, the level of utility players gain from non-agreement – influences to their favour the equilibrium result: intuitively, the higher is the disagreement payoff relative to that of the others, the stronger is his bargaining position (Adams et al, 1996; Richards and Singh, 1996; Simon et al, 2006).
- *Result 3.* There are gains to be made when the negotiating parties are able to trade-off different variables. This is in line with an important result of the

theory on multiple issue bargaining – and the related theory on issue linkage (see, for instance, Carraro and Siniscalco, 1997; Katsoulacos, 1997; and, more recently, Alesina et al, 2001). So, for instance, the LRBs may increase their utility by lowering their demands on their winter share in exchange for a higher summer share, as they prefer to be closer to their ideal allocation in summer. When external factors limit players' flexibility to trade-off between the two policy dimensions of the water sharing agreement, all negotiating parties suffer. The negotiator who is now constrained in his flexibility to make concessions is able to extract a higher share of the resource for himself. Yet, he is not necessarily better off. This is shown in Figure 13.2, which compares the utilities that players derive from the self-enforcing equilibrium allocations in the baseline case and in the case where the preferences of ENEL are more similar to those of the other players. Another interesting result seems to emerge from Figure 13.2: most of the players are worse off under this scenario with the restricted scope for trade.

Even though these results are sensitive to the choice of parameters' values, they do nonetheless provide an important insight. This simulation provides formal support to the empirical evidence that, when players have similar preferences, the scope for gains from trade is reduced, and the allocation process in a purely competitive bargaining situation is generally more difficult. As a consequence, players are likely

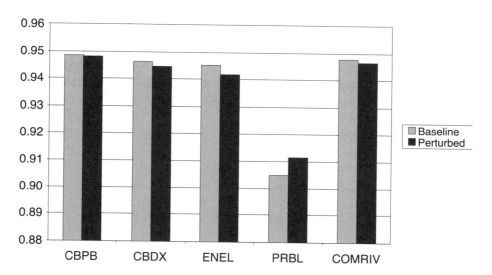

**Figure 13.2** *Varying the relative importance of the negotiated variables: comparing equilibrium utilities*

to be worse off in this situation, as the distance between their ideal allocation and the allocation agreed upon is, on average, larger.

- *Result 4.* When players take into account the fact that water availability cannot be predicted with certainty, their bargaining strategies are significantly affected.[13] All players bargain harder under their stochastic strategy as compared to their deterministic strategy in the final round of the negotiation, since they attempt to secure for themselves a higher share of the resource to hedge against water scarcity conditions. However, as in equilibrium, the sharing agreement must satisfy the total quantity constraint, the emerging self-enforcing agreement favours only some players – generally those who are stronger.
- *Result 5.* The explicit introduction of uncertainty in players' strategies does have an additional cost, which affects all players and, on average, it is more difficult for a self-enforcing equilibrium agreement to emerge. Players' offers do not always converge and, for the purpose of this application, on several occasions no agreement is found. This problem is exacerbated when players have different risk attitudes: in this case, not only is agreement more difficult to achieve, but also more pessimistic players bargain much harder to extract a larger share of the resource, and weaker players bear even more the burden of risk.

## Assessing players' strategies and allocation rules in the face of uncertainty

The value added of the proposed approach to explore the water allocation problem lies in its ability to provide useful information to policy makers. The previous exercises have provided some intuitions as to what features of preferences and contexts can affect negotiation strategies and outcomes, with a view of identifying potential entry points for water managers to influence the strategic behaviour of resource users. In these final exercises, we will explore two aspects that may be of more direct interest for policy making: the individual efficiency of accounting for uncertainty in negotiating water policy; and the individual and overall welfare implications of different sharing rules, when players are left to negotiate among themselves. These issues are of particular relevance in the face of climate change, which is expected to increase uncertainty over water availability.

### Ex post efficiency

What happens to players' utility level, when they follow a deterministic versus a stochastic bargaining strategy, in the face of a certain realization of water availability? To assess whether players are better off by taking into account uncertain water supply as compared to the case when they consider water quantity as known, we assume that uncertainty over the quantity of water available in the summer is resolved after an agreement over water allocation has been achieved, and we

**Table 13.2** *Summary of results – ex post assessment*

|  | Change in utilities | | |
|---|---|---|---|
|  | Low | Medium | High |
| CBPB | ++ | ++ | —— |
| CBDX | ++ | ++ | — |
| ENEL | — | — | + |
| PRBL | +++ | +++ | ++ |
| COMRIV | — | — | + |

compute players' utilities as derived from their equilibrium agreement shares under both the deterministic and stochastic strategies. We do so for three realizations of the resource: low, medium and high, and compute players' quantities and the resulting utility.

Table 13.2 summarizes the results of the ex post comparison of players' strategies. The three columns report the qualitative changes in players' utility under three realizations of winter water: low, when water is scarce; medium, when summer water constraint is still binding, but not as tight as in the previous case; and high, when the summer water constraint would not be binding on players.

The results of our simulations suggest that uncertainty will not affect players' strategies and their payoffs in the same manner. Stronger players will, on average, benefit by explicitly taking into account uncertain water supply in their negotiation strategies in a wide range of situations and, in particular, when water is scarce and there are likely to be acute conflicts. On the other hand, players with a lower bargaining effectiveness will be worse off in these situations, as the more aggressive bargaining strategies of the stronger players lead to an equilibrium outcome that is unfavourable to them. Interestingly, the results are robust for a wide range of values for summer water flow, indicating that accounting for uncertainty is a winning bargaining strategy for the stronger players. Only when the quantity of water available is above the observed average, and the availability constraint is not binding, would these players experience a reduction in utility. However, this result is likely to be strongly driven by the fact that, by construction, negotiators suffer a loss of utility from allocations that are above their ideal point. In real life, an agreed water share above the ideal point of a player is unlikely to negatively impact his utility, as the excess water can simply be left in the river – benefiting, in addition, downstream players. This is true for a quantity of water below a critical value, after which floods and related damages may occur.

In conclusion, and with all the necessary caveats, the results of these simulation exercises seem to indicate that accounting for uncertainty in water availability when bargaining over how to share the resource does influence to a significant extent players' strategies. The sharing agreement that emerges as a self-enforcing equilibrium from the non-cooperative negotiation will be even more skewed in favour of the players with stronger political influence: thus, if the observed trend in decreasing river water will continue in the future, we can expect the weaker players to bear a larger share of the burden of water scarcity.

## Proportional versus fixed allocation

A final exercise, which may be of interest for policy making, is the comparison between fixed and proportional allocation rules. A fixed rule allocates a fixed quantity of water to players, in an exogenously specified order: thus, the needs of the priority user are satisfied first, and the residual water is allocated to other uses. Fixed upstream distribution gives priority to the upstream users, while fixed downstream distribution prioritizes downstream users. On the other hand, a proportional rule allocates a share of the resource to the users, which, however, need not be the same.

In this section, we will compute the utilities of individual players and the overall welfare (in a utilitarian sense, this will be computed as the sum of players' utilities) under the two different sharing rules: fixed (downstream) and proportional allocation, when players account for uncertainty in water supply.

In particular, we use the multilateral bargaining model to mimic a negotiation process in which the five players haggle over how to allocate a fixed quantity of resource, rather than share. The underlying parameters of the model are the same as the baseline stochastic simulation exercise discussed above. We then assume, as in the previous exercise, three realizations of water availability, and compute the utilities that each player derives from the equilibrium allocation agreement under the two allocation rules. For conditions of severe and medium drought, when the sum of the equilibrium quantities allocated to each player under the fixed sharing rule exceeds the total quantity of water available, we impose a reduction in players' allocations as implemented in reality by the RBA (and as detailed in Decision 4/2001). Key to our result is the assumption of a fixed downstream allocation: it is in fact the case that, in situations of water shortage, all the reservoirs are managed by ENEL in such a way as to ensure that downstream (agricultural) needs are satisfied. We thus assume that the Province of Belluno loses all of its allocation. Furthermore, we reduce the allocation to the downstream municipalities (the residual flow) to the emergency minimum water flow (ADB, 2001).

The results indicate that, under fixed downstream allocation, the downstream players who have priority – namely the two LRBs – and ENEL are able to achieve a higher welfare compared to the equilibrium agreement under the proportional allocation rule. This is despite the fact that these two players are the stronger bargainers, and are still able to extract a relatively large allocation, in equilibrium, following their stochastic proportional allocation strategy. The welfare gains of these two players, however, come at the expense of the weaker players. These results are represented graphically in Figure 13.3, where each bar represents the change in players' welfare under a fixed versus a proportional sharing rule, for two drought conditions. Figure 13.3 shows us another insight: the gains to the three stronger

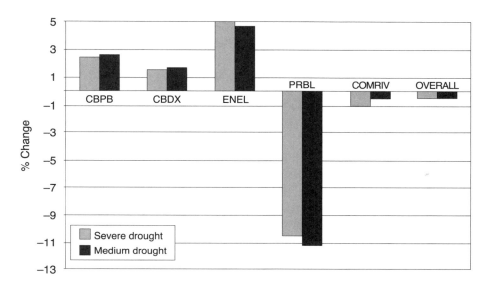

**Figure 13.3** *Comparing individual and social welfare under different water allocation rules – quantities versus shares*

players are not sufficient to offset the losses suffered by the upstream player and the downstream, weak player. That is, overall welfare is higher, with uncertain water supply, under a proportional versus a fixed quantity allocation rule.

These results are in line with previous findings, and reflect the fact that, under a proportional allocation, the risks of water shortage are shared more equally among players. They are robust for a wide range of parameters for the underlying probability distribution – both in terms of changing means and spread.

## Conclusions and policy implications

The model presented in this chapter is a stylized and simplified representation of the complex problems of the Piave River Basin. The offer–counter-offer procedure mimicked in this work may be a realistic enough representation of an actual negotiation process and the parameters of players' utility functions have been estimated with the direct input of players, thus adding more realism to the application of non-cooperative bargaining theory to explore real life negotiation processes. The model proposed simulates the process of negotiation among multiple players, who have to decide how to share a surplus of fixed size. Players have welfare that depend on the share of the surplus that they can secure for themselves – with different negotiated variables having different importance for each player, thus generating space for trade-offs among them. Furthermore, players have varying access probabilities, which signal the relative strength at the bargaining table and thus influence the equilibrium agreement.

The results of these simulation exercises mimic – in behaviour though not necessarily in real quantities – the observed situation in the Piave River Basin. In particular, when we consider a fixed downstream allocation rule, we are able to replicate the current water sharing arrangement and ranking of water users' welfare.

Our results conform with intuition, and provide some useful insights, based on formal models, as to which factors influence to a significant extent players' strategies and, as a consequence, the resulting agreement. Our simulations support the findings of similar applications of non-cooperative, multilateral, multiple issues bargaining models – and thus strengthen the argument for the potential usefulness of this approach in exploring allocation problems. The benefits deriving from implementing similar simulation models are primarily in a new ability to predict negotiation strategies and outcomes, and alter the rules, incentives and structures in such a way as to obtain a more desirable agreement – be it a 'fairer' or a more efficient one, and in the ability of the approach to help find a politically and socially acceptable compromise.

The approach highlights important sources of bargaining power, which can be used by resource management authorities as leverage points to ensure a more desired state of affairs, or easier implementation of proposed legislations. So, for instance, we find that players' acceptance regions are embedded in the structure of the game, thus their bargaining strength is determined not only by their political influence, but also by this parameter. If the managing authority can credibly threaten to impose a default policy in case the parties reach no agreement, it may be able to influence players' strategies and, as a consequence, the equilibrium emerging as a self-enforcing policy from the non-cooperative negotiations. In particular, if the default policy yields differentiated advantages to a subset of players, the resulting equilibrium agreement will favour this subset.

According to our results, investing resources to reduce uncertainty, as well as to disseminate knowledge evenly among resource users, is a winning strategy. This facilitates cooperation among negotiators. In the specific case of water sharing, the managing authority could reduce water availability uncertainty through more targeted reservoir management, or through the building of additional reservoirs to hedge against water flow fluctuations. This is indeed in line with the finding of the literature (see for instance Tsur, 1990; Tsur and Graham-Tomasi, 1991) on the stabilization role of groundwater – even though the use of groundwater may be limited to buffering temporary situations of surface water scarcity, rather than being a long-term solution to the problem. The role of information and knowledge dissemination is even more important when we consider climate change, which is expected to have important implications for water variability.

Sharing rules that spread the risks associated with an unforeseen reduction of the resource are better, from a social point of view, than rules favouring specific resource users. Yet, a switch from fixed downstream to proportional allocation necessarily implies a significant change in individual welfare allocation, with some water users losing their privileges. A policy to induce the players who would be worse off to accept the change for the benefit of society as a whole would therefore be needed: financial resources for upgrading irrigation infrastructure could be made available, extension services to encourage farmers switching to less water-intensive crops, or funds for building additional water reservoirs.

The modelling framework proposed in this paper is particularly relevant for environmental applications. Natural resource and environmental management problems are often complex, they interest several parties at the same time, and are characterized by different, often conflicting, objectives. Negotiated decision making can be viewed as a multiparty decision making activity: through strategies and movements, actors (players) try to achieve an agreement that is acceptable to all parties, and maximize their own satisfaction. As such, the model can be seen as an applied simulation tool that can be used by decision makers as a 'negotiation-support-tool'.

Our model can be helpful in exploring conflicts and opportunities in the management of natural resources that are shared among different users, for example at the national level, where there is a regulatory body that can impose a solution and enforce it, and thus the threat of enforcing a policy that is not negotiated among players is real. In addition, given that the equilibrium requires the consensus of all negotiating parties, our model is consistent with a situation in which a regulatory authority wishes to identify policies that could be acceptable to its regulated subjects. Furthermore, it can prove a useful catalyst to initiate and manage dialogue among resource users with conflicting interests.

Our non-cooperative bargaining approach can be useful to explore individuals' decisions and strategic incentives in the case of managing common pool resources, in particular whenever the use of a natural resource by one individual diminishes the quality available to others, but limitation of access and policy enforcement can be difficult. A similar framework has been used by Pinto and Harrison (2003) to model trade negotiation over environmental policies to abate carbon dioxide. In the field of climate change, one could envisage an application of our model to endogenize climate policy burden sharing rules – whereby players participating in an agreement negotiate over how to share emissions' allowances first, and then optimize their strategies, given the (negotiated) emission constraints. In the case of the EU, for instance, the model could be used to explore the implications of bargaining for the allocation of allowances under the EU Emission Trading Scheme. Uncertainty may concern the global target to be achieved in the future (that depends on decisions taken at the UNFCCC level). The non-cooperative multilateral, multiple issues bargaining model proposed in this chapter could also be fruitfully applied to exploring strategic incentives for more general trade negotiations, such as WTO rounds on agricultural subsidies.

## Notes

1   See, for instance, the UN World Water Development Report (UNESCO, 2003), UNEP's Global Environment Outlook 3 (UNEP, 2003), the numerous reports by the World Water Council (www.worldwatercouncil.org), and references therein.
2   In economics, utility is a measure of the relative satisfaction from consumption of goods. Economic agents are then assumed to act rationally, choosing their strategies as to maximize their utility. A utility function is a mathematical formulation to represent utilities, which allows the ranking of different consumption levels or combination of bundle of goods – water in our case. In game theory, a utility function for a player assigns a number for every possible outcome of a game – payoffs. The higher the number, the more preferred is the outcome.
3   Unanimity is often the decision rule when the focus is on natural resources management or environmental policies at the national or local level, rather than at the international level. Even at the international level, however, 22 out of the 122 multilateral environmental agreements provided by the Center for International Earth Science Information Network, require a unanimous decision. Furthermore, the requirement of unanimous consent is consistent with a situation in which an environmental regulatory authority wishes to identify the policy space that can be acceptable to all its regulated subjects: a policy belonging to such an acceptable space would be less likely to cause conflict and certainly easier to enforce.
4   River Basin Authorities (RBAs) are established at the level of the river basin (Law No. 183/1989). RBAs are self-governing bodies with responsibility for soil protection, reclamation of water resources, use and management of water heritage, and environmental protection. The operative instrument of the RBAs is the River Basin Management Plan, an area plan for the water sector, which gives guidance for the conservation, protection and improvement of soil and for the correct use of soil and water resources. The River Basin Plan is above other territorial development plans, such as waste management, urban development, water and wastewater use, providing a framework within which all activities influencing the river basin should be located. RBAs also provide guidelines concerning the issue of concession for large and small diversions, and on water saving in agriculture.
5   NetSyMoD is the result of several years of experience in the field of participatory modelling and planning within the Natural Resource Management Programme at Fondizione Eni Enrico Mattei (FEEM). NetSyMoD has been designed as a flexible and comprehensive methodological framework, using a combination of methods and support tools, and aimed at facilitating the involvement of actors or experts in decision making processes. For further information, see www.netsymod.eu.
6   The field work benefited from the EC funded project ISIIMM (Institutional and Social Innovations in Irrigation Mediterranean Management). See www.isiimm.agropolis. org/. The ISIIMM experience with local actors is summarized in Sgobbi et al (2007). Interactions with local actors took two main forms: either individual, face-to-face

interviews or group decision making processes, and had the main objective of validating and fine-tuning the preliminary findings, allowing for a more precise specification of the model constitutional structure. This included the identification of key players and key issues, as well as the information necessary to construct their preference functions, and the constraints in the Piave River basin.

7   It should be stressed that at no stage during the bargaining game do players compare their utility with that of the other players: at each round, each player only compares his utility with the utility he can expect at the following round. Utility functions thus do not need to be comparable. Furthermore, only local preferences are required, so that players' preferences need not be completely described.

8   In order to reduce potential interview biases and an excessively subjective and mediated preference elicitation process, the interviews were structured around a questionnaire, with a mix of open-ended and close-ended questions: the former are particularly useful in interviews, as they leave space for the respondent to freely describe his experience with respect to the issue. Close-ended questions, on the other hand, are less problematic both analytically, and psychologically, as they minimize biases in responses. A mix of the two is therefore likely to provide more information, with in-built reliability checks and balances.

9   Although in reality players may decide to invest in their bargaining power – and hence the vector of access probability would be determined endogenously in the game – in this model we will not consider this possibility.

10  Wetherell et al (1994) provide a useful definition of Social Network Analysis: 'Most broadly, social network analysis (i) conceptualizes social structures as a network with ties connecting members and channelling resources, (ii) focuses on the characteristics of ties rather than on the characteristics of the individual members, and (iii) views communities as "personal communities", that is, as networks of individual relations that people foster, maintain, and use in the course of their daily lives' (p645). For a more detailed review of Social Network Analysis and its techniques, see Wasserman and Faust (1994).

11  With the resources available for this study, it was not possible to design a system to estimate players' risk attitude, and we have thus relied on existing literature on the subject. In this application, players are therefore assumed to have the same degree of aversion to risk, which is constant and independent of the space variables (Romer, 1996).

12  Note, however, that the availability of water in the summer period will necessarily depend on winter use, through the regulation of the water reservoirs, and the accumulation of water in the reservoirs themselves – a function of rainfall level, as well as water release patterns. To simplify matters, we will assume that these variations are captured by the stochastic variability of summer water flow.

13  To explore the impact of introducing uncertainty over the realization of 'summer water', we solve the game computationally for 300 simulations. In each simulation, the games are identical with the exception of the realized quantity of water that is available for consumption in the summer period. We compare the equilibrium agreement with the baseline case, in which players do not take into account in their strategic choices the possibility of uncertainty in the summer water flow of the river.

# References

Adams, G., G. Rausser, and T. Simon (1996) 'Modeling multilateral negotiations: An application to California Water Policy', *Journal of Economic Behavior and Organization*, vol 30, pp97–111

ADB (1998) *Progetto di Piano Stralcio per la gestione delle risorse idriche del fiume Piave*, Autorità di Bacino dei fiumi Isonzo, Tagliamento, Livenza, Piave, Brenta-Bacchiglione, Venice

ADB (2001) *Annex A of Decision N. 4/2001: Adozione delle misure di salvaguardia relative al Piano stralcio per la gestione delle risorse idriche*, Autorità di Bacino dei fiumi Isonzo, Tagliamento, Livenza, Piave, Brenta-Bacchiglione, Venice

Alesina, A., I. Angeloni and F. Etro (2001) 'The political economy of unions', NBER Working Paper

Baruffi, F., M. Ferla and A. Rusconi (2002) 'Autorità di bacino dei fiumi Isonzo, Tagliamento, Livenza, Piave, Brenta-Bacchiglione: Management of the water resources of the Piave River amid conflict and planning', 2nd International Conference on New Trends in Water and Environmental Engineering for Safety and Life: Eco-compatible solutions for aquatic environments, Capri, Italy, 24–28 June

Breslin, J. W., E. Siskind and L. Susskin (1990) *Nine Case Studies in International Environmental Negotiation*, Harvard Law School, Cambridge, MA

Breslin, J. W., E. J. Dolin and L. F. Susskind (1992) *International Environmental Treaty Making*, Harvard Law School, Cambridge, MA

Carlini, F. and A. Sulis (2000) 'Programmazione dinamica stocastica e politiche di gestione in un sistema multi-invaso: il bacino del Piave', Unpublished thesis, Fac. di ingegneria, Laurea in ingegneria per l'ambiente e il territorio, Politecnico di Milano, Milano

Carraro, C. and A. Sgobbi (2007) 'A stochastic multiple players multi-issues bargaining model for the Piave River Basin', FEEM Nota di Lavoro 101.07, CEPR Discussion Paper 6585 and CESifo Working Paper N. 2178

Carraro, C. and A. Sgobbi (2008) 'Modelling negotiated decision making in environmental and natural resource management: A multilateral, multiple issues, non-cooperative, bargaining model with uncertainty', *Automatica*, vol 44, no 6, pp1488–1503

Carraro, C. and D. Siniscalco (1997) 'R&D cooperation and the stability of international environmental agreements', in C. Carraro (ed), *International Environmental Negotiations: Strategic Policy Issues*, E. Elgar, Cheltenham

Churchman, D. (1995) *Negotiation: Process, tactics, theory*, University Press of America, Inc., Lanham, MD

Conca, K. and G. D. Dabelko, (eds) (2002) *Environmental Peacemaking*, Woodrow Wilson Center Press, Washington, DC

Dalla Valle, F. and I. Saccardo (1996) *Caratterizzazione idrologica del Piave*, ENEL, Venezia

Fisher, R., W. Ury and B. Patton (1991) *Getting to Yes: Negotiating Agreement Without Giving In*, Penguin Books, New York

Franzin, R., M. Fiori and S. Reolon (2000) *Il conflitto dell'acqua: il caso Piave*, ISBREC, Belluno

Gleick, P. H. (2000) 'The changing water paradigm: A look at twenty-first century water resources development', *Water International*, vol 25, pp127–138

Katsoulacos, Y. (1997) 'R&D spillovers, R&D cooperation, innovation and international environmental agreements', in C. Carraro (ed), *International Environmental Negotiations: Strategic Policy Issues*, Edward Elgar, Cheltenham

Muthoo, A. (1999) *Bargaining Theory with Applications*, Cambridge University Press, Cambridge

Pinto, L. M. and G. W. Harrison (2003) 'Multilateral negotiations over climate change policy', *Journal of Policy Modeling*, vol 25, no 9, pp911–930

Raskin, P., G. Gallopín, P. Gutman, A. Hammond and R. Swar (1998) 'Bending the curve: Toward global sustainability', PoleStar Series Report, 8, Stockholm Environment Institute, Stockholm

Rausser, G. and L. Simon (1992) 'A non cooperative model of collective decision making: A multilateral bargaining approach', Department of Agricultural and Resource Economics, University of California, Berkeley

Richards, A. and N. Singh (1996) 'Two level negotiations in bargaining over water', Department of Economics Working Paper, University of California, Santa Cruz

Romer, D. (1996) *Advanced Macroeconomics*, McGraw Hill, New York

Roy, B. and J. Figueira (2002) 'Determining the weights of criteria in the ELECTRE type methods with a revised Simos' procedure', *European Journal of Operational Research*, vol 139, pp317–326

Rubinstein, A. (1982) 'Perfect equilibrium in a bargaining model', *Econometrica*, vol 50, pp97–110

Sgobbi, A., A. Fassio and C. Giupponi, (2007) 'Modellistica partecipativa come tecnica di analisi delle prospettive dell'agricoltura irrigua nel trevigiano', in C. C. Giupponi and A. Fassio (eds), *Agricoltura e acqua: modelli per una gestione sostenibile. Il caso della riorganizzazione irrigua nel Trevigiano*, Il Mulino, Bologna

Simon, L., R. Goodhue, G. Rausser, S. Thoyer, S. Morardet and P. Rio (2006) 'Structure and power in multilateral negotiations: An application to French water policy', paper presented at 6th Meeting of Game Theory and Practice, Zaragoza, Spain, 10–12 July

Ståhl, I. (1972) *Bargaining Theory*, Stockholm School of Economics, Stockholm

Thoyer, S., S. Morardet, P. Rio, L. Simon, R. Goodhue and G. Rausser (2001) 'A bargaining model to simulate negotiations between water users', *Journal of Artificial Societies and Social Simulation*, vol 4, www.soc.surrey.ac.uk/JASSS/4/2/6.html

Tsur, Y. (1990) 'The stabilization role of groundwater when surface water supplies are uncertain: The implications for groundwater development', *Water Resources Research*, vol 26, pp811–818

Tsur, Y. and T. Graham-Tomasi (1991) 'The buffer value of groundwater with stochastic surface water supplies', *Journal of Environmental Economics and Management*, vol 21, pp201–224

UNEP (2003) *Global Environment Outlook 2003*, Earthscan, London

UNESCO (2003) *The UN World Water Development Report: Water for People, Water for Life*, UN World Water Assessment Programme, New York

Wasserman, S. and K. Faust (1994) *Social Network Analysis: Methods and Applications*, Cambridge University Press, New York

Wetherell, C., A. Plakans and B. Wellman (1994) 'Social networks, kinships and community in Eastern Europe', *Journal of Interdisciplinary History*, vol 24, pp639–663

Yoffe, S., G. Fiske, M. Giordano, K. Larson, K. Stahl and A. T. Wolf (2004) 'Geography of international water conflict and cooperation: Data sets and applications', *Water Resources Research*, vol 40, pp1–12

# 14

# Strategic Behaviour in Water Policy Negotiations: Lessons from California

*Rachael E. Goodhue, Leo K. Simon and Susan E. Stratton*

Much of water policy is determined through negotiation in most societies. In some instances, the negotiations occur as part of the legislative process. In others, the negotiations take place under the auspices of one or more regulatory agencies, or in an effort to obtain a settlement to a court case. Depending on the context, the negotiations may involve stakeholders, regulators, other policymakers or some combination thereof.

The objective of this chapter is to use bargaining theory to examine the role of strategic behaviour in the design of negotiations regarding water resources. We begin from the perspective that negotiation participants are *strategic*: each stakeholder group takes into account what it expects other participants to do when it makes decisions. When designing a negotiation process, bargaining theory provides insights regarding how the structure of a negotiation process among strategic participants influences whether it will be successful and, if successful, the characteristics of a successful outcome.

We anchor our discussion of bargaining theory, negotiation processes and strategy by using as an illustration a simplified version of a current policy debate regarding investment in California's water system through issuance of a state bond. Water-related infrastructure is an important policy concern for California. Its population is growing, as is the value of its (mostly irrigated) agricultural production, leading to an increased demand for water. Global climate change is projected to reduce the ability of California's existing infrastructure to capture the Sierra Nevada snowmelt, reducing available supplies. Due in no small part to the federal Endangered Species Act (ESA), environmental uses are becoming increasingly important, reducing the water available for urban and agricultural uses given current supply sources and technology. In the current debate, the size of the bond issue, the allocation of the resulting funds across various prospective uses, and the sharing of the financial burden between California taxpayers (via the bond) and water users (via fees) are all topics of negotiation.

This chapter has several sections. First, we provide a non-technical introduction to non-cooperative bargaining theory. The next section provides background information on the California water system and the current debate. We then structure a highly simplified version of the current debate in terms of non-cooperative bargaining theory. Using this framework, the next section examines negotiation structure features identified by bargaining theory as important determinants of the strategic incentives facing participants and the ultimate negotiated outcome. The final section discusses policy implications and concludes.

## A bargaining theory primer

Bargaining theory is a subset of game theory, which is used by economists, political scientists and other researchers to examine strategic interactions. Consequently, it is a natural approach for modelling negotiations regarding water policy, whether they involve stakeholders, politicians, government agencies or some combination thereof. In this section, we offer a general description of the elements required to define a negotiation structure, and discuss different game theoretic approaches to solving a bargaining problem. Later in the chapter, we describe how to fit a particular policy problem into a negotiation structure.

### Fundamental elements of a bargaining model

A bargaining process involves a group of stakeholders debating different policy options. Each stakeholder has a set of concerns or goals they care about and some belief about what will happen if they fail to participate in the bargaining. Game theory puts this general description in specific mathematical terms. The stakeholders involved in the bargaining are described as the *set of players*. These players negotiate over a set of policy choices known as the *issue space*. If they fail to reach an agreement, a *default outcome* is realized. A series of utility or *payoff functions* describe how happy each stakeholder is with any specific policy choice (including the default).

Identifying the players, the issue space, the default outcome and the payoff functions is the first step in building any game theoretic model of a bargaining process. In principle, this seems like a straightforward task. However, as we discuss in later sections, placing a real world policy problem in this stylized setting offers many challenges.

## Solution concepts for bargaining models

Once an analyst has described the basic structure of the bargaining, the next step is to describe how a bargaining solution is reached. The multitude of different bargaining models in the literature reflects uncertainty among game theorists about how best to model the complicated real world interactions that lead to a bargaining solution. Broadly speaking, game theorists take two different approaches to identifying bargaining solutions. Following Osborne and Rubinstein (1990), we refer to these approaches as the *axiomatic* and *strategic* approaches.[1]

Nash (1950) is the quintessential example of the axiomatic approach to bargaining. Nash begins with the basic structure of a bargaining game, including players, policy choices, a default and a set of payoff functions. He hypothesizes that the solution to such a problem should satisfy several specific properties, or axioms. He then proves mathematically that only one possible solution satisfies all of these axioms, and concludes that this solution is the expected outcome of the bargaining process, known as the 'Nash bargaining solution'.

While Nash's work remains the most frequently used axiomatic bargaining model, authors have developed alternative ones. These models differ primarily in their specification of the axioms that a solution must satisfy. Notable examples include Kalai and Smorodinsky (1975) and Roth (1977, 1979).[2]

Strategic models of bargaining focus on describing the actual bargaining process. Rubinstein's (1982) model of alternating offers in a game with two players provides the template for most of these models. In each round, one player makes an offer to her opponent. If the opponent accepts, the game ends. Otherwise, the game moves to the next round and the opponent makes an offer. Play continues indefinitely, with players alternating offers. Rubinstein shows that there is a unique subgame perfect equilibrium.

The literature contains a wide variety of variations on the basic Rubinstein approach. Osborne and Rubinstein (1990) describe several two-player variations. Multi-player models include Krishna and Serrano (1995) and Rausser and Simon (1999).

While axiomatic and strategic models take a very different approach to modelling bargaining, the differences are not as large as they seem. In particular, Binmore et al (1986) demonstrate Nash's axiomatic model and Rubinstein's strategic alternating offer model produce the same solution, and Krishna and Serrano (1996) link their multi-player strategic model to axiomatic models.

Both axiomatic and strategic bargaining models have been used to analyse water policy negotiations. Parrachino et al (2006) and Dinar et al (1992) provide extensive reviews of applications of axiomatic bargaining models to water policy issues. Simon et al (2007) review the much smaller literature applying strategic bargaining models to water policy issues, as do Carraro et al (2007).

Here, we wish to identify the broad insights that game theory has to offer. Instead of implementing a specific model, we examine the class of processes that require unanimous agreement. If any player can veto a proposal, it must be at least as well off under the final agreement than it would be if bargaining failed. In other words, any game theoretic model of a unanimous bargaining process necessarily requires that the final solution be a 'Pareto improvement' on the default.

# California water: Background and current debate

California, the third largest state in the US based on land area, is geographically very diverse. Annual rainfall varies from less than 4 inches (10cm) in the southeast, near Death Valley, to more than 140 inches (335cm) on the northwest coast. Overall, in an average rainfall year about half of all the water obtained through precipitation and inflows is a dedicated supply allocated across urban, agricultural and environmental uses (DWR, 2005).

## Geography of California water

California includes ten hydrological regions. Figure 14.1 shows the ten regions and reports their natural and man-made inflows and outflows in an average year. The economic geography of California's water system is driven by a mismatch between the location of water supply and the location of water consumption. As shown in Figure 14.1 (Plate 5), most of California's water comes from the north, while most of its water use is in the south of the state. Figure 14.2 (Plate 6) includes the major federal, state and local water projects that transport water supply to water demand.

Our discussion of the current policy debate will focus on the water transferred from the Sacramento River basin to the Tulare Lake (southern San Joaquin Valley) and South Coast (Los Angeles and San Diego) hydrological regions. Snowmelt from the mountains drains into the Sacramento River. Major above-ground water storage facilities on the Sacramento and its tributaries capture much of this water and release it throughout the year. Water releases flow down the Sacramento River system into the Delta where the Sacramento joins the San Joaquin River, which flows northward. Water releases intended for uses in southern California are pumped out of the Delta into canals associated with federal and state water projects. Water remaining in the Delta flows into the San Francisco Bay. Of the diversions we examine, 6.7 million acre-feet are diverted upstream from the Delta by Sacramento Valley users, who are primarily agricultural. Another 5.4 million acre-feet are conveyed through the Delta to southern California through the state and federal water projects (Lund et al, 2007). An additional 4.0 million acre-feet are diverted from the San Joaquin River system. In total, about two-thirds of Californians depend on the Delta for their domestic water use. Delta water also irrigates roughly 5 million acres of cropland – about 57 per cent of the state's total irrigated cropland.

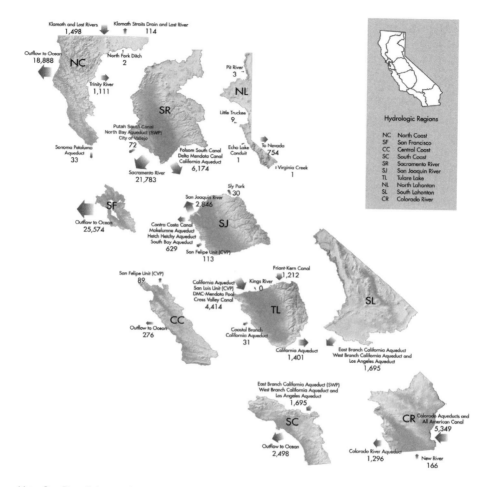

*Note:* See Plate 5 for a colour version.
*Source:* DWR, 2005

**Figure 14.1** *California hydrologic regions and regional water flows*

The use of the Delta as a key component of California's water system creates a number of policy concerns. Because the Bay-Delta is an environmentally sensitive ecosystem that is home to endangered and threatened species, environmental water uses have become increasingly important. Because water supplies for southern California are conveyed through the Delta, there are major concerns regarding Delta infrastructure. The Delta comprises primarily man-made 'islands' below sea level, many of which are surrounded by levees deemed incapable of withstanding the major earthquake predicted for the San Francisco region in the next few decades. If enough levees collapse, the Pacific Ocean will rush into the Delta and fresh water supplies for southern California will be lost. These concerns are

Note: See Plate 5 for a colour version.
Source: DWR, 2005

**Figure 14.2** *California rivers and water facilities*

important in the ongoing debate over a water infrastructure bond issue that we will examine in our analysis, as they were in the seminal Three Way Negotiations among stakeholders in the early 1990s, and the 1994 formation of the California Bay-Delta Program (CALFED), a group of state and federal agencies charged with managing the Bay-Delta.

## Current water policy debate

In early 2007 Republican Governor Schwarzenegger introduced a $4.5 billion bond measure that provided funds for conservation, underground storage, environmental enhancement in the Bay-Delta and elsewhere, and above-ground storage and conveyance. In California, all bond issues must be approved by voters. A bond issue may qualify directly for inclusion on the ballot through the initiative process or it may follow the standard process of approval by the legislature and the governor in order to qualify.

Schwarzenegger's initial proposal (carried by Senator Dave Cogdill) failed to be approved by the legislature for inclusion on the state ballot during the regular legislative session. The legislature met in a special session called by the governor in the fall of 2007 to attempt to pass a bond proposal to put on the February 2008 ballot for voter approval. Democrats and Republicans proposed competing bills but neither passed.[3] In November, Governor Schwarzenegger called for a joint effort to come up with a consensus proposal to put to the voters. The governor and various interest groups are preparing initiatives for the fall 2008 election through an alternative process, even while negotiations are ongoing.

One of the primary points of disagreement among interest groups regards the future role of additional dams and above-ground water conveyance in California's water system. Since Californian voters rejected a major water infrastructure project, the Peripheral Canal, in a 1982 referendum, proposals for the construction of additional major infrastructure designed to increase water supplies from surface water has been largely absent from the policy debate. Water allocation issues have been addressed through conservation measures, rationing and some use of water markets. Broadly speaking, Schwarzenegger, the Republicans in the legislature and water users who depend on water conveyed through the Delta support the development of additional infrastructure, while the Democrats and environmental interests do not.[4]

A closely related policy action that is playing a role in the ongoing negotiations is a recent judicial ruling resulting from a lawsuit filed to protect the delta smelt by Earthjustice, on behalf of the National Resources Defense Council, Friends of the River, California Trout, The Bay Institute and Baykeeper. US District Court Judge Oliver Wanger in Fresno ruled that, under the ESA, the state and federal agencies' Delta operation plan's assessment of its biological impacts (the Long-Term Central Valley Project Operations Criteria and Plan Biological Opinion) did not consider sufficiently the possibility of harm to the delta smelt. Until a new plan is developed (by the court-ordered deadline of 15 September 2008), he ruled, protection of the delta smelt required that the water pumps that convey water to southern California had to be shut down when environmental conditions suggested the smelt would be harmed by additional pumping. The new management plan is expected to include reductions in pumping as well.

The plan implemented by the judge's order is estimated to reduce water deliveries from the Delta by roughly a third in an average year. One study, sponsored by the Western Growers Association, estimated a decrease in agricultural revenues of as much as $294 million (Schultz, 2007c). The Metropolitan Water District (MWD), a consortium of cities and water districts providing drinking water to much of urban Southern California, began an advertising campaign encouraging urban users to conserve water. As of 1 January 2008, MWD had not ruled out instituting mandatory water use cuts.

## A brief note on other California water policy issues

Other sources of California water have had their own policy controversies and associated negotiations.[5] Historically, California exceeded its allocated portion of water captured by the Hoover Dam on the Colorado River. As Nevada and Arizona grew, they wished to use their allocated water, requiring California to reduce its use. Disagreements among California beneficiaries prompted the federal government to impose a deadline for Californians to negotiate a solution. In response, the Quantification Settlement Agreement was signed in 2003.[6] The San Joaquin River watershed has been another source of controversy. Although it carries much less water than the Sacramento, it is also used as a water supply. Friant Dam caused the San Joaquin to dry up entirely for part of each year. In 2006, there was a negotiated settlement to a lawsuit brought by environmental groups against the Friant Water Users Authority. Federal legislation implementing the settlement is currently in committee. Notably, one of the dams under discussion in the current state bond issue debate would be in the San Joaquin watershed.

## Analysed issues and stakeholders

We represent the current water policy debate discussed above in terms of negotiations over the funds spent on four types of water-related expenses (new dam construction and conveyance infrastructure, Delta restoration and other environmental purposes, agricultural water supply infrastructure improvement, and urban water conservation measures); how paying these expenses is allocated among three types of water users (urban users and two types of agricultural users) via fees and taxpayers (via the bond issue); and how the available water supply is allocated among the three types of fee-paying users and environmental purposes. We include five negotiating parties: the three types of water users subject to fees, taxpayers and environmentalists.

# Model

To illustrate the main points in this chapter, we sketch the components of a bargaining model representing negotiations over a California water bond measure to be placed on the November 2008 state ballot, proposed in fall 2008 to the California voters for approval. To confine this problem within manageable bounds, we limit our attention to issues relating to water flowing from the Sacramento River into the Delta, and then pumped from the Delta on to southern California. We ignore issues associated with water flowing from the San Joaquin River into the

Delta and with other watersheds. Even so, the outline we present here is far too simple to adequately reflect the complexity of the issues considered. Moreover, we do not propose a specific structure for the bargaining model; the ingredients we discuss below – players, issue space, model outputs, technology, etc. – are necessary components of a diverse array of different kinds of models. Our intention is to provide a minimal amount of detail, in order to anchor and illustrate the discussion of the interactions between negotiation structure and strategic behaviour.

## Players

In the Californian water policy debate, the most visible actors have been the governor and the leaders of both parties in the state legislature. We view these politicians merely as conduits through which stakeholders transmit their influence on the political process. Accordingly we exclude them from our model, and focus instead on the stakeholder groups themselves. We have selected five stakeholder groups to be seated at our bargaining table. Our first and second groups are agricultural interests, distinguished by the nature of their water rights: *senior agricultural right-holders*, denoted by senAg, and *junior agricultural right-holders*, denoted by junAg. Having assumed away the San Joaquin River, we can now say that farmers belong to either the former or the latter group, depending on whether they have access to water *before* or *after* it passes through the Delta. Consequently, junior agricultural rights-holders, but not senior ones, have been adversely impacted by the restrictions on using the Delta pumps imposed to protect the delta smelt. The third group is *urban water users*, in particular, the cities of Los Angeles and San Diego and some cities in the Bay Area. We denote this group by urban. Fourth is a composite of *environmental interests*, denoted by enviro. The last and the most diffuse group, denoted by taxpr, is a residual one, consisting of those *taxpayers* whose interests are at most peripherally aligned with any of the other five groups.

Clearly, this selection implies a considerable degree of aggregation of diverse preferences. Most obviously, it is a gross over-simplification to include a single group representing environmental interests. There are many active groups in the debate, ranging from narrowly focused groups like the California Sportfishing Protection Alliance, a state-wide organization that 'fights for water for fish' (CSPA, 2001) to the broadly focused Nature Conservancy, 'a wide conservation organization working around the world to protect ecologically important lands and waters for nature and people' (Nature Conservancy, 2008). While it would be interesting to explore the consequences of disaggregating our set of players into a larger number of more homogeneous groups, it seems appropriate to begin with a coarse partition of the many participants in the debate.[7]

Notably, we omit a representative of the managers of California's many water districts from the player list. These districts actively participate in water policy debates, and undoubtedly constitute one of the better organized stakeholder groups. We do not include them because it is very difficult to disentangle their preferences from those of their constituents, which are not necessarily aligned: urban and agricultural districts tended to take opposing sides in the October 2007 special session of the California Legislature. The urban MWD supported Perata's Democratic bill, while many Central Valley agricultural agencies supported Cogdill's Republican bill (Schultz, 2007b).

An alternative approach to modelling water authorities would be to consolidate the interests of water managers and taxpayers into a single social welfare-maximizing player, such as the state Department of Water Resources (DWR). DWR's mission is 'To manage the water resources of California in cooperation with other agencies, to benefit the State's people, and to protect, restore and enhance the natural and human environments' (DWR, 2008). We do not follow this approach for a number of reasons, most critically that the strategic goals operationalizing its mission do not necessarily correspond to social welfare maximization. For example, Strategic Planning Goal 2 is to 'Plan, design, construct, operate and maintain the State Water Project to achieve maximum flexibility, safety and reliability' (DWR, 2008). Another important reason we do not include DWR as a player in our model is that while it provides technical information to policy makers and stakeholders, its approval is not required for the bond measure we examine.

## Issues on the bargaining table

We consider three broad classes of issues: a list of categories for which dollar expenditures are earmarked, a scheme for sharing the burden of funding the earmarked expenditures, and an allocation of available water among competing uses. In reality, a bond proposal would not explicitly treat the third item as a distinct negotiating item. Rather, these allocations would be implicit in the details of each proposal. For clarity, we abstract from these details, and assume that the allocations are negotiated directly.

To construct our list of expenditure items, we draw from the proposals recently tabled in Sacramento by the Republicans and Democrats. We identify four categories. First, *dam construction and conveyance infrastructure* is a major component of the Republican proposal. Let $\$_{dam}$ denote the proposed expenditure on this item. The remaining categories are all emphasized in the Democratic proposal, although we have reorganized them to align closely with our model. The second item, *delta restoration* ($\$_{delta}$) is geared towards 'increasing the effective water in the Delta available for the fish'. This is a broad class of expenditures, ranging from specific items like screens that prevent fish from being entrained in the Delta, to general items such as 'restoration of the Bay-Delta ecosystem'. The third expenditure category is *agricultural water supply infrastructure improvement* ($\$_{infra}$), water use efficiency and water conservation measures intended to increase the ratio of effective to applied water in agriculture. The last category is *urban water conservation measures* ($\$_{conserv}$), such as desalinization and water recycling. Let $\underline{\$}$ denote the four-vector of dollar expenditures and $\sum \$$ denote their sum.

Responsibility for funding $\sum\$$ is allocated among four of these five composite players. The only player exempted from sharing the financial burden is the one (the environmentalists) that represents a public rather than private interest. We denote by $\omega_i$ the fraction of $\sum\$$ that is the responsibility assigned to the $i$ th player. The amount $\omega_{taxpr} \sum\$$ is the magnitude of the bond issue. The remainder of the funds are paid by agricultural and urban water users. Summarizing, our players will negotiate to select a vector $\omega = (\omega_{senAg}, \omega_{junAg}, \omega_{urban}, \omega_{taxpr})$ from the set of admissible burden allocations.

Let $W_0$ denote the total available supply of 'applied' water flows per annum. $W_0$ will depend on annual precipitation, releases from existing storage, and $\$_{dam}$, the amount spent on new dam and conveyance infrastructure construction. (Other expenditure types will increase effective water but not applied water.) In reality, obviously, $W_0$ is a random variable with a high variance, but for simplicity we ignore this complication. As a crude approximation to reality, we will assume that senAg has first claim on all available water, up to its entire legal allocation, denoted by $\overline{W}_{senAg}$. The water that remains after senior right-holders claims have been satisfied, $(W_0 - \overline{W}_{senAg})$, is available to junAg, urban and enviro. Summarizing, the players will negotiate to select a vector $\lambda = (\lambda_{junAg}, \lambda_{urban}, \lambda_{enviro})$ from the set of admissible water allocations. To recapitulate, the objective of the negotiation is to select a triple ($\$$, $\omega$, $\lambda$). We call each such triple a *bargaining proposal*.

## Outputs of the model

For each bargaining proposal, players' payoffs can be computed from the *model outputs*. The specification of these outputs is very stylized; many, indeed most, of the outputs that are important in reality are either aggregated into broad, abstract categories or omitted altogether. The first two outputs are the dollar values of *agricultural production* by senAg and junAg. The third is the percentage rate of *urban development* in areas dependent on water conveyed through the Delta. The fourth and fifth outputs are scalars regarding the delta smelt, representing its *population* and the species' *survival probability*.[8] The sixth output is the vector of *financial obligations* ($\omega_{senAg}\sum\$, \omega_{junAg}\sum\$, \omega_{urban}\sum\$, \omega_{taxpr}\sum\$$) implied by the negotiated expenditures and burden shares.

For the purposes of this chapter, there is no need to declare specific functional forms for the technologies that relate bargaining proposals to model outputs. Instead, we will simply identify the variables which determine each output, and the qualitative nature of the dependence. For $i$ = junAg, senAg, we assume that the agricultural production of farmers of type $i$ depends on the amount of effective water available to farmer $i$. The difference between effective and applied water is determined by the degree of water efficiency, which is an increasing function of $\$_{infra}$, the funds earmarked for agricultural water supply infrastructure improvement. Applied water to farmers is an increasing function of $\$_{dam}$ and, for junAg, the fraction $\lambda_{junAg}$ of total water allocated to this group from the residual water supply.

Likewise, urban development depends on the effective water available for urban uses, which depends positively on $\$_{dam}$, $\$_{conserv}$ and $\lambda_{urban}$. The fish population depends positively on effective water available for environmental uses, determined by $\$_{dam}$ and $\$_{delta}$; and on the share of water allocated to the fish, $\lambda_{enviro}$. It depends negatively on the quantity of water pumped through the Delta. Thus, holding the water allocation vector $\lambda$ constant, the net effect of an increase in $\$_{dam}$ is ambiguous, since this will increase the amounts of water available to both the fish and to users south of the Delta. The next output, the survival probability for the smelt, is the component of the model about which least is known or is knowable. While its determinants are identical to the determinants of the fish population, the science relating the smelt population to its probability of survival is in its infancy. The final output is the vector of financial obligations, which is specified explicitly as part of each bargaining proposal.

## Players' payoffs

Players' payoffs depend on components of bargaining proposals both directly (in particular, each one's payoff is negatively related to its share of the financial burden) and indirectly through the impact of these proposals on model outputs. Each farmer group's payoff increases with the value of its agricultural output. Urban's payoff increases with urban development. Enviro cares about the welfare of the fish – its payoff increases with the fish population and its survival probability – and the preservation of wilderness areas. This creates an ambiguity with respect to $\$_{dam}$: to the extent that it benefits the fish, it increases enviro's payoff; to the extent that it degrades wilderness areas, it decreases it. Since most environmental groups adamantly oppose dam construction, we assume that the latter effect dominates. Taxpr's utility is the most difficult to define because the group is so amorphous. Past voting behaviour indicates that it is willing to incur some additional financial obligation in support of vaguely articulated objectives such as 'bolster(ing) the state's water system and keep(ing) our economy strong' (Russo, 2007b). In the absence of more information, we assume that taxpr's payoff is concave increasing in the total cost, $\sum \$$, of the bargaining proposal, and declining in its share of the burden, implying an upper bound on the cost–benefit trade-off that it will be willing to accept.

This completes our description of the structural framework. Obviously, what we have built so far is no more than a skeleton onto which 'flesh' would have to be superimposed before it can be applied in practice. The 'flesh' in this context is a set of functional forms with specific numerical parameters that will enable us to map each bargaining proposal to a vector of model outputs and then, in turn, to a payoff vector.

## Structure and strategy

Bargaining theory provides a number of lessons regarding how negotiation structure affects the chances that the negotiation concludes successfully. We discuss three specific structural features of a negotiation and illustrate their role in the context of the current California water policy debate: the issue space (what's on the table), the definition of a successful negotiation, and the default outcome (what occurs if the negotiation fails). While the current debate is not a *policy maker-designed* negotiation, it illustrates the importance of these structural features, and allows us to draw lessons for cases where policy makers are able to design the structure of a stakeholder negotiation.

### Choosing what's on the table

The first element of negotiation structure is the topic of the negotiation. Defining the issue space is a more subtle question than it may first appear to be. In his initial 2007 proposal, Schwarzenegger included above-ground storage and conveyance. In terms of our framework, Schwarzenegger expanded the issue space by adding a fourth category – dam and conveyance construction – to an initial list of only three. Agricultural and urban business interests (and many Republican members of the California Legislature) strongly support above-ground storage and conveyance, while many members of the Democratic Party in the legislature and many environmental interests strongly oppose it. Many of the other measures in the bond proposal, such as funds for environmental restoration, are broadly supported by groups opposing above-ground storage. In response to criticism regarding the inclusion of above-ground storage in the bond measure, DWR director Lester Snow responded 'We have everything on the table from groundwater to conservation to waste water recycling' (Young, 2007). By broadening the set of alternatives, the governor provided space for negotiation and compromise, as recognized by members of both political parties.

> If we're going to get any kind of agreement, everybody's got to give a little,' said Assemblyman John Laird, D-Santa Cruz, leader of a group of Assembly Democrats working on solutions to the state's water problems. (Thompson, 2007)

> No surface storage, no deal,' said GOP leader Mike Villines of Fresno. 'The idea that we let millions of acre feet of water every year run to the ocean totally wasted is insanity.' ... But how many dams and what proportion of the cost should be paid by the state are open to negotiation, Villines said at a state Capitol news conference with an ornamental fish pond as a backdrop. (Thompson, 2007)

Several months after the failure of the governor's initial proposal, Assemblyman Laird replied the following to an interviewer who asked him for his prognosis regarding the likelihood of a negotiated agreement:

> *I really believe it's too important to not come to some sort of an agreement. But if this is about dams, and nothing else, it won't happen. And if this is about water cleanup and conservation and leaving out the Delta, it won't happen. It's going to have to see where there's a place in the middle to give everyone involved the comfort level to move ahead.*
> *Assemblyman John Laird* (Goldmacher, 2007)

These comments illustrate two lessons regarding negotiation design and negotiation success. First, a broad issue space provides more room for negotiation, and increases the likelihood that a solution can be reached that makes all parties better off. Second, every valued variable should be negotiated. In this case, if above-ground storage is excluded from the water policy negotiations, then the relative costs and benefits of other policy options for stakeholders will almost certainly be distorted.

This lesson was documented for an earlier California water policy debate. Adams et al (1996) examined the effect of excluding above-ground storage from the Three Way Agreement among agricultural, urban and environmental stakeholders in the early 1990s. They found that when spending on above-ground storage was sufficiently limited all parties were made worse off, *including* environmentalists. Consistent with this finding, stakeholders indicated that the agreement to consider all major issues simultaneously was important for the negotiations to progress.

## Defining the terms of success

The definition of success is a critical component of negotiation structure. The obvious way to define success is as a binary variable: either it is achieved or it is not. That is, a negotiation is successful if and only if it results in an agreement. In the context of the current California bond issue debate, negotiations will be considered successful in this sense if they result in a bond issue proposal that is approved by both houses of the legislature, the governor and the voters. In complex, multi-issue negotiations such as the one we consider there is a second, more subtle definition: success is located on a continuum defined by the degree of implementability of the negotiated outcome. A negotiation is completely successful if it results in an agreement that requires no further wrangling regarding policy: the only decisions remaining are administrative ones. A partially successful negotiation is one that leaves policy details unresolved, to be negotiated at a later date; the more important and numerous are these details, the lower is the degree of success. Just as the issue space is a choice variable for those designing a negotiation, so is the target degree of success. The first choice regards the scope and number of issues that are placed on the bargaining table; the second regards how 'finely partitioned' are the decisions negotiated concerning these issues.

To illustrate, consider the model presented above. We considered four categories of expenditure. In the preceding section, we discussed Schwarzenegger's role in expanding the issue space, by adding the fourth category – dam and conveyance construction – to a list that initially had only three. We modelled our players as bargaining over a four-vector of dollar expenditures, one for each of these categories. Since these categories were quite broad, however, an agreement on a particular four-vector $ would be classified as only partially successful. Many policy-level allocation questions would remain unresolved, and further negotiations would be necessary before the components of $ could actually be put to work. We could, however, have set the bar for success even lower. For example, our players negotiated over distinct expenditure levels $_{infra}$ and $_{cnsrv}$ earmarked for increasing the ratios of effective to applied water in, respectively, agricultural and urban areas; we could have partitioned the expenditure allocation less finely, and had them negotiate a single expenditure level that would cover both of these objectives. Agricultural and urban interests would then have had to fight over the distribution of the aggregate earmark at a later date, before any money could be spent.

The decision over how high to set the bar for success is an extremely delicate one. Some negotiators would be extremely reluctant to agree to a coarse partition of a total bond allocation, in the absence of any guarantee that in the subsequent negotiations over the fine details, their particular interests would be funded to their satisfaction. Others might prefer to seek a general, imprecisely specified agreement as a first step, and be happy to postpone the difficult fights over details until later. In the current bond issue debate, we can observe Senate President Pro Tem Perata becoming more willing over the past year to lower the bar for success. On 25 January 2007, Perata said the following in response to the introduction of Schwarzenegger's proposal: 'We don't believe new dams at this point are needed. They cost billions of dollars and they take years, in fact decades, to build' (Lin, 2007). In contrast, his October 2007 proposal included funding for regional water supply projects that would be allocated based on bids by local water agencies, and could be used for above-ground storage. We interpret the following October 2007 quote as a clear request for a coarser partition of the expenditure allocation:

> 'We're not against dams,' said Perata spokeswoman Lynda Gledhill. 'We just feel they should compete with other projects.' (Thompson, 2007)

Perata's motivation for taking this stance is made clear by the following quote:

> There seems to be broad consensus on some very basic principles. If you look at Cogdill's bill and Perata's bill, there's a lot of commonality there. So, maybe we should just build on the commonality and wait for another day to fight over the differences. Senate President Pro Tem Don Perata (quoted in Russo, 2007a)

On the other hand, agricultural water users consider the partition proposed by Perata far too coarse. Steve Patricio, chairman of Western Growers, said the following regarding the need to specify surface storage and conveyance in the bond agreement:

> *'For the last 12 years, the grower community has supported and voted for bond proposals with all of our friends, and they always said the same thing: "Take care of this and next time we'll get you some storage. Take care of this and next time we'll get you some storage." There is no next time,' Patricio said. 'Storage has to be part of the solution and conveyance, too.'* (Krauter, 2007)

In our language, Patricio is arguing that the partition of the allocation must be sufficiently fine that negotiators will be obliged to link specific funds to specific activities.

As the CALFED negotiations illustrate there is no guarantee that requiring agreement over funding will increase the likelihood of a successful negotiation. CALFED has as part of its mandate that stakeholders must pay for private benefits that they obtain from CALFED projects. Howitt (2007) identifies this requirement as one of the pressures that is limiting the scope of CALFED to advance policy successfully.

Although the CALFED definition of success was imposed as a constraint on its players while in the current negotiations, the definition of success is endogenous, the two cases provide a common lesson regarding the design of stakeholder negotiation processes: there is a trade-off when specifying the precision of the negotiated agreement. The finer the partition of the issue space that must be specified in the agreement, the more likely that it can be implemented without further decision making by policy makers or others. On the other hand, the finer the partition, the more difficult it may be to achieve a successful negotiated outcome.

## Defining the consequences of failure

One way to induce parties to agree to negotiate and to do so successfully is to threaten a bad outcome if the negotiation fails. In game theoretic terms, a default outcome that has significant negative consequences for a stakeholder group will increase the number of negotiation proposals that it will accept in order to avoid the consequences of failure. In California water policy, some commentators have simplified this point into the statement that a drought-induced crisis is required in order to make major policy changes.

While a crisis undoubtedly provides an impetus for negotiation, stating that a drought is required for major water policy advancement is a simplification. 'Crises' are not necessarily exogenous, but can be induced by policy makers or judges. Whether or not these actions facilitate successful negotiation depends on how they alter the implications of failure for negotiating parties. In the current debate, the recent judicial decision regarding the ESA and the delta smelt altered the default payoff for major interest groups. Practically speaking, the ruling implied that the requirements of the ESA preceded any water delivery obligations for urban or agricultural uses. Clearly this decision made agricultural and urban users worse off, regardless of the outcome of any negotiations involving water allocation and the Delta. Senator Perata stated that the ruling 'is so far-reaching it could have such a deleterious effect on the state's economy ... that everything has to be looked at and a compromise has to ensue' (Schultz, 2007c). This comment on the consequences of the ruling for negotiations suggests that the change in the default outcome will increase the incentives to reach a negotiated agreement. On the other hand, the ruling benefited environmentalists, particularly those whose primary concern was the Delta. These stakeholders were guaranteed the primacy of ESA requirements over all other uses. Consequently, these stakeholders have an improved default outcome. '"It's better than what there was before," said Trent Orr, an attorney with the environmental group Earthjustice, which was party to the suit' (Conaughton, 2007).

In the context of the current bond issue debate, the funds devoted to environmental restoration in the Bay-Delta can be interpreted as a way of addressing the relatively strong position of environmentalists concerned with that ecosystem and the delta smelt. Because their default outcome is more desirable, in order to obtain their support any proposal must be more attractive than would have been the case previously.

As this example illustrates, changes in the default outcome due to an external crisis or government action do not necessarily enhance the chances of a successful negotiating outcome. One lesson for policy makers is that in order for a change to increase the likelihood of success, it must make some or all of the negotiating parties worse off. If one or more are made better off, they will have less of an incentive to negotiate, as in the case of environmentalists and the judicial ruling regarding the ESA and the delta smelt. On the other hand, urban and agricultural water users' default payoffs declined as a result of the ruling, which increases their incentive to negotiate.

## Conclusion and policy implications

Strategic behaviour plays an important role in water policy negotiations. In this chapter, we consider how bargaining theory can aid policy makers in addressing the implications of strategic behaviour when designing stakeholder negotiations. To conclude, we link our findings regarding structural features of negotiations to the strategic positions of each stakeholder group and the progress of the bond issue negotiations to date, and examine the future policy implications.

First, the definition of the issue space is a critical determinant of the success of the negotiation and the nature of its outcome. While sometimes the contribution of economics to policy debates is summarized as 'everything has a price', this statement ignores the fact that strategic players may assign different values to an item. Bargaining theory demonstrates that the more accurate statement is 'every negotiated variable has a value, and every relevant variable should be negotiated'. Relevance is based on technical relationships *and* on the preferences of negotiating parties. In order for a negotiation to succeed, there must be technically feasible outcomes that Pareto dominate the default. Consequently, narrow definitions of policy problems that limit the issue space are likely to inhibit the search for a solution. Schwarzenegger's decision to widen the issue space by adding dams and other infrastructure was intended to increase the likelihood that a negotiated outcome would be better than the default outcome for stakeholders favouring increased water supplies, including many of the groups represented by our player junAg.

In the case of water and other common property resources, the statement 'everything has a price' omits another consideration: in the presence of a market failure, there is no reason to expect that the price determined by the market will be the socially optimal price. The market price will not necessarily incorporate all of the costs and benefits incurred by members of society as the result of a given allocation of water. Government action is one solution for correcting market failure. However, in many instances, the information needed to implement optimal regulation can be costly or impossible to obtain, due to strategic behaviour by stakeholder groups and other considerations. Consequently, there is an important role for collective action, such as stakeholder negotiations. In California the market does not incorporate the value of water used for environmental purposes for the most part, although there are small but well-publicized exceptions. As mentioned earlier, judicial decisions regarding the enforcement of the ESA have played an important role in California water policy, and altered the default payoffs of stakeholder groups involved in the current negotiation. However, the lack of flexibility in the current enforcement of these judicial decisions, in turn may lead to larger losses incurred by other stakeholders than would be necessary in the context of a negotiated solution involving all stakeholders.

Second, the impacts of structural features of a negotiation process on its success are interconnected. The impact of the definition of the issue space is tied to the definition of success. While a broad issue space that follows the principle of including all relevant variables increases the likelihood that there will be technically feasible outcomes that Pareto dominate the default outcome, agreement over all of the details regarding each variable increases the negotiation's complexity. A coarser definition of success that does not require agreement on every detail may be easier to meet. However, the definition must be fine enough that all participants feel that their payoffs from agreeing to the negotiated solution are greater than their payoffs from the default outcome. In the current debate, this is well illustrated by the disagreements regarding whether funds for dams and above-ground storage must be specified in the negotiated bond measure, or whether these items can be included in a broader group of funded expenditure classes, with the allocation of the funds among classes to be determined later.

Third, the role of the default outcome is both technical and political. In the default outcome, technical relationships determine the value of relevant variables. Stakeholder preferences over these default values drive their willingness to compromise: the less attractive is the default outcome for a given stakeholder group, the more likely it is that there will be alternative outcomes that it will consider to be an improvement. For example, because urban water users tend to be willing to pay high prices for water purchased from senAg users and other sources, they would likely be more willing to endorse a negotiated solution that involved statewide water markets than would junAg users, whose willingness to pay for water tends to be lower than urban users'.

At this writing (March, 2008), negotiations are still ongoing to develop a water bond proposal to place on the November, 2008, state ballot. Consequently, we cannot link our analysis to the ultimate negotiation outcome. However, we can identify implications for future policy of the strategic positions of the players and the evolution of the negotiations to date. First, environmental interests have strengthened their strategic position through the use of the courts. Given this success, they have an incentive to continue to pursue legal action, as illustrated by their early 2008 lawsuit regarding the enforcement of the ESA for the case of the longfin smelt in the Bay-Delta. A ruling in this case further reduced water exports from the Delta. Consequently, other interests should be more willing to come to the bargaining table when there is a possibility that negotiations can replace court actions or aid in a legal settlement. Because government agencies are the legal defendants in these lawsuits, they have an incentive to design regulations (perhaps using stakeholder negotiations) that are less likely to be challenged by environmentalists. Second, in February, 2008, it came to light that Governor Schwarzenegger had hired a small team of specialists to examine the logistics and environmental implications of a new above-ground conveyance system that would route water exports around the Delta. The immediate political reaction when this action was discovered was predicated on the assumption that Schwarzenegger simply wanted to push through a new version of the Peripheral Canal. However, a more strategic interpretation of the governor's action was that he wanted to alter the state of knowledge regarding technical relationships associated with above-ground conveyance. Changes in technical relationships will in general be expected to alter stakeholder groups' payoffs from potential solutions, and may facilitate a negotiated outcome, even if it is not one that ultimately involves above-ground conveyance around the Delta. In this instance, as is often the case, the strategic implications of an action are broader than the immediate political interpretations. Bargaining theory provides a means for modelling negotiations that enables the strategic behaviour of negotiators to be analysed.

## Acknowledgements

The authors thank Ariel Dinar, Jose Albiac, Kelly Grogan and Rich Sexton for their comments on this chapter. They thank Richard Howitt for graciously sharing his knowledge of California water policy, and for stimulating discussions of the relationship between economic-engineering optimization modelling and bargaining-theoretic modelling.

## Notes

1   Game theorists often refer to the two approaches as *cooperative* and *non-cooperative* bargaining theory. As the game theoretic understanding of the distinction between cooperative and non-cooperative solutions is not always consistent with the lay person's understanding, we use the *axiomatic* and *strategic* labels to avoid confusion.
2   See Osborne and Rubinstein (1990) for an overview of axiomatic bargaining theory or Thomson and Lensberg (1989) for a more detailed description.
3   See Keene and Martinson (2007) for a good summary of the proposals.
4   There are exceptions to this broad rule. Notably, Lund et al (2007) document that some environmental groups are willing to consider a peripheral canal to route water exports to southern California around the Delta.
5   California water policy is complex, and our summary here quite abbreviated. Howitt and Sunding (2003) discuss water infrastructure and recent policy, including water markets. Rausser and Stratton (2007) for a discussion on groundwater policy.
6   Some observers point to the Quantification Settlement Agreement as one of the causes of the current environmental problems in the Bay-Delta. Southern California has increased its use of water from northern California since its use of Colorado River water has been curtailed, although the increase has not been on a one-to-one basis.
7   See Goodhue et al (2008) for a discussion of the disaggregation of stakeholder groups.
8   Obviously, the delta smelt is just one species whose survival is in jeopardy; we treat it as a metaphor for all the others.

## References

Adams, G., G. C. Rausser and L. K. Simon (1996) 'Modeling multilateral negotiations: An application to California water policy', *Journal of Economic Behavior and Organization*, vol 30, pp97–111

Bazerman, M. (1983) 'Negotiator judgment: A critical look at the rationality assumption', *American Behavioral Scientist*, vol 27, no 2, pp211–228

Binmore, K., A. Rubinstein and A. Wolinsky (1986) 'The Nash bargaining solution in economic modeling', *RAND Journal of Economics*, vol 17, no 2, pp176–188

California Sportfishing Protection Alliance (CSPA) (2001), 'California Sportfishing Protection Alliance', http://users.rcn.com/ccate/CSPAPagerev0.html (accessed 16 January 2008)

Carraro, C., C. Marchiori and A. Sgobbi (2007) 'Negotiating on water: Insights from non-cooperative bargaining theory', *Environment and Development Economics*, vol 12, no 2, pp329–349

Conaughton, G. (2007) 'Judge's ruling could cut water supplies by 30 percent', *North County Times*, 1 September, www.nctimes.com/articles/ 2007/09/01/news/top_stories/ 22_09_448_31_07.txt (accessed 7 January 2008)

Department of Water Resources (DWR), State of California (2005) *California Water Plan Update 2005*, Department of Water Resources Bulletin 160-05, Sacramento, CA, www.waterplan.water.ca.gov/ (accessed 8 January 2008)

Department of Water Resources (DWR), State of California (2008) 'Missions and goals', www.water.ca.gov/ (accessed 17 February 2008)

Dinar, A., A. Ratner and D. Yaron (1992) 'Evaluating cooperative game theory in water resources', *Theory and Decision*, vol 32, pp1–20

Goldmacher, S. (2007) 'Laird talks water in Q & A', *Capital Alert, Sacramento Bee*, www.sacbee.com/1066/story/392067.html (accessed 7 January 2008)

Goodhue, R. E., G. C. Rausser, L. K. Simon and S. Thoyer (2008) 'Multilateral negotiations over the allocation of water resources: The strategic importance of bargaining structure', in A. Dinar, J. Albiac and J. Sanchez-Soriano (eds), *Game Theory for Policymaking in Natural Resources and the Environment*, Routledge Press, London

Howitt, R. E. (2007) 'Delta dilemmas: Reconciling water-supply reliability and environmental goals', *ARE Update*, Giannini Foundation of Agricultural Economics, University of California, vol 10, no 4, pp1–4, www.agecon.ucdavis.edu/extension/ update/articles/v10n4_1.pdf (accessed 24 December 2007)

Howitt, R. E. and D. Sunding (2003) 'Water infrastructure and water allocation in California', in J. Siebert (ed.), *California Agriculture: Dimensions and Issues*, http:// giannini.ucop.edu/CalAgBook/Chap7.pdf (accessed 24 December 2007)

Kalai, E. and M. Smorodinsky (1975) 'Other solutions to Nash's bargaining problem', *Econometrica*, vol 43, no 3, pp513–518

Keene, K. and C. Martinson (2007) 'Water: Agriculture and natural resources', *California Counties Legislative Bulletin*, vol 107, no 28, www.imakenews.com/csac/ e_article000928717.cfm?x=b11,0, (accessed 7 January 2008)

Krauter, B. (2007) 'Group pushes $11.7B bond', *Capital Press*, 14 December, www. capitalpress.com/main.asp?Search=1&ArticleID=37614&SectionID=67&SubSectio nID=616&S=1 (accessed 6 January 2008)

Krishna, V. and R. Serrano (1995) 'Perfect equilibria of a model of n-person noncooperative bargaining', *International Journal of Game Theory*, vol 24, no 3, pp259–272

Krishna, V. and R. Serrano (1996), 'Multilateral bargaining', *Review of Economic Studies*, vol 63, no 1, pp61–80

Lin, J. (2007) 'Democrats dubious on pair of dams', *Sacramento Bee*, 26 January, http://parkwayblog.blogspot.com/2007/01/dam-dubiousness.html (accessed 9 January 2008)

Lund, J., E. Hanak, W. Fleenor, R. Howitt, J. Mount and P. Moyle (2007) *Envisioning Futures for the Sacramento–San Joaquin Delta*, Public Policy Institute of California, San Francisco

Nash, J. F. (1950), 'The bargaining problem', *Econometrica*, vol 18, pp155–162

Nature Conservancy (2008) 'About us: Learn more about the Nature Conservancy', www.nature.org/aboutus/index.html?src=home (accessed 16 January 2008)

Osborne, M. J. and A. Rubinstein (1990) *Bargaining and Markets*, Harcourt Brace Jovanovich, Academic Press, Economic Theory, Econometrics, and Mathematical Economics series, San Diego, London, Sydney and Toronto

Parrachino, I., A. Dinar and P. Patrone (2006) 'Cooperative game theory and its application to natural, environmental, and water resource issues: 3. Application to water resources', paper presented at the Sixth Meeting on Game Theory and Practice, Zaragoza, July 2006

Rausser, G. C. and L. K. Simon (1999) 'A non-cooperative model of collective decision-making: A multilateral bargaining approach', Working Paper Number 620, Department of Agricultural and Resource Economics, University of California, Berkeley

Rausser, G. C. and S. E. Stratton (2007) 'The political economy of groundwater management in California', Department of Agricultural and Resource Economics, University of California, Working paper

Roth, A. (1977) 'Individual rationality and Nash's solution to the bargaining problem', *Mathematics of Operations Research*, vol 2, no 1, pp64–65

Roth, A. (1979) *Axiomatic Models of Bargaining*, Springer-Verlag, New York

Rubinstein, A. (1982) 'Perfect equilibrium in a bargaining model', *Econometrica*, vol 50, no 1, pp97–110

Russo, F. D. (2007a) 'Drama this afternoon in the California capitol – Senate to vote on water bond; Governor Schwarzenegger to speak on health care special session at press conference', *California Progress Report*, 9 October, www.californiaprogressreport.com/2007/10/drama_this_afte.htm (accessed 5 January 2008),

Russo, F. D. (2007b) 'Perata to Schwarzenegger: You've failed twice this week on California's water crisis', *California Progress Report*, 17 October, www.californiaprogressreport.com/2007/10/perata_to_schwa_2.html (accessed 16 January 2008).

Samuelson, L. (2005) 'Economic theory and experimental economics', *Journal of Economic Literature*, vol 43, no 1, pp65–107

Schultz, E. J. (2007a) 'Ballot delay for water bond?' *Sacramento Bee*, 22 November, www.sacbee.com/111/story/510716.html (accessed 7 January 2008)

Schultz, E. J. (2007b) 'Dueling, multibillion-dollar water bonds may hit ballot', *Sacramento Bee*, 9 October, www.sacbee.com/111/v-print/story/421759.html (accessed 16 January 2008)

Schultz, E. J. (2007c) 'Water plans back on tap', *Fresno Bee*, 6 September, www.pwmag.com/industry-news-print.asp?sectionID=760&articleID=568399 (accessed 7 January 2008)

Simon, L. K., R. E. Goodhue, G. C. Rausser, S. Thoyer, S. Morardet and P. Rio (2007) *Structure and Power in Multilateral Negotiations: An Application to French Water Policy*, Giannini Foundation Research Monograph 47, University of California, Oakland, CA, http://giannini.ucop.edu/ Monographs/47_Water.pdf (accessed 24 December 2007)

Thompson, D. (2007) 'Republicans pledge no dams, no water deal as negotiators huddle', *North County Times*, 4 October, http://nctimes.com/articles/ 2007/10/04/news/state/6_01_2410_3_07.txt (accessed 24 December 2007)

Thomson, W. and T. Lensberg (1989) *Axiomatic Theory of Bargaining with a Variable Number of Agents*, Cambridge University Press, Cambridge

Young, S. (2007) 'Governor's plans for dams rejected', *Oakland Tribune*, 25 April, http://findarticles.com/p/articles/mi_qn4176/is_20070425/ai_n19034066 (accessed 24 December 2007)

# 15
# Strategic Behaviour in Transboundary Water and Environmental Management

*George B. Frisvold*

How can an understanding of strategic behaviour and game theory improve transboundary environmental management? Game theory can improve environmental management in two ways. First, one can use game theory to explain and evaluate past conflicts, negotiations and outcomes. Second, insights from game theory can improve the design of current transboundary environmental policies. The first point deals with ex post policy assessments, the second with ex ante policy guidance. In this second area, game theory applications have yet to reach their full potential.

This chapter draws on the history of US–Mexico border environmental management to illustrate the power of game theory as an analytic tool, but also to discuss why it has proven so difficult to apply insights from game theory to improve border environmental management. Here, I paint with a broad brush, neither delving into technical details of game theory, nor discussing border institutions in detail. This does not mean institutions are unimportant. Quite the contrary, game theory provides a richer understanding of just how important institutions are. The goal, though, is to use the history of US–Mexico border environmental management to derive general lessons for transboundary environmental management.

Developments in the strategic bargaining approach have made game theory more applicable to policy evaluation and design. The strategic bargaining approach may be contrasted with the axiomatic approach to bargaining. The axiomatic approach relies on developing a set of convincing properties that a bargaining solution would (or should) have. The next step is to show that such a solution is possible and (even better) unique. The focus is the mathematical properties of the solution and the approach abstracts from the actual bargaining process itself. The attributes of the players, the institutional rules of the game, or the precise bargaining environment are implicit. There is an emphasis on the economic efficiency of outcomes. Often (as in the classic case of bilateral monopoly) the distribution of payoffs to bargaining parties is discussed all too vaguely in terms of differences in bargaining power.[1] But, what determines differences in bargaining power?

Under the strategic bargaining approach, in contrast, the bargaining process itself, along with the attributes of bargaining parties are examined explicitly (Binmore et al, 1986). A party's bargaining power is enhanced by patience (the ability to wait for a negotiated payoff) and the speed at which it can make and respond to offers, while bargaining power is weakened by aversion to risk.

The strategic approach has features that make it naturally more appealing. While also being highly mathematical, it is couched in the language of playing games (in terms of players and moves), which is more accessible than discussion of axioms. The explicit consideration of the environment and rules of negotiations highlight how much institutions matter. Consideration of repeated games further illustrates the importance of reputation, reciprocity, institutional memory and the history of negotiations.[2] Analysis of trade assuming perfectly competitive markets focuses on the efficiency of market-based outcomes (often downplaying distributional consequences). Game theory, in contrast, is well equipped to explore issues such as the exercise of monopoly or monopsony power, distribution and often highly asymmetric benefits of exchange. By dealing with these aspects of the 'dark side' of exchange, game theory can overcome some of distrust other social scientists, environmental advocates or environmental managers have of economics and reliance on market-oriented outcomes. Staff members of international environmental institutions or academics who serve on their advisory committees are rarely economists (at least this is true in the US). So, improving communication with these groups is an important challenge.

The next section provides a brief overview of US–Mexico border environmental problems and institutions and presents examples where game theory proves useful in evaluating transboundary environmental management. The main themes include:

- the role of issue linkage and side payments;
- cost sharing rules for binational environmental projects;
- the potential for institutions to transform negotiations from one-shot prisoners' dilemma games to repeated games with greater scope for cooperative solutions.[3]

Table 15.1 summarizes some important institutional arrangements that have developed to address transboundary water issues on the US–Mexico border. Each institutional arrangement arose to deal with rather specific water management issues. Negotiated solutions required the countries to make specific decisions about how to finance and implement projects. One can view negotiation, in turn, in terms of different applications of game theory to find solutions. Finally, I highlight key aspects of negotiated solutions.

The third part of the chapter discusses the potential to apply game theory insights to ex ante policy design, rather than merely for ex post policy analysis. Game theory can provide guidance to better-designed multilateral funding programmes and facilitate issue linkage to resolve multi-faceted environmental conflicts. Yet, several constraints remain. Binational environmental programmes are often reactive and crisis-driven, which preclude negotiation that is more sophisticated. This also makes it difficult to complete sophisticated game theoretic modelling exercises quickly enough to inform policy decisions ex ante.

## Environmental management on the US–Mexico border

'¡Pobre Mexico! ¡Tan lejos de Dios y tan cerca de los Estados Unidos!'
(*Poor Mexico, so far from God and so close to the United States*)

— Porfirio Diaz, President of Mexico

Despite periodic conflicts, the US and Mexico actually have a long history of agreements and cooperation over water resources. The 1889 Convention on Boundary Waters established the International Boundary Commission (IBC) – the world's first binational agency to govern a river. The IBC was tasked with resolving disputes over the Rio Grande. A 1933 treaty authorized the first joint water infrastructure project between the countries, a canal and flood control project overseen by the IBC. It also authorized the construction of a dam for the Caballo Reservoir in New Mexico to capture irrigation water for both countries. The costs of the dam project were allocated based on the relative value of agricultural assets on lands served by the project and relative benefits of flood control on each side of the border. The US was to bear 88 per cent of project costs and Mexico 12 per cent.

This agreement set three important precedents. First, it formally recognized that investments made by one country could benefit the other and that joint project development could be more cost-effective than unilateral action. Second, it allocated project costs based on the share of expected benefits accruing to each country. Third, benefits were estimated using a simple approach requiring limited data that was available, easily understandable and verifiable for both countries.

In the 1940s, Mexico successfully linked negotiations over allocations of the Rio Grande and the Colorado River. Mexico desired an assured allocation of Colorado River water where it was the downstream riparian. But, the US held to the Harmon Doctrine asserting sovereignty over waters flowing within

Table 15.1 Institutional arrangements for transboundary water management on the U.S.–Mexico border water management

| Year | Institutional arrangement | Water management issues | Strategic problem | Game theory applications | Negotiated solutions |
|---|---|---|---|---|---|
| 1933 | Convention between the US and Mexico for Rectification of the Rio Grande | Rio Grande flood control project Construction and cost apportionment of the Caballo Dam in New Mexico | Determining sites for flood control infrastructure Cost apportionment of Caballo Dam in New Mexico | Two-party cooperative game Flood control benefits measured in terms of relative productivity of agricultural assets at risk | Infrastructure chosen to minimize cost of flood control objectives regardless of location Costs apportioned based on relative benefits of flood control to each country |
| 1944 | Treaty between the US and Mexico Relating to the Waters of the Colorado and Tijuana Rivers, and of the Rio Grande (called 1944 Water Treaty) | US diversions of the Colorado River impose costs on Mexico Mexican diversions of the Rio Bravo/Rio Grande tributaries impose costs on the US | Because of downstream positions, Mexico had weak bargaining power in Colorado dispute and US had weak bargaining power in Rio Grande/Rio Brave dispute | Interconnected game | Allocations of waters of Colorado and Rio Grande/Rio Bravo Rivers determined simultaneously |
| 1944 | 1944 Water Treaty | Border faced numerous transboundary water issues | Binational projects often more cost-effective than unilateral action Each new transboundary water project required a new binational treaty | Shift from a one-shot to repeated game negotiation structure | Scope of International Boundary and Water Commission (IBWC) expanded. US and Mexican Sections of IBWC given authority to negotiate terms of projects Repeated game nature of negotiations facilitates cooperation with over 300 agreements (called Minutes) approved to date |

| Year | Event | Issue | Game | Outcome |
|---|---|---|---|---|
| 1973 | Minute 242 Permanent and definitive solution to the international problem of the salinity of the Colorado River | Salinity of Colorado River water reaching Mexico from the US | Salinity harmed agricultural productivity of Mexico's Mexicali Valley | Two-party cooperative game with side payments | Minimum salinity standard established for transboundary water flows<br>Groundwater pumping limits established around San Luis on Colorado River<br>Side payments: US to provide assistance for Mexicali Valley rehabilitation and to assist Mexico in securing financing for rehabilitation |
| 1974 | Colorado River Basin Salinity Control Act | Compliance with Minute 242 (above), controlling Colorado River Salinity | Achieving salinity control by individual US states to comply with terms of US–Mexico binational agreement | Multi-level game with top level between US and Mexico and next level between US federal government and states | US constructs Yuma desalinization plant to bind commitment to control salinity<br>Authorizes payments for land fallowing and improve irrigation infrastructure to control salinity |
| 1958– | Various IBWC minutes | Border wastewater treatment | Developing sites for treatment facilities<br>Cost apportionment of facilities<br>Property right determination over treated effluent | Sequential bargaining game with asymmetries: US down stream facing external costs; Mexico has fewer financial resources; Public health crises make US an 'impatient' player | Countries site facilities to minimize cost of treatment objectives<br>Costs initially apportioned based on relative project benefits to each country<br>US adopts then abandons equal cost sharing rule after it leads to inferior solutions |

its borders. The US position changed, however, when Mexico began diverting water from tributaries of the Rio Grande, reducing water available for irrigation in southern Texas. Mexico insisted on tying negotiations over allocation of the Rio Grande to allocation of the Colorado. Ragland (1995) examined this process as an interconnected game.[4] In an isolated game allocating the Colorado, Mexico could expect little or no assurance from the US. By linking games, however, Mexico was able to achieve a greater water allocation than would otherwise be possible. The resulting 1944 Treaty between the US and Mexico Relating to the Waters of the Colorado and Tijuana Rivers, and of the Rio Grande (known as the 1944 Water Treaty) allocated Mexico 1.5 million acre-feet of Colorado River water per year, while Texas was to receive an annual average of 350,000 acre-feet from the Rio Grande.

The 1944 Water Treaty also changed the name of the IBC to the International Boundary and Water Commission (IBWC) and elevated its role, placing it in charge of 'settlement of all disputes' arising from the treaty and stating the commission 'shall in all respects have the status of an international body'. The IBWC was thus given authority as the primary vehicle for settling water disputes and coordinating water projects on the US–Mexico border. The IBWC is made up of US and Mexican Sections, with each section required to be led by a licensed engineer. The jurisdiction of the IBWC is specific and narrow. It extends only to water issues that are fundamentally binational. The IBWC may address water sanitation problems, through projects mutually agreed upon by the two nations. These agreements are called 'Minutes'. The commission is primarily a technical agency, focusing on scientific appraisals and engineering solutions to water management problems. Although the commission's jurisdiction is limited in scope, on US–Mexico border water issues, its authority supercedes the claims of other domestic agencies. To alter the jurisdiction or authority of the commission would require a new treaty approved by both governments.

This new mandate for the IBWC was a significant event in the history of transboundary water and environmental management. Prior to 1944, the 1889 Convention was extended numerous times (1895, 1896, 1897, 1898, 1899 and 1900) and separate treaties on border water issues were signed in 1906 and 1933. This meant that any border water settlement or project required a separate treaty and a two-thirds majority in the US Senate for passage. By allowing disputes to be resolved and projects to be planned via 'Minutes', it allowed negotiations between the two sections of the IBWC to take on the character of repeated games (where cooperative outcomes are more likely). Over 300 Minutes have been approved to date. The role of Congress (in both countries) was thus scaled back to approving funding for proposed projects in up-or-down votes. The IBWC has received praise for its ability to find cooperative solutions to border water problems and for its sheer longevity as a bilateral negotiation institution (Mumme, 1993; Szekely, 1993a). The IBWC has been the only permanent institution, conducting bilateral negotiations and planning of any kind, between the US and Mexico.

Since 1944, the border population has increased twelvefold, placing stress on the region's water treatment infrastructure. The IBWC's attention has been drawn increasingly toward water quality problems, particularly the salinity of Colorado River water reaching Mexico and the treatment of wastewater from rapidly growing Mexican border cities. Agreements to finance, construct and operate border water infrastructure has taken the form of binding commitments. This allows the sections of the IBWC to negotiate in a two-party cooperative game setting. Indeed both countries have insisted on agreements with more binding provisions. The US has insisted that joint wastewater treatment facilities be constructed and operated on the US side of the border in part to maintain control over pollution control operations. In 1973, the countries agreed to Minute 242, which established salinity standards for Colorado River water reaching Mexico. The Minute required the US to construct a large desalinization plant in Yuma, Arizona. The plant has proven uneconomical to operate and the US meets its treaty obligations by diverting irrigation drainage water. Yet, the plant serves as a backstop – a commitment – to meet Minute 242's obligation.

Frisvold and Caswell (2000) have examined negotiations over wastewater treatment projects between US and Mexican Sections of the IBWC as a Nash bargaining game.[5] The Nash solution has several desirable features. The outcome is Pareto efficient. For two agents bargaining over the division of treatment effort to meet a drinking water quality standard, the Nash solution guarantees that the standard is achieved at the least cost (Frisvold and Caswell, 1995). Finally, despite its simplicity, the Nash solution can closely approximate solutions to more sophisticated non-cooperative games (Binmore et al, 1986).[6]

Untreated sewage is a major transboundary externality, as polluted water flows northward from Mexican to American cities. At one time, the city of Nuevo Laredo deposited 24 million gallons per day (mgd) of raw sewage into the Rio Grande (Johnstone, 1995). In Tijuana, over 10mgd of untreated sewage, combined with industrial waste, flow into the Tijuana River and San Diego (IBWC, Minute 283, 1990; Johnstone, 1995). Flows of sewage into the ocean have led to frequent beach closures in San Diego (Ganster, 1996). The New River – flowing north from the Mexicali Valley, through the Imperial Valley and into the Salton Sea – has the dubious distinction of being one of the most polluted rivers in the US (Kishel, 1993; Johnstone, 1995; Ganster, 1996). The Nogales Wash, a tributary of the Santa Cruz River, flows through the twin cities of Nogales, Sonora, and Nogales, Arizona. During summer rains, there have been raw sewage flows into the Wash and through neighbourhoods on both sides of the border (Ingram and White, 1993; Varady et al, 1995). Giardia and cryptosporidium have been detected in the Wash and the aquifer serving as the primary water source for both cities (Varady and Mack, 1995).

Frisvold and Caswell (2000) consider the problem of the US attempting to meet water pollution control standards at least cost. IBWC engineers frequently make recommendations about the location and scale of waste collection and treatment systems based on the principle of minimizing cost to achieve particular objectives, such as compliance with environmental laws. Often, this can be achieved via a joint wastewater collection and treatment project that requires investment and can provide benefits to both countries. Once the US and Mexican sections agree on the least-cost project, the problem simplifies to allocating project costs between the two countries.

The IBWC negotiated construction of the first joint US–Mexico sewage treatment facility in 1951 to serve the border cities of Nogales, Arizona, and Nogales, Sonora. The IBWC recommended apportioning costs in proportion to benefits (Mumme, 1993). This follows the precedent of the 1933 Treaty. The downstream position of the US, combined with its greater willingness to pay for water sanitation meant that the US would derive relatively larger benefits from the project. The US therefore assumed a higher share of the project costs. This policy of apportioning costs in proportion to benefits was used repeatedly as a guideline in subsequent negotiations over wastewater treatment (Mumme, 1993). Sharing costs in proportion to benefits is certainly consistent with a Nash solution. In 1984, however, the Reagan administration adopted the position that the Mexican government should finance half the cost of jointly developed pollution control projects (Mumme, 1993).

The Nash bargaining approach can be used to examine negotiated outcomes of pollution control projects in three border metro areas: San Diego–Tijuana, Calexico–Mexicali and Laredo–Nuevo Laredo in response to this equal cost-sharing rule. Frisvold and Caswell (2000) argue that the equal cost-sharing rule fundamentally changes the nature of the game. Instead of choosing how to share costs, given optimal project size and scope, the problem becomes one of negotiating over project scale subject to the equal cost-sharing constraint. Requiring joint projects to be equally funded will generate efficient solutions only in limited and unlikely cases. The equal cost-sharing rule can discourage cooperation on projects where both total benefits and the US' share of the benefits are large (Frisvold and Caswell, 2000). A likely outcome is that Mexico will not cooperate, but rather unilaterally construct projects, ignoring transboundary impacts on the US. This is exactly what occurred.

The equal cost rule impeded a cooperative solution to border sanitation problems in San Diego–Tijuana (Mumme, 1993). In the 1980s, IBWC engineers recommended a gravity-flow collection system, with the main treatment plant located in San Diego. The objective of this system was to eliminate uncontrolled sewage flows into the Tijuana River and San Diego. Mexico balked at paying half of the estimated $730 million project cost. Instead, Mexico acted unilaterally, building a smaller, less expensive, self-financed system in Tijuana (IBWC, Minute 270, 1985). Rapid growth in Tijuana soon outstripped the capacity of the first of two facilities built and Mexico developed plans to construct a secondary treatment plant at the Rio Almar. US engineers, however, considered the proposed plant 'suboptimal and less reliable as a mechanism of managing Tijuana's growing sewage production' (Mumme, 1993, p117).

In 1990, the IBWC agreed to pursue the larger joint sewage collection and treatment project along the lines originally proposed, a gravity-flow system with the treatment facility sited in San Diego (IBWC, Minute 283, 1990). Under Minute 283, the US abandoned equal cost sharing:

> *The cost corresponding to Mexico shall be in an amount ... equal to that which would have been used in the construction, operation and maintenance of the treatment plant planned for the Rio Almar.*
> (IBWC, Minute 283, 1990)

Minute 283 improves on the earlier non-cooperative outcome. The US Section believed the scale and location of facilities would allow it to comply with domestic water quality standards cost-effectively. The Mexican government would incur no greater costs than those associated with its disagreement point, yet would derive benefits from the more efficient, larger system. Fernandez (2006) notes that this allocation strategy is consistent with a Chander–Tulkens (1992) allocation rule in that Mexico's costs were no greater than its costs under non-cooperation.[7]

The equal cost constraint also affected Minute 274 (IBWC, 1987b), *Joint Project for Improvement of the Quality of the Waters of the New River at Calexico, CA–Mexicali, BC*. The principal engineers were asked to develop plans for a jointly funded project to improve the waters of the New River 'Utilizing funds to be provided in equal parts by the Governments of the United States and Mexico' (IBWC, Minute 274, 1987b). The result was a small project that the engineers conceded was, 'but a small part of the total works required for solution of the border sanitation problem' (IBWC, 1987a). The engineers also noted some project features were abandoned because they fell outside of Mexico's budget constraint. Subsequent Minutes regarding the New River have dropped language about equal cost sharing.

In 1997, the commission signed Minute 297, apportioning the costs of a wastewater treatment project for the Rio Grande at Laredo–Nuevo Laredo. Here, the externalities of untreated wastewater affect the two countries more symmetrically. The project expanded collection and treatment capacity in Nuevo Laredo, Mexico. The project's goal was to prevent discharges of untreated sewage into the Rio Grande and to have discharges from new treatment facilities conform to US water quality standards. US standards are higher than standards required by Mexican law. The US agreed to pay Mexico for the incremental cost of operating and maintaining the project to meet the higher US effluent standard. The US Section believed expanding facilities in Nuevo Laredo was a more cost-effective way to meet US standards than to unilaterally build infrastructure in the US. The US, in turn, compensated Mexico for its incremental costs of meeting the higher

US standard. Again, this conforms to a Chander–Tulkens (1992) cost-sharing rule.

These examples illustrate game theoretic models can be used to assess cost-sharing rules, not just abstractly, but for actual projects. It also illustrates how politically imposed constraints on bargaining parameters can thwart cooperation and lead to less desirable outcomes. This type of analysis can be quite simple, as it is in Frisvold and Caswell (2000). Indeed, in an introductory environmental economics course, I use a simple graph from Field and Field's (2006) *Environmental Economics: An Introduction* in a homework assignment that has students evaluate cost-sharing rules for transboundary pollution control (Figure 15.1). Students are asked to show (and generally succeed!) how an equal cost-sharing rule under circumstances like those on the US–Mexico border will lead to suboptimal treatment plant scale. They also derive cost-sharing rules that could lead to cooperative financing of the optimally scaled facility.

## From programme assessment to policy design

Many externalities on the US–Mexico border, when viewed in isolation, are unidirectional (the lining of the All-American Canal, Colorado River salinity, sewage flows from Mexican to US cities).[9] Because of national sovereignty, dealing with such transboundary externalities must take the form of Coasian bargaining.[10] Two means of addressing unidirectional externalities are side payments and using an interconnected games approach to link issues for negotiation.[11] Table 15.2 summarizes important border institutions with the potential to encourage side payments and linked negotiations.

### Institutionalizing side payments

In 1994, as side agreements to the North American Free Trade Agreement (NAFTA), the US and Mexico established the Border Environmental Cooperation Commission (BECC) and the North American Development Bank (NADBank). The NADBank arranges financing of border water and other environmental issues that the BECC must certify, based on environmental, technical and financial criteria. In 2006, the organizations were merged with a common board of directors. A goal of BECC/NADBank is to address market failures that are at the centre of border environmental problems. Firms located on the border have not had to pay the full social costs of their production and release of industrial wastes into water bodies. While the IBWC has focused on responding to border sanitation problems after they arise, its mandate and organization structure is not designed to address problems of market failures and incentive problems that lead to water pollution crises in the first place.

A second problem has to do with the provision of water infrastructure needed to support the rapidly growing workforce on the border. Historically, firms have

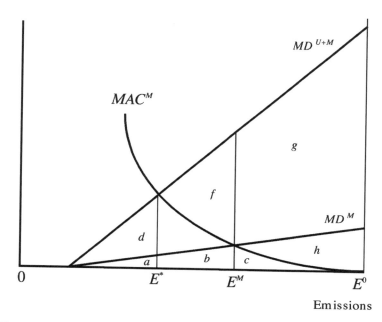

Note: Wastewater emissions from Mexico border cities cause environmental damage to Mexico and the US. Line $MD^M$ shows marginal damages (the damage from each additional unit of pollution) to Mexico. Line $MD^{U+M}$ shows combined marginal damages to both countries. Acting unilaterally, Mexico's net benefits from damage reduction are greatest if it operates a plant to reduce emissions from $E^0$ to $E^M$. Pollution control benefits are $h + c$ and abatement costs are $c$, so Mexico's net benefits are $h$. Reducing emissions to $E^M$ reduces damage to the US by $g$. Net benefits to both countries are highest if Mexico operated a larger plant and reduced emissions to $E^*$. US environmental benefits would increase by $e + f$ and Mexican environmental benefits would increase by $b$. Abatement costs, however, would increase by $e + b$, more than Mexico's benefits. To induce Mexico to reduce emissions from $E^M$ to $E^*$, the US could offer abatement cost sharing of $e$ or greater. The US would be willing to pay up to $e + f$, so the countries have room to negotiate a mutually beneficial deal with cost-share payments between $e$ and $e + f$. Under equal cost sharing, the US could not offer more than $\frac{1}{2}(e + b)$. But this is lower than $e$ (the minimum acceptable offer to Mexico) if $b < e$, as it is in the figure.[8] So, Mexico will unilaterally operate a plant too small to achieve the highest net benefits for both countries. The figure illustrates how insistence on equal cost sharing can discourage cooperative solutions and lead to pollution control projects of inadequate scope.

**Figure 15.1** *An equal cost-sharing rule can discourage cooperative solutions to transboundary pollution problems*

not paid much in the way of user fees or taxes to finance safe drinking water or sewer systems for the growing workforce. Local municipalities pay only a fraction of the cost of water treatment infrastructure. The US federal government's willingness to bail out border cities is an understandable response to immediate health concerns. However, because cities are not internalizing the full costs of border growth, population and sewage growth has outstripped local infrastructure (Ingram and White, 1993; Udall Center, 1993; Johnstone, 1995).

Table 15.2 Border institutions and their potential roles in facilitating cooperative transboundary water management solutions

| Year | Institutional arrangement | Water management issue | Strategic problem | Game theory application | Solution |
|---|---|---|---|---|---|
| 1983 | La Paz Agreement (Border XXI and Border 2012) | Problems of economic development and pollution control in border regions divided among multiple federal and state resource management agencies in each country | Jurisdictional fragmentation increases transactions costs of negotiations | Identify issues for interconnected game Identify institutional partners to implement joint programmes | Increase knowledge base for negotiations Establish common knowledge base for negotiations Improve coordination between sister agencies in each country |
| 1990 | Southwest Consortium for Environmental Research and Policy (SCERP) | Multiple environmental problems linked to water quantity and quality issues | Need to increase knowledge of cross-media effects | Identify issues for interconnected game | Increase knowledge base for negotiations Establish common knowledge base for negotiations |
| 1992 | The Good Neighbor Environmental Board (GNEB) | Multiple economic and institutional problems linked to water quantity and quality issues | Provides policy recommendations to US regarding transboundary management | Identify issues for interconnected game Multi-level game with top level between US and Mexico and next level between US federal government and states | Increase knowledge base for negotiations Establish common knowledge base for negotiations Facilitate cooperative state-level response to federal binational negotiations |
| 1993 | Border Environment Cooperation Commission (BECC) and the North American Development Bank (NADBank) | Many US–Mexico border water pollution problems are unidirectional Pollution from Mexico makes it difficult for US jurisdictions to comply with US environmental laws | Financing needed for border pollution control infrastructure 'Victim pays' solutions often politically unattractive | Comparison of non-cooperative and cooperative solutions Side payments key for improving solutions | BECC/NADBank institutionalize side payments For US, financing Mexican pollution control is often more cost-effective than domestic control Work of Fernandez (2005) suggests increase in funding for side payments could improve outcomes |

Yet, border cities are limited in their abilities to self-finance water infrastructure (Hinojosa-Ojeda, 1999). Because of the risks associated with these investments, it is difficult to obtain long-term financing through international markets. In addition, Mexico's legal system limits the ability of local governments to issue bonds against user fees or real estate taxes.

The NADBank's purpose is to help border communities with long-term funding of water and solid waste projects. Capitalized by both the Mexican and US governments, NADBank can secure financing at lower commercial rates than would otherwise be possible for border communities. The bank also uses its funds to leverage other private loans and grants that local entities may not otherwise be able to secure. The NADBank is not a grant-giving agency (although it does help administer an Environmental Protection Agency (EPA) grants programme). Water projects must be able to repay loans, raising funds through user fees or other mechanisms.

The BECC certification criteria include human health and environment, technical feasibility, financial feasibility and project management, community participation, and sustainable development. Along with certifying projects for funding, the BECC provides technical assistance for local entities developing projects. In addition, it analyses environmental and financial aspects of projects and helps arrange public financing for projects (EPA et al, 1998).

In its first two years, the BECC failed to secure NADBank funding for any of its certified projects. NADBank (1998a) identified five constraints limiting project development: (i) insufficient community resources for high cost projects; (ii) a lack of master plans and inadequate proposal preparation; (iii) limited financial, administrative and commercial capabilities of local water agencies; (iv) inadequate revenue for the sound operation of existing services and resistance to raising user fees; and (v) lack of private sector involvement in environmental projects.

To address these constraints, the EPA and NADBank established the Border Environmental Infrastructure Fund (BEIF) (NADBank, 1998b). The fund receives and administers grants that may be combined with loans or loan guarantees. Grants may support municipal infrastructure, drinking water treatment plants and treated water distribution systems. Funds may be used to allow user fees to be phased in over time. In its approval process, BECC gives preference to projects addressing transboundary pollution, while the BEIF (funded through appropriations to the USEPA) funding criteria state that projects must have a US interest and that priority will be given to projects that benefit both countries (NADBank, 1998b).

The NADBank also established a Project Development Program (PDP) to provide technical assistance to communities and utilities to help finance the costs involved in preparing projects for construction. The aim of the programme is to aid communities that lack the expertise or financial resources needed to plan and design infrastructure projects. By the end of 2006, NADBank had provided over $835 million in grants and loans for 97 environmental infrastructure projects in the US–Mexico border region.

The BECC/NADBank system has institutionalized a way for the US to provide Mexico with side payments for transboundary pollution control. Fernandez (2005) has examined the role of side payments in reducing sediment pollution emanating from Tijuana, Mexico, and affecting San Diego, California. The analysis considered outcomes under a Nash non-cooperative equilibrium versus a cooperative game solution following alternative cost-sharing rules.[12] Because benefits accrue largely to the US, side payments are a crucial part of cooperative solutions. The level of side payments will vary, depending on whether cost allocation is determined by Shapley value, Chander–Tulkens or is based on Article V of the Helsinki Rules.[13] Currently, the US provides Mexico with side payments via the BEIF and PDP programs, but Fernandez's (2005) results suggest that payment levels would increase under a cooperative solution. The study is interesting in that it brings a wealth of empirical data to bear on the problem and ties recommendations to specific funding mechanisms and institutions.

## Interconnected games and issue linkage

Bennett et al (1997) note that game theoretic solutions to unidirectional externalities tend toward victim pays outcomes and Fernandez's (2005) results support this.[14] Bennett et al (1997) find victim pays regimes unsatisfactory because they run counter to the polluter pays principle accepted in the international community and because countries may wish to avoid appearing to be weak negotiators.

An alternative to side payments or accepting externalities is to link negotiation issues. For example, Mexico obtained a greater allocation of Colorado River water, where it was the downstream country, by linking Colorado River negotiations to negotiations over allocation of Lower Rio Grande waters, where it was the upstream country.

In interconnected games, negotiations over separate issues are joined in a repeated game. Each country's action in one game is conditional on the outcome of another. This allows for equilibrium solutions not attainable in isolated games and may yield higher joint payoffs. Interlinked solutions may also avoid side payments when isolated solutions do not (Folmer et al, 1994; Bennett et al, 1997).

Bennett et al (1997) and Ragland (1995) discuss how the interconnected game approach can identify issues for linkage simply by identifying issues with payoffs of the same order of magnitude and where the games have asymmetric prisoner's dilemma structure.[15] One might use their approach as a low-cost method of screening issues for potential linkage. To be policy-relevant in international water negotiations, Dinar and Dinar (2003, p1287) recommend that, 'economists should develop models that do not rely on sophisticated approaches, which necessitate accurate data that is probably as scarce as the water in the basin they are investigating'. Identifying 'same order of magnitude' payoffs would appear to follow this advice.

Linking water negotiations with other water or environmental issues may be attractive to Mexico. While the US has entered into agreements involving side payments, Mexico is less able to do so. Kishel (1993) has suggested linking negotiations over the lining of the All-American Canal to issues such as construction of a Yuma–Mexicali pipeline, groundwater banking, rights to treatment plant effluent, and transfer of water conservation technology. The canal diverts 3.5 million acre-feet (MAF) of Colorado River water to farmers in California's Imperial Valley. Because the unlined canal is built on sandy soils, 0.2 MAF of diverted water seeps into the ground annually. The US plans to line part of the canal to reduce seepage. This, however, would reduce recharge and raise the salinity of the Mesa San Luis aquifer supplying groundwater to Mexican farmers in the Mexicali Valley (La Rue, 1999). Mumme and Lybecker (2006) also suggest that issue linkage might be a useful approach for resolving the All-American Canal controversy.

The EPA's Border 2012 Program could become a vehicle for identifying issues amenable to linked negotiations. The US and Mexico signed the Border 2012: US–Mexico Environmental Program (Border 2012) in 2003. Border 2012 seeks to reduce water, solid waste, hazardous materials and air pollution, and to improve environmental health in both countries within 100km of the border. A key element of Border 2012 is coordination between the USEPA and its Mexican counterpart, SEMARNAT, as well as collection and sharing of data and information by various workgroups. Workgroups are organized into both regional and topical workgroups. The different Border 2012 Workgroups could help supply information for this screening process.

Other groups could also facilitate issue linkage. The Good Neighbor Environmental Board (GNEB) was created in 1992 as a US federal advisory committee to advise the president and the Congress about environmental and infrastructure issues in US border states. The GNEB does not carry out specific programmes, but provides policy recommendations. Its board members include government officials from Border States, interest group representatives and academics. Another group is the Southwest Consortium for Environmental Research and Policy (SCERP), a collaboration of five US and five Mexican universities located in all ten border states. Funded by the US Congress since 1990, SCERP addresses US–Mexico border environmental issues and is tasked to 'initiate a comprehensive analysis of possible solutions to acute air, water and hazardous waste problems that plague the United States–Mexico border region'.

## Policy lessons and remaining challenges

What broader lessons can the history of US–Mexico environmental negotiations provide countries facing problems of transboundary water and environmental management? All countries have certain unique environments, institutions and histories of conflict and cooperation with their neighbours. Yet, I believe there are three areas where game theory insights can improve the design of transboundary water and environmental policies. First, negotiating and planning institutions could be structured to operate under repeated game rather than one-shot rules. Much of the success of the International Boundary and Water Commission (IBWC) is owed to the ability of negotiators to develop a history and rules of cooperation.

Second, given many types of unidirectional externalities, side payments will often be unavoidable as a way of finding cooperative solutions. Yet, paying one's neighbours not to pollute is a 'victim pays' outcome that countries may find unattractive politically. US–Mexico border infrastructure financing projects in the post-NAFTA area have institutionalized methods of making side payments. In the US, states facing federal water and air quality mandates have found that investments in pollution control in Mexico are more cost-effective ways to meet these mandates than domestic regulation. Projects on both sides of the border supported by the IBWC, BECC, NADBank and EPA's Border Environmental Infrastructure Fund appear to have overcome resistance to side payments by reducing regulatory costs at the state level.

Finally, issue linkage in negotiation is an alternative to side payments. The theory of interconnected games suggests that issue linkage has the potential to achieve superior outcomes. Linked negotiated settlements over allocation of the Colorado and Rio Grande/Rio Bravo Rivers demonstrate that this approach is more than just a theory.

Challenges remain, however. One challenge is to develop methods that have modest data requirements and allow for timely model development. It is one thing to develop sophisticated game theoretic models to examine negotiations after the fact. It is quite another thing to develop models that are literally useful for finding cooperative solutions when negotiations are taking place. A second challenge is for economists to begin a more active dialogue with the social scientists in other disciplines and environmental scientists who participate more actively in environmental advisory committees and resource management agencies. Finally, an important precursor to greater transboundary cooperation between governments is greater international scientific collaboration in social and environmental sciences.

## Notes

1 In a bilateral monopoly, a market has a single buyer and a single seller. Each must strategically consider the actions of the other in reaching an agreement on negotiation terms (usually the price and quantity of a good exchanged).
2 In game theory, there is an important distinction between static, or one-shot games, and games that actors play over and over. A repeated game allows players' strategies to depend on past moves. Actions in one period affect a player's reputation in later periods. Players may face later rewards (or retribution) for their actions.

3   An example of the classic prisoner's dilemma follows. Two suspects are arrested for a robbery. The police have insufficient evidence for a conviction without confessions, but place the suspects in separate cells so they cannot communicate. The police tell each suspect the same thing. If you both confess, you both get two years in prison. If you do not confess, but your partner does, he goes free and you get five years. If you confess, but your partner does not, you go free and your partner gets five years. If neither confesses, both will be charged on a lesser count and sentenced to six months. The outcome for each prisoner ranges from freedom to five years in prison, depending on what one's partner does. Not knowing what the partner will do, each prisoner has an incentive to confess to minimize his or her sentence. The result, however, is that both confess and are each sentenced to two years, while their best outcome is if neither confessed, each getting six months. The prisoner's dilemma illustrates a general class of problems where actors would be better off cooperating, but have individual incentives not to cooperate. As a result, actors do not cooperate and are worse off for it.

4   In an interconnected game, players can take a position in one game (or one set of negotiations) in response to another player's action in a separate game (negotiation). This allows countries to bargain over a wider set of issues and may allow a country to accept 'losses' in one set of negotiations in exchange for greater concessions or gains in other negotiations.

5   The Nash bargaining game is a two-player game where parties agree over negotiation terms where failure to agree gives each player a fixed payoff known as a threat point. For example, parties can expect a certain set of benefits if an agreement is reached and a set of lesser benefits (or even losses) if it is not. Nash's (1953) solution to the bargaining problem has the feature that no other solution can make both parties better off.

6   In a cooperative game, players can discuss strategies and make binding commitments. In contrast, in non-cooperative games, players cannot explicitly coordinate their strategies and make binding commitments. Any cooperation must be self-enforcing.

7   In a cost-sharing rule discussed by Chander and Tulkens (1992), the cost savings a country receives from cooperation is at least equal to what the country would achieve under non-cooperation.

8   One might argue equal cost sharing could be applied to Mexico's total costs, not just incremental costs of expansion so that the United States could offer more, $½(e + b + c)$. Yet, there could be cases where $(b + c) < e$ and negotiations would still break down.

9   In economics, an externality is an impact on any party not involved in a given economic transaction. Pollution is a classic example of a negative externality. For example, a power plant generating and selling electricity can generate air pollution that harms others downwind. This harm is external to the market consideration of the energy producer or consumers.

10  Because of national sovereignty, one country cannot unilaterally impose its environmental laws on another. Economist and Nobel Laureate Ronald Coase (1960) proposed that under certain conditions (clearly defined property rights, low bargaining costs, absence of wealth or income effects) then parties could (i) bargain

to achieve an efficient level of pollution reduction; and (ii) that the outcome did not depend on whether the polluter had the right to pollute or those affected had the right to be free of pollution.

11  Side payments are monetary or in-kind transfers that can be used to give a party incentive to undertake costly actions as part of an agreement. In environmental agreements, in-kind transfers can take the form of technical assistance or access to technology.

12  Nash (1950, 1951) proposed a solution concept to a game where no player can increase their own benefits by unilaterally changing their strategy. A Nash equilibrium exists if each player is making the best decision he or she can, accounting for decisions of other players. A Nash equilibrium is not necessarily the best solution for all the players. They could be able to increase all their payoffs if they could agree on a coordinated strategy. In contrast, the cooperative solution maximizes the cumulative payoff to all players.

13  In a cooperative game, the Shapley Value awards gains to players in proportion to their marginal contribution to overall gains (Shapley, 1953). Fernandez notes, 'The Helsinki Rule formulated by the International Law Association [Cano, 1989] suggests, "reasonable and equitable sharing" of environmental protection according to several criteria. The criteria can include: land area, hydrological share, population, and practicability of compensation among other items.'

14  Under the victim pays principle, an entity or entities harmed by pollution pay the polluter to reduce or stop pollution. In contrast, dating back to the Trail Smelter Case of 1941, international law has affirmed a polluter-pays principle, where polluters are responsible for compensating for environmental damage. The polluter pays principle was reaffirmed in 1972 by the declaration at the Stockholm Conference.

15  In asymmetric prisoner's dilemma games, outcomes to individual players differ. Because payoffs differ, the dilemma is more complicated than the choice between cooperation and non-cooperation. The timing and sequence of actions can affect outcomes more. Asymmetries reduce cooperation rates in repeated games. Players can have difficulty even agreeing what constitutes a desirable outcome and dilemmas require more complicated negotiations (Murnighan, 1991; Beckenkamp et al, 2007).

# References

Beckenkamp, M., H. Hennig-Schmidt and F. P. Maier-Rigaud (2007) 'Cooperation in symmetric and asymmetric Prisoner's Dilemma games', (March). MPI Collective Goods Preprint No. 2006/25, Max Planck Institute for Research on Public Goods, Bonn

Bennett, L., S. Ragland and P. Yolles (1997) 'Facilitating international agreements through an interconnected game approach: The case of river basins', in D. Parker and Y. Tsur (eds), *Decentralization and Coordination of Water Resource Management*, Kluwer Academic Publishers, Boston, pp61–85

Binmore, K., A. Rubinstein and A. Wolinsky (1986) 'The Nash bargaining solution in economic modeling', *Rand Journal of Economics*, vol 17, pp176–188

Cano, G. (1989) 'The development of the Law of Water Resources and the work of the International Law Commission', *Water International*, vol 14, pp167–171

Chander, P. and H. Tulkens (1992) 'Theoretical foundations of negotiations and cost sharing in transfrontier pollution problems', *European Economic Review*, vol 36, no 2–3, pp388–399

Coase, R. H. (1960) 'The problem of social cost', *Journal of Law and Economics*, vol 3, pp1–44

Dinar, S. and A. Dinar (2003) 'Recent developments in the literature on conflict, negotiation and cooperation over shared international freshwater', *Natural Resources Journal*, vol 43 (Fall), pp1217–1287

Environmental Protection Agency (EPA) (1998) *Colonias Facts*, Office of International Affairs, USEPA, Washington, DC

Environmental Protection Agency (EPA), BECC and NADBank (1998) *Promoting Environmental Infrastructure on the U.S. – Mexico Border*, Office of International Affairs, USEPA, Washington, DC

Fernandez, L. (2005) 'Coastal watershed management across an international border in the Tijuana River watershed', *Water Resources Research*, vol 41, W05003

Fernandez, L. (2006) 'Transboundary water management along the U.S.–Mexico Border', in R.-U. Goetz and D. Berga (eds), *Frontiers in Water Resource Economics*, Springer, New York, pp153–176

Field, B. and M. Field (2006) *Environmental Economics: An Introduction*, 4th edn, McGraw-Hill Irwin, New York

Folmer, H. P., P. van Mouche and S. Ragland (1994) 'Interconnected games and international environmental problems', *Environmental and Resource Economics*, vol 3, pp313–335

Frisvold, G. and M. Caswell (1995) 'A bargaining model of water quality and quantity', in A. Dinar and E. Loehman (eds), *Water Quantity / Quality Disputes and their Resolution*, Praeger Press, Westport, CT, pp399–407

Frisvold, G. and M. Caswell (2000) 'Transboundary water management: Game-theoretic lessons for projects on the U.S.–Mexico Border', *Agricultural Economics*, vol 24, pp101–111

Ganster, P. (1996) 'Environmental issues of the California-Baja California border region', Border Environment Research Reports, No. 1, Southwest Center for Environmental Research and Policy, San Diego

Hinojosa-Ojeda, R. (1999) 'From NAFTA debate to democratic and sustainable integration: Potential implications of the North American Development Bank', mimeo, School of Public Policy and Social Research, University of California, Los Angeles

Ingram, H. and D. White (1993) 'International boundary and water commission: An institutional mismatch for resolving transboundary water problems', *Natural Resources Journal*, vol 33, pp153–176

International Boundary and Water Commission (IBWC) (1967) *Enlargement of the International Facilities for the Treatment of Nogales, AZ and Nogales, Sonora Sewage*, Minute 227. El Paso, TX, 5 September

IBWC (1973) *Permanent and Definitive Solution to the International Problem of the Salinity of the Colorado River*, Minute 242. Mexico, DF, 30 August

IBWC (1985) *Recommendations for the First Stage Treatment and Disposal Facilities for the Solution of the Border Sanitation Problem at San Diego, California – Tijuana, Baja California*, Minute 270. Ciudad Juarez, 30 April

IBWC (1987a) *Joint Report of Principal Engineers Proposing Technical Basis for Jointly Funded New River Water Quality Improvement Measures in Mexicali, Baja California*, Ciudad Juarez, Mexico, 9 February

IBWC (1987b) *Joint Project for the Improvement of the Quality of the Waters of the New River in Calexico, California – Mexicali, Baja California*, Minute 274. Ciudad Juarez, Mexico, 15 April

IBWC (1990) *Conceptual Plan for the International Solution to the Border Sanitation Problem in San Diego, California / Tijuana, Baja California*, Minute 283. El Paso, TX, 2 July

IBWC (1995) *Facilities Planning Program for the Solution of Border Sanitation Problems*, Minute 294. El Paso, TX, 24 November

IBWC (1997) *Operations and Maintenance Program and Distribution of its Costs for the International Project to Improve the Quality of the Waters of the Rio Grande at Laredo, Texas – Nuevo Laredo, Tamaulipas*, Minute 297. Ciudad Juarez, Mexico, 31 May

Johnstone, N. (1995) 'International trade, transfrontier pollution, and environmental cooperation: A case study of the Mexican–American border region', *Natural Resources Journal*, vol 35, pp33–62

Kishel, J. (1993) 'Lining the all-American canal: Legal problems and physical solutions', *Natural Resources Journal*, vol 33, pp697–726

La Rue, S. (1999) 'U.S. water-saving plan roils Mexican officials: Lined canal would harm Mexican farmers they say', *San Diego Union-Tribune*, ppA3, 28 February

Mumme, S. (1992) 'New directions in United States–Mexican transboundary environmental management: A critique of current proposals', *Natural Resources Journal*, vol 32, pp539–562

Mumme, S. (1993) 'Innovation and reform in transboundary resource management: A critical look at the international boundary and water commission, United States and Mexico', *Natural Resources Journal*, vol 33, pp93–120

Mumme, S. P. and D. Lybecker (2006) 'The All-American Canal: Perspectives on the possibility of reaching a bilateral agreement', in V. S. Munguía (ed.), *The U.S.–Mexican Border Environment: Lining the All-American Canal: Competition or Cooperation for the Water in the U.S.–Mexican Border?* SCERP Monograph Series, no. 13, San Diego State University Press, San Diego

Murnighan, J. K. (1991) 'Cooperation when you know your outcomes will differ', *Simulation and Gaming*, vol 22, pp463–475

Nash, J. (1950) 'Equilibrium points in N-person games', *Proceedings of the National Academy of Sciences*, vol 36, pp48–49

Nash, J. (1951) 'Non-cooperative games', *The Annals of Mathematics*, vol 54, pp286–295

Nash, J. (1953) 'Two-person cooperative games', *Econometrica*, vol 21, pp128–140

North American Development Bank (1998a) *Current Status and Outlook*, NADBank, San Antonio, TX

NADBank (1998b) *Border Environmental Infrastructure Fund*, NADBank, San Antonio, TX

Ragland, S. (1995) 'International Environmental Externalities and Interconnected Games', PhD Dissertation, University of Colorado, Boulder

Rubinstein, A. (1982) 'Perfect equilibrium in a bargaining model', *Econometrica*, vol 50, pp97–110

Shapley, L. (1953) 'A value for N-person games', in H. Kuhn and A. Tucker (eds), *Contributions to the Theory of Games*, II, *Annals of Mathematical Studies*, vol 28, Princeton University Press, Princeton

Szekely, A. (1992) 'Establishing a region for ecological cooperation in North America', *Natural Resources Journal*, vol 32, pp563–622

Szekely, A. (1993a) 'Emerging boundary environmental challenges and institutional issues: Mexico and the United States', *Natural Resources Journal*, vol 33, pp33–46

Szekely, A. (1993b) 'How to accommodate an uncertain future into institutional responsiveness and planning: The case of Mexico and the United States', *Natural Resources Journal*, vol 33, pp397–403

Udall Center for Studies in Public Policy (1993) *State of the U.S. – Mexico Border Environment: Report of the U.S. EPA U.S.–Mexico Border Environmental Plan Public Advisory Committee*, Udall Center, University of Arizona, Tucson, AZ

Varady, R. G. and M. D. Mack (1995) 'Transboundary water resources and public health in the U.S.–Mexico border region', *Journal of Environmental Health*, vol 57, pp8–14

Varady, R. G., H. Ingram and L. Milich (1995) 'The Sonoran Pimería Alta: Shared environmental problems and challenges', *Journal of the Southwest*, vol 37, pp102–122

Varady, R. G., D. Colnic, R. Merideth and T. Sprouse (1996) 'The U.S.–Mexico Border Environment Cooperation Commission: Collected perspectives on the first two years', *Journal of Borderlands Studies*, vol 11, pp89–119

# 16

# Climate Change and International Water: The Role of Strategic Alliances in Resource Allocation

*Ariel Dinar*

Scientists have gained a great deal of understanding on how the climate is changing. They are now confident that the 'global average net effect of climate since 1750 has been one of warming' (IPCC, 2007, p3). Most atmospheric scientists concur that '[A]t continental, regional and ocean basin scales, numerous long-term changes in climate have been observed. These include changes in arctic temperatures and ice, widespread changes in precipitation amounts, ocean salinity, wind patterns and aspects of extreme weather including droughts, heavy precipitation, heat waves and the intensity of tropical cyclones' (IPCC, 2007, p7).

The Fourth Assessment Report (IPCC, 2007, pp1–10) is much more assertive regarding climatic results. It suggests that 'Warming of the climate system is unequivocal, as is now evident from observations of increases in global average air and ocean temperatures, widespread melting of snow and ice, and rising of global average sea levels. The 100-year linear trend (1906–2005) of 0.74 [0.56–0.92]°C is larger than the corresponding trend of 0.6 [0.4–0.8]°C (1901–2000) given in the Third Assessment Report.'

These higher world temperatures are expected to increase the hydrological cycle activity leading to a general change in precipitation patterns and increase in evapotranspiration. 'There is high confidence that by mid-century, annual river runoff and water availability are projected to increase at high latitudes (and in some tropical wet areas) and decrease in some dry regions in the mid-latitudes and tropics. There is also high confidence that many semi-arid areas (e.g. Mediterranean basin, western US, southern Africa and northeast Brazil) will suffer a decrease in water resources due to climate change' (IPCC, 2007, p8). The Fourth Assessment Report further verifies the findings from the Third Assessment Report that states: 'One major implication of climate change for agreements between competing users (within a region or upstream versus downstream) is that allocating rights in absolute terms may lead to further disputes in years to come when the total absolute amount of water available may be different' (IPCC, 2001, section 4.7.3). Climate change is expected to increase heat, reduce/increase precipitation, and also increase water supply variability both intra- and inter-annually.

Some experts emphasize that climate change can lead to conflict between states who share international bodies of water because of the possibility of dwindling water supplies (Gleditsch et al, 2007). On the other hand, some experts suggest that further exacerbation in the water situation may even open the door to new water allocation opportunities between these riparians (ESCAP, 1997). Game theory, a mathematical-economic approach used for strategic decision making, is a tool that allows basin riparians to address climate change consequences in the water sector and assess the viability of these various potential arrangements.

The purpose of this chapter is to demonstrate the use of game theory to set and identify options for cooperation. These options are needed in case two or more riparian countries face increased variability in water supply within a shared river basin that might jeopardize their existing water agreements. Finding a partner riparian with which to share the risk of a variable water supply is a strategic decision. In what follows, I will demonstrate that under increased variability of water supply, the cooperative approach, in the form of investment in infrastructure, may be preferred over any individual solution. Furthermore, I will also argue that under variable water supply conditions, partial coalition cooperation may be preferred to the grand coalition.

The second section reviews the scientific basis for the climate–hydrology nexus that affects the flow regime in river basins. The next section develops an analytical framework to help assess the possibility of addressing the variability in the water supply via unilateral or joint arrangements, both through investment in infrastructure and by institutional measures. The subsequent section summarizes several relevant basin cooperation cases where some, if not all, basin riparians may or may not have been able to identify strategic alliances and pull themselves away from the unsustainable situation created by water supply variability. A more illustrative example is then provided, where sub-basin and basin-wide arrangements are compared and the chapter concludes by emphasizing the lessons learned and highlighting areas still open for further research.

# Climate–hydrology nexus and river basin water regimes

The hydrology of river basins is sensitive to changes in climatic conditions. Anthropogenic-induced climate change is expected to affect water resource cycles significantly. However, the stochastic nature of the changes in the water cycle is uncertain. As a result, much of the work by hydrologists, planners, engineers and economists has been brought to the global forefront in an attempt to assess the vulnerability of water supply systems to climate change and variability (Frederick et al, 1997; Frederick, 2002; Miller and Yates, 2005; Smith and Mendelsohn, 2006).

## Water runoff effects

A useful explanation of the scientific interaction between climate change and the hydrological cycle can be found in Miller and Yates (2005). They suggest that global climate change is expected to modify the hydrologic cycle by affecting the amount, intensity and temporal distribution of precipitation. Warmer temperatures will affect the amount of winter precipitation in the form of rain or snow, the amount stored as snow and ice, and its melting dynamics. Long-term climatic trends could trigger vegetation changes that would alter a region's water balance. In forest areas, the combination of warmer temperatures and drying soils caused by snow melting earlier than usual or longer droughts can lead to more frequent and extensive wildfires. When this occurs, land cover and watershed runoff characteristics may change quickly and dramatically as wildfires reduce forest cover and thereby alter the runoff response. Less dramatic but equally important, changes in runoff can affect the transpiration of plants, altered by changes in soil moisture availability, as well as plant responses to elevated $CO_2$ concentrations. In addition, changes in the quantity and quality of water percolating to groundwater storage will result in changes in aquifer levels and quality, in base flows entering surface streams, and in seepage losses from surface water bodies to the groundwater system (Miller and Yates, 2005, p37).

A comprehensive assessment of the stock of water hydrology–climate studies from around the world is provided in IPCC (1996a, 1996b) and IPCC (2001). The findings in IPCC (2001) suggest that:

> In general, the patterns found are consistent with those identified for precipitation: Runoff tends to increase where precipitation has increased and decrease where it has fallen over the past few years. Flows have increased in recent years in many parts of the United States, for example, with the greatest increases in low flows... Variations in flow from year to year have been found to be much more strongly related to precipitation changes than to temperature changes... There are

304  *Interaction between Policy and Strategy*

> *some more subtle patterns, however. In large parts of eastern Europe, European Russia, central Canada ..., and California ..., a major – and unprecedented – shift in streamflow from spring to winter has been associated not only with a change in precipitation totals but more particularly with a rise in temperature: Precipitation has fallen as rain, rather than snow, and therefore has reached rivers more rapidly than before. In cold regions, such as northern Siberia and northern Canada, a recent increase in temperature has had little effect on flow timing because precipitation continues to fall as snow.* (IPCC, 2001, section 4.3.6.1)

However, the IPCC (2001) concludes that

> *it is very difficult to identify trends in the available hydrological data, for several reasons. Records tend to be short, and many data sets come from catchments with a long history of human intervention. Variability over time in hydrological behaviour is very high, particularly in drier environments, and detection of any signal is difficult. Variability arising from low-frequency climatic rhythms is increasingly recognized, and researchers looking for trends need to correct for these patterns. Finally, land-use and other changes are continuing in many catchments, with effects that may outweigh any climatic trends.* (IPCC, 2001, section 4.3.6.1)

Specifically, not all river basins are affected by climate in the same way. Differences have been observed both within a given country or even a state, such as in Miller et al (2006), who studied six basins in Central-Northern California. While the trend of the impact of the various future climate scenarios on the six water systems is similar, it is evident that the six basins differ in their level of sensitivity to the same expected changes in temperature and precipitation.

A comparison between five international river basins (the Nile, Zambezi, Indus, Mekong and Uruguay) in Riebsame et al (2002) suggests that basins in drier regions (e.g. Nile, Zambezi) would be most hydrologically sensitive to the climate change scenarios that were used in the simulation. Hydrological sensitivity of the Indus and Uruguay basins is described as moderate and that of the Mekong is described as low. The adaptation scenarios that have been considered in the basins include mainly investment in larger storage and adjustments to allocation regimes. However, because these two adaptation interventions are associated with transboundary property rights, the authors correctly identify that climate change could likely lead to either cooperation or conflict among the basin riparians.

Arora and Boer (2001) analysed, using simulations, 23 basins, among them 12 that are international. Applying one climate change scenario they can simulate future mean annual discharges and mean annual floods in 2100. Findings suggest that rivers in middle to high latitude are expected to face between +67 and −16 per cent change in mean annual discharge and between +68 and −28 per cent change in mean annual flood. On the other hand, rivers in tropical and low latitudes are expected to face between +5 and −79 per cent change in mean annual discharge and between +26 and −74 per cent change in mean annual flood. These findings necessitate a serious consideration of water management adaptation, including a possible adjustment of infrastructure. A recent study (Palmer et al, 2008) evaluated globally the future (2050, A2 Scenario) impact of climate change on the discharge of major dammed rivers. The findings are in agreement with Arora and Boer (2001), but much more comprehensive in coverage. They then evaluate a set of river basin management strategies (Bernhardt et al, 2005) to propose a range of interventions that may mitigate the future impact of climate change and man-made development on river flow.

Similar findings are suggested by Milly et al (2005), namely increase of runoff (10–40 per cent) by 2050 in high latitude basins in north America and Europe, and in certain low latitude basins such as La Plata and basins in western Africa. A decrease in runoff between 10–30 per cent is expected in basins in southern Europe, the Middle East and basins in mid-latitude western regions of north America and southern Africa.

This chapter will focus on three international basins. A closer look at three basins addressed in this chapter, the Ganges, Jordan and Aral Sea basins,[2] suggest three different long-term trends of water flow. The Ganges shows a decline in water flow over the last 50 years (yet has the lowest inter-annual variability among the three – coefficient of variation of 0.16). The Jordan water flow, over the last 35 years, has neither increased nor decreased (yet has the highest inter-annual variability among the three – coefficient of variation of 0.37) and the Aral Sea flow has increased over the last 93 years (with a moderate inter-annual variability – coefficient of variation of 0.21).

## Economic impacts of water supply variability

Increased flow variability entails more extreme and frequent events of drought (although with no pattern of excessive and deficient flows). Documented economic damages (both from excessive and deficient flows) include direct and indirect effects (from deficient flows) of loss of yields in irrigated agriculture, hydropower generation, fisheries, biodiversity, loss of industrial production and loss of agricultural production from water quality degradation. Losses from excessive flows include flood damages to crops, roads, other infrastructure, water quality, etc.

An estimated cost (Mogaka et al, 2006) of the 1997/1998 El Niño floods and the 1999/2000 La Niña drought in Kenya suggests $870 million in damage from floods (infrastructure, public health, loss of crops) and $2.8 billion in damage from drought in loss of crops, livestock, forest fires, fisheries, hydropower, industrial production and water supply. Of the $2.8 billion, $2.3 are from loss of hydropower generation (including loss of manufacturing production). Brown and Lall (2006) observe in a simple model that national GDP is negatively correlated with rainfall variability. They arrive at the conclusion that storage infrastructure would mitigate the negative impacts of rainfall variability (both intra-annual and inter-annual) on water supply, food production and economic growth.

However, Quiggin and Horowitz (2003) arrive at some quite convincing arguments against the tendentious conclusion that investment in storage infrastructure is the solution to inter-temporal water availability. They argue that the value of dams and reservoirs, as well as irrigation systems and hydropower facilities, is a function of present climate (precipitation, evaporation and growing conditions) in the basin. Because all of these parameters will be affected, in an unpredictable way, as climate changes, the locations of these infrastructures has to be changed and the investment costly. There is no evidence that the future distribution of rainfall, and thus of runoff, from future climate change would be any more or less suitable for the production of irrigated crops or hydropower than the present distribution. Therefore, it could be that existing dams may require costly redesign or replacement, making investment in infrastructure, as an adaptation intervention, prohibitively expensive.

Are existing treaties among basin riparians resilient to water supply variability? As can presently be seen from water supply variability impacts, some treaties do and some do not exhibit resilience. A recent crisis in the Aral Sea, in spite of an existing agreement to deal with water supply variability, is presented in Box 16.1.

---

**BOX 16.1 TAJIK ENERGY CRISIS DEEPENS AS UZBEKISTAN CUTS DOWN NATURAL GAS SUPPLIES**

Tajikistan has experienced the coldest winter in 25 years. The country has plunged deeper into a winter energy crisis. In January 2008, neighbouring Uzbekistan cut down, by one-third, natural gas supplies due to a US$7 million debt (Bishkek Treaty 1998).

The energy crisis forced many residents to go without heat. The national gas company, TajikGaz, will only continue to supply natural gas from emergency reservoirs, to strategic facilities and those enterprises and private consumers who have no unpaid gas bills. Only a handful of key enterprises, as well as dairies and bakeries, have been spared from the cut. Power is also maintained in hospitals and schools wherever possible; rural households are provided with only three hours of electricity daily, while districts in the capital Dushanbe, are subjected to rolling blackouts. Power shortages have been caused by a sharp drop in water levels at the Nurek reservoir that powers an important hydroelectric plant.

*Source:* The Associated Press, 2008

The following analytical framework will provide an understanding of the relationship between riparians under variable water supply situations and the options they face.

## Analytical framework

So far I have discussed the links between climate change and water resources, and their likely impacts. To better understand the concept of strategic alliance, I will develop a simple analytical framework. Assume a basin that is shared by $N$ riparian states. Each state has different water resources it may use on its territory, in addition to the shared basin. The water basin is allocated between the $N$ riparian states, based on an existing treaty that was previously signed between these states. As is the case in most treaties, water is allocated in a fixed proportion between the riparians (Wolf et al, 1999; Kilgour and Dinar, 2001). Water is used for joint projects (e.g. hydropower production, environmental flows), and/or used unilaterally on each riparian territory (e.g. for irrigation, hydropower, urban supply). For the purpose of this discussion it is not important how water is used beyond the allocation stage. For simplicity assume that only annual flows are the subject of the allocation. No consideration of monthly or weekly flows are given in the basin treaty (this information is actually very important in real life treaty allocations).

Once a riparian state is faced with a given allocation, investments (infrastructure and domestic allocations among sectors) are made and the entire water system is designed to meet these allocations. Changes to the original basin allocation are difficult to accommodate by the riparian states in the short run because they necessitate altering fixed infrastructure assets and regulations, which is associated with high cost. Therefore, flow variability may pose harm to the shared basin riparian states. This will be discussed in the following subsection.

## A deterministic world

Assume that annual flow in the shared basin is $F$ (m³/year) and that the treaty allocates it in full between the $N$ riparian states (environmental flows are not assigned any allocation). Since treaties refer to long-term annual flows, $F = \overline{F}$, where $\overline{F}$ is the long-term mean annual flow in the shared basin. Let $f_i$ be the annual allocation of water in the shared basin to riparian $i$, $i \in N$; $\sum_{i \in N} f_i \leq F$. Each riparian then allocates the water internally/domestically among competing uses, using their own criteria. Let $f_{ij}$, $j=1, 2, ..., J$; $J=\{\text{hydropower, irrigation, drinking, ...,}\}$, be the internal use of state $i$'s allocation from the shared basin, with $\sum_{j=1}^{J} f_{ij} \leq f_i; \forall i \in N$. Assume that water production functions for each use are known in each riparian state. Each riparian state has a payoff function from its internal use of the shared basin allocation, given the treaty parameters that are based on the long-term mean flow $v_i = \sum_{j=1}^{J} h_j^i(f_{ij} | \overline{F})$, $\forall i \in N$.

Assume further that each riparian state has also other sources of, say, capital ($x$) and water ($w$) that are outside the shared basin and are used for economic activities in regions other than the shared basin. The production functions of these resources are also known, and the state's payoff function is $u_i = \sum_{j=1}^{J} k_j^i(x_{ij}^d, w_{ij}^d)$, $\forall i \in N$. A state is a rationale decision maker and maximizes its resources. Therefore, a state payoff ($S$) is:

$$S_i = Max \sum_{j=1}^{J} \{h_j^i(f_{ij}|\overline{F}) + k_j^i(x_{ij}^d, w_{ij}^d)\}, \forall i \in N \quad (1)$$

Subject to:

$$\sum_{j=1}^{J} f_{ij} \leq f_i; \forall i \in N, \quad (1a)$$

$$\sum_{j=1}^{J} x_{ij}^d \leq X_i^d, \forall i \in N, \quad (1b)$$

$$\sum_{j=1}^{J} w_{ij}^d \leq W_i^d, \forall i \in N. \quad (1c)$$

For simplicity assume that only these constraints are considered in the optimization problem of state $i$.

I argue that the basin riparian states have incentives to cooperate. The treaty among the riparian states is one type of cooperation. It takes place in the shared basin through agreement on a formula to allocate the flow in that basin between them. The basin-wide profit $B$ is:

$$B = Max \sum_{i \in N} \sum_{j=1}^{J} \{h_j^i(f_{ij}|\overline{F}) + k_j^i(x_{ij}^d, w_{ij}^d)\} \quad (2)$$

Subject to:

$$\sum_{i=1}^{N} f_i \leq F \quad (2a)$$

$$\sum_{j=1}^{J} f_{ij}^d \leq f_i; \forall i \in N \quad (2b)$$

$$\sum_{j=1}^{J} x_{ij}^d \leq X_i^d, \forall i \in N \quad (2c)$$

$$\sum_{j=1}^{J} w_{ij}^d \leq W_i^d, \forall i \in N. \quad (2d)$$

For simplicity assume that only these constraints are effective in the optimization problem of state $i$ and various cooperative agreements between subsets of states in the shared basin.

The model in (1)–(2d) suggests that $\overline{F}$, $X$ and $W$ are the resources that affect the potential payoff in the basin. Remember that $\overline{F}$ is a joint resource while $X$ and $W$ are resources owned individually by each riparian state. In the case where the parameter in the basis of the treaty is the shared basin flow, $\overline{F}$, then, in most cases, if not in all known treaties (see below under 'Actual management options'), the riparian states only cooperate over the resources in the shared basin – in our case the flow of the shared basin $\overline{F}$.

Referring to the set of $N$ riparian state, let the individual coalition comprised of each riparian state optimizing its payoff without cooperation be $\{i\}$, $\forall i \in N$. The payoff to coalition $\{i\}$ is given by the solution of (2)–(2c) as $v(\{i\}) = z_i(\overline{F}, X_i, W_i); \forall i \in N$. In addition, for any coalition of states $n \subseteq N$, where $n$ includes a subset of all the riparians $N$ of the shared basin, one has $v(\{n\}) = z_n(\overline{F}, \sum_{k \in n \subseteq N} X_k, \sum_{k \in n \subseteq N} W_k)$, and finally, for the grand coalition of the shared river riparians, $N$, one obtains $v(\{N\}) = z_N(\overline{F}, \sum_{k \in N} X_k, \sum_{k \in N} W_k)$.

## Introducing flow variability

Assume that flow in the domestic basins is deterministic,[3] and that the flow in the international basin is variable. (The analysis is similar in the case that all basins face variability, but is easier to demonstrate with only one basin facing variable flow.) Let the flow variability be represented by a departure from the annual mean, $\overline{F}$. I measure variability by dividing actual flow by mean annual flow $\theta = F/\overline{F}$. If $\theta = F/\overline{F} > 1$ then the flow exceeds mean annual values and leads to damage or loss from floods and from not being able to capture all water. If $\theta = F/\overline{F} < 1$ then the flow is below mean and there is damage from crop loss, energy underproduction etc. Therefore, $h$ is a quadratic function in $\theta$.

The basin-wide profit $B$ is:

$$B = Max \sum_{i \in N} \sum_{j=1}^{J} \left\{ h_j^i(f_{ij} \mid \theta \overline{F}) + k_j^i(x_{ij}^d, w_{ij}^d) \right\} \quad (3)$$

Subject to:

$$\sum_{i=1}^{N} f_i \leq \theta F \quad (3a)$$

$$\sum_{j=1}^{J} f_{ij}^d \leq f_i; \forall i \in N \quad (3b)$$

$$\sum_{j=1}^{J} x_{ij}^d \leq X_i^d, \forall i \in N \quad (3c)$$

$$\sum_{j=1}^{J} w_{ij}^d \leq W_i^d, \forall i \in N, \quad (3d)$$

and various cooperative agreements between subsets of states in the shared basin.

As it appears in the deterministic case, the payoff to coalition $\{i\}$ is given by the solution of (3)–(3c) as $v(\{i\}) = s_i(\theta \overline{F}, X_i, W_i); \forall i \in N$. In addition, for any coalition of states $n \subseteq N$, where $n$ includes a subset of all the riparians $N$ of the shared basin, one has $v(\{n\}) = z_n(\overline{F}, \sum_{k \in n \subseteq N} X_k, \sum_{k \in n \subseteq N} W_k)$, and finally, for the grand coalition of the shared river riparians $N$, one obtains $v(\{N\}) = z_N(\overline{F}, \sum_{k \in N} X_k, \sum_{k \in N} W_k)$. Again, $s_{\bullet}$ is a quadratic function of $\theta$.

I claim (based on Just and Netanyahu, 1998) that because of the basin-level externalities steaming from the variation in water supply, $\left|\frac{\partial s_n}{\partial \theta}\right| \leq \left|\frac{\partial s_N}{\partial \theta}\right|$. This will push the basin riparians to seek solutions to the situations resulting from water supply variations in partial coalitions rather than the grand coalition. They must then rely on resources that exist outside the basin that may be subject to variable water supply conditions too. In the next section I provide examples from three actual basins, the Ganges, the Jordan and the Aral Sea.

## Actual management options

While the economic impacts associated with the anecdotal information in Box 16.1 are not available, I anticipate a variety of impacts resulting from water supply variability. Some impacts may be devastating (Kahn, 2005) and others costly (e.g. Gleick and Adams, 2000). In the case of domestic basins, possible adjustments to management of the water supply variability include very specific options such as modifications to operational regimes of existing water systems, new supply options and demand management (Gleick and Adams, 2000). In the case of international basins, the set of options may be greater, although its implementation could be more complicated and costly.

To demonstrate an actual range of possible options with which riparians respond to water supply variability I will refer to three cases, the Ganges–Brahmaputra–Meghna Basin (with focus on the Farakka Barage) shared by Bangladesh and India, the Jordan Basin (1994 treaty) shared by Israel and Jordan (for the sake of this chapter I focus only on the signatories to the 1994 treaty) and the Aral Sea Basin (regulated by several bi- and multilateral treaties minus a basin-wide treaty) between Kazakhstan, Kyrgyz Republic, Tajikistan, Turkmenistan and Uzbekistan.

### Ganges–Brahmaputra–Meghna Basin

The Farakka Barrage is a contentious point in the hydro politics of India and Bangladesh over the Ganges–Brahmaputra–Meghna water. Until 1996 the two countries regulated the allocation of flow at Farakka between themselves, based on interim agreements (in 1975, 1977 and 1985) that were based on fixed amounts. These fixed amount allocations, coupled with a high level of flow variability (Figure 16.1), and the asymmetric information on river flows upstream to the Farakka, led to devastating floods[4] and droughts affecting areas mainly downstream to the Farakka. The 1996 treaty that was signed between India and Bangladesh (for a period of 30 years) marks a major change in the allocation schemes compared to prior agreements. In 1996 differential percentage allocations in the dry (lean) season were based on trigger flow values. The specific allocations were supposed to give Bangladesh a priority (augmentation) as a downstream riparian in event of low flows. This is a very sophisticated treaty that is both equitable and adjustable to hydrological situations.

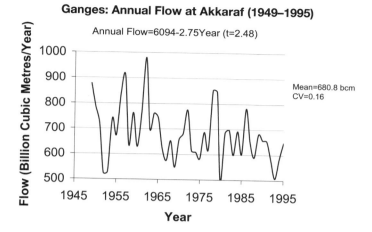

**Figure 16.1** *Annual flow in the Ganges at Akkaraf*

However, the 1996 treaty did not prevent the upstream riparian – India – from diverting water from the Farakka. This lead to a slow and steady decline of flows to Farakka (Nishat and Faisal, 2000). Comparison of the performance of the 1977 agreement with that of the 1996 treaty over the flows of the Ganges at Farakka suggests that the 1996 treaty has not been good for Bangladesh. Because the 1996 treaty was signed for a period of 30 years, its modification now is probably not feasible. Instead, a strategic alliance between Bangladesh and India on other issues would be much more useful. There is still a possibility for collaboration on issues of bank erosion, salt-water intrusion and ecological damages. 'In fact, the major damage that Bangladesh has encountered through the common river [system with India] is not flood, but gradual distruction of the ecosystem of the south western part' (Nishat and Faisal, 2000, p300). The goal of the Gorai Restoration Project is to allow water to flow during the lean season through the Gorai in Bangladesh towards southern Bangladesh and to prevent an increase in salinity in parts of the river that are left without water. Having India participate in funding the dredging of the river would be considered highly cooperative, compensate for low flow and be just as important as augmenting the flow at Farakka.

## Jordan Basin

The 1994 peace treaty between Jordan and Israel provided for water sharing and alleviation of water shortages in the Jordan River, shared between these two riparians. Recognizing the relative advantage, experience and other water resources available to each, the treaty also includes clauses. Several of these clauses provide for strategic cooperation, such as an exchange of groundwater pumping rights for an equivalent desalinized amount, increased operational storage and utilization of existing storage to capture water flows otherwise lost (Haddadin, 2000, p281).

312  *Interaction between Policy and Strategy*

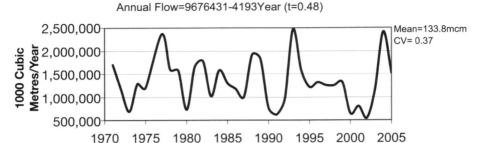

**Figure 16.2** *Annual flow in the Jordan at Lake Terenik*

Figure 16.2 shows that water flow entering Lake Tenerik, the main natural reservoir of the Jordan River, is highly variable, with an annual coefficient variation of 37 per cent. Droughts in the region are frequent and as a result the parties have not been able to deliver their treaty quota to each other. In an analysis of future scenarios, Dinar (2004, p220) asserts that 'no matter what the final allocation Jordan is entitled to under the Treaty the Kingdom would still have a water deficit. Non traditional means of augmenting supply will be needed.'

To address the natural variability of water flow and future needs, several strategies have been considered. These include wastewater reclamation in both Israel and Jordan, large scale seawater reclamation facilities in the Mediterranean and the Red Sea, and desalinization of fossil brackish groundwater in the Israel and Jordan inland areas (Jordan Valley, Arava and Hisban area (Dinar, 2004)). In addition to unilateral projects that should augment the Basin water supply, there are also joint projects such as the Red–Dead Sea project that will pump water from the Red Sea for desalinization. Brine would then be discharged to the Dead Sea and the desalinized water would be supplied to Jerusalem, Hebron and Amman.

## Aral Sea basin

The Aral Sea Basin, shared by five asymmetric states, demonstrates the ability of weak, small upstream states to contemplate strategic arrangements for water and water-related resource exchange. Two upstream states, the Kyrgyz Republic and Tajikistan, have succeeded in convincing the three downstream states to buy their hydropower. This is produced in the summer when water is released for irrigation in the downstream states irrigated agriculture. Still, the two upstream states face difficulties producing electricity during the winter due to low levels of water in the reservoirs.

In the 1992 Bishkek Agreement, supplies of oil, coal and natural gas are exchanged during the winter between the upstream and downstream states for water being released during the spring and summer. The problem is during drier years; when upstream states release less water, they receive less fuel. During wet years, when downstream states irrigate less, they often return less fuel in the winter (Dinar, 2005, p151). The evidence from the 2007/2008 winter (Box 16.1) demonstrates the vulnerability of the Bishkek Treaty and its sensitivity to climate variability. Therefore, widening the scope of possible cooperation in this basin may be needed to accommodate water supply variability. The next section illustrates the structure of cooperation, using a basin model.

## More-than-an illustrative example

To illustrate the importance of strategic alliances in the presence of flow variability, I will refer to a simplified version of the Aral Sea Basin. A compressed version of the Aral Sea Basin is modelled after the Lara River Basin model.[5] The features of the model are also explained in Dinar et al (2007, Annex 3).

The Lara Basin model includes three riparian states, A, B and C. It evaluates payoff for single coalitions {A}, {B} and {C}. It allows evaluating the payoff of various regional arrangements between sub-coalitions {A, B}, {A, C} and {B, C}, and the payoff for the grand coalition of riparian states {A, B, C}. To save space I will skip showing the model equations and will focus solely on the description of the geography, hydrology and economics of the model, as well as on the various regional arrangements that are modelled in this chapter. For simplicity, the environmental damage to the Aral Sea is not considered. Only the amount of flow that reaches the Aral Sea is reported.[6] The three riparian states have different water economies and also vary in size and economic power.

## Description of the riparian states and their inter-links

State A is an upstream riparian, where the majority of flow in the basin is generated by snow and glacier melt runoff from its territory. It has a small economy, using the river water mainly for hydropower generation (Reservoir A in Figure 16.3) for heat in the winter. Excess energy production can be sold to downstream states. Experiencing increasing population and energy demand, it is concerned about meeting winter energy needs once the demand exceeds the hydroelectric capacity of Reservoir A. Thus, State A would negotiate with States B and C over transfers (cash payment or equivalent energy sources – electricity, natural gas or fossil fuels) for releasing water for irrigation during this period. Clearly, water supply variability is detrimental to State A's economy.

State B is a midstream riparian with a major agricultural economy and a large demand for irrigation water. Its irrigated agriculture depends on the multi-year

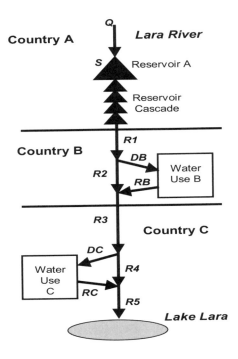

**Figure 16.3** *The Lara River basin geography and hydrology*

storage capacity of Reservoir A to supply the demand for irrigation water during the summer months for drier-than-average years. Thus, State B is interested in negotiating with State A for an appropriate storage and release regime for the reservoir to meet its irrigation water, which may be reciprocal to the needs of State A. State B has energy resources (e.g. coal) with which to make fuel payments to State A in the winter in compensation for the irrigation season releases. Under low to normal flow conditions in the river, State B has the ability to divert all of the water out of the river and use it for irrigation, thus leaving no water in the river for State C. However, under high flow conditions, due to capacity constraints, some unused flow passes downstream to State C.

State C is a downstream riparian (the Aral Sea is not included in the model). It also has an agricultural economy with an associated irrigation water demand. State C diverts water from dedicated releases out of Reservoir A, when they are allowed to bypass State B territory, as well as agricultural return flows from State B to the river. State C would like to negotiate with States A and B for adequate flows to supply its irrigation demand. State C also has energy resources (e.g. natural gas) to make fuel payments to State A in the winter in compensation for the irrigation season releases.

All water that is not diverted by State C and agricultural return flow goes to Lara Lake, which is downstream of State C where the river terminates. In this chapter I will not address the sensitivity of Lake Lara to the quantity and quality of the water that enters its area. There are also physical links between the states that cause the actions of State A to affect countries B and C, and actions of B to affect C. Under Coalition {A}, State A releases sufficient water to cover only its internal power demands. States B and C receive the resulting water from these power releases – named 'residual' water. Water flow in excess of Reservoir A's capacity is spilled and made available to States B and C. State B can affect the amount of water available to State C by diverting more water from the river and by changing water regimes to release less return flow (the portion of water that State C applies on its fields that returns to the river downstream of C). State C cannot affect any of the riparians with any of its actions.

## State of nature and regional arrangements among riparian states

Figure 16.4 suggests that the flow in the Lara Basin is variable and has an upslope trend. Inter-annual variation is significant with a coefficient of variation (CV) equal to 21 perent. The (actual) 100-year data used to plot the flow fluctuation in Figure 16.4 suggests five climatic scenarios: extreme dry, dry, average, wet and extreme wet, with corresponding values of 6525, 8900, 11,900, 14,900 and 20,725 million cubic metres (MCM) per year (probability distribution of these values are 0.01, 0.15, 0.67, 0.13 and 0.03 respectively). In addition, a trend analysis suggests that in 2020 the wet climate (14,900) will become the average. Population increase is also included in the 2020 average scenario.

What are the possible arrangements each riparian state should undertake in order to increase its payoff under various climatic/flow scenarios? I will distinguish between energy swap arrangements and unilateral investments that have both

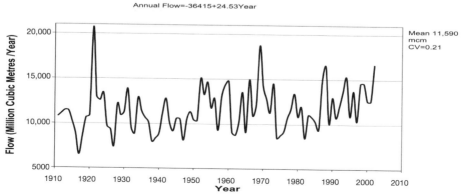

**Figure 16.4** *Annual flow in the Aral Sea at Lugotkot*

Table 16.1 Regional arrangements

| | | Energy swap {A} | {B} | {C} | {AB} | {AC} | {BC} | {ABC} |
|---|---|---|---|---|---|---|---|---|
| No investment | | No energy swap agreement; No investment | No energy swap agreement; No investment | No energy swap agreement; No investment | Energy swap between A and B | Energy swap between A and C | No energy swap agreement | Energy swap between A, B and C |
| Investment in water infrastructure | {A} | No energy swap agreement; A invests in reservoir improvements | No energy swap agreement; A invests in reservoir improvements | No energy swap agreement; A invests in reservoir improvements | Energy swap between A and B; A invests in reservoir improvements | Energy swap between A and C; A invests in reservoir improvements | No energy swap agreement; A invests in reservoir improvements | Energy swap between A, B and C; A invests in reservoir improvements |
| | {B} | No energy swap agreement; B invests in irrigation infrastructure | No energy swap agreement; B invests in irrigation infrastructure | No energy swap agreement; B invests in irrigation infrastructure | Energy swap between A and B; B invests in irrigation infrastructure | Energy swap between A and C; B invests in irrigation infrastructure | No energy swap agreement; B invests in irrigation infrastructure | Energy swap between A, B and C; B invests in irrigation infrastructure |
| | {C} | No energy swap agreement; C invests in irrigation infrastructure | No energy swap agreement; C invests in irrigation infrastructure | No energy swap agreement; C invests in irrigation infrastructure | Energy swap between A and B; C invests in irrigation infrastructure | Energy swap between A and C; C invests in irrigation infrastructure | No energy swap agreement; C invests in irrigation infrastructure | Energy swap between A, B and C; C invests in irrigation infrastructure |
| | {AB} | No energy swap agreement; A and B invest in water infrastructure | No energy swap agreement; A and B invest in water infrastructure | No energy swap agreement; A and B invest in water infrastructure | Energy swap between A and B; A and B invest in water infrastructure | Energy swap between A and C; A and B invest in water infrastructure | No energy swap agreement; A and B invest in water infrastructure | Energy swap between A, B and C; A and B invest in water infrastructure |

Investment in water infrastructure

| | | | | | | | |
|---|---|---|---|---|---|---|---|
| {AC} | No energy swap agreement; B and C invest in water infrastructure | No energy swap agreement; A and C invest in water infrastructure | No energy swap agreement; A and C invest in water infrastructure | Energy swap between A and B; A and C invest in water infrastructure | Energy swap between A and C; A and C invest in water infrastructure | No energy swap agreement; A and C invest in water infrastructure | Energy swap between A, B and C; A and C invest in water infrastructure |
| {BC} | No energy swap agreement; B and C invest in water infrastructure | No energy swap agreement; B and C invest in water infrastructure | No energy swap agreement; B and C invest in water infrastructure | Energy swap between A and B; B and C invest in water infrastructure | Energy swap between A and C; B and C invest in water infrastructure | No energy swap agreement; B and C invest in water infrastructure | Energy swap between A, B and C; B and C invest in water infrastructure |
| {ABC} | No energy swap agreement; A, B and C invest in water infrastructure | No energy swap agreement; A, B and C invest in water infrastructure | No energy swap agreement; A, B and C invest in water infrastructure | Energy swap between A and B; A, B and C invest in water infrastructure | Energy swap between A and C; A, B and C invest in water infrastructure | No energy swap agreement; A, B and C invest in water infrastructure | Energy swap between A B and C; A, B and C invest in water infrastructure |

positive and negative externalities on the rest of the riparian States. First, State A can establish energy swap agreements solely with State B or State C, and a multilateral agreement with both. These sub-basin and basin-wide agreements can take place irrespective of any other regional or unilateral arrangement between any of the riparian states. Second, State A can invest in increasing the capacity of Reservoir A, and both State B and C can invest in improving its irrigation system. These investments can be done for any combination of riparian states.[7]

The energy swap arrangement for States B and C guarantees adequate water supply for irrigation in exchange for fuel supply to State A, using the existing infrastructure. The investment programme of the water systems in the three riparian states allows them to become even more efficient and to cope with situations of harsh climates (low flow). Investment in State A allows larger storage in high flow years. The various arrangements are summarized in Table 16.1. All parameter values and model results, except for payoffs, are not reported in the chapter (they can be obtained upon request). These hypotheses will be checked using the Lara Basin Model. One objective is to verify the hypothesis that under extreme climate situations there is an increased likelihood for strategic alliances, which prefer sub-basin coalitions rather than the grand coalition.

## Selected results of the Lara Basin arrangements under flow variability scenarios

As previously indicated, the results presented will include only payoffs to the states from various regional arrangements under the various climate scenarios. All parameters and model equations can be found in Dinar et al (2007).

Table 16.2 presents coalitional values under various climates without investments. Only energy swap arrangements are considered. Table 16.3 presents coalitional values under the 2020 flow estimate (and population values). Therefore, the 'No investment' column in Table 16.3 pertains to the same results under the 14,900MCM flow in Table 16.2.

**Table 16.2** *Coalitional payoffs for various climates without investments in water infrastructure (US$ billion)*

| Climate/flow (MCM/year) | 6525 | 8900 | 11,900 | 14,900 | 20,725 |
|---|---|---|---|---|---|
| Coalition {A} | 472 | 611 | 785 | 786 | 756 |
| Coalition {B} | 67 | 74 | 83 | 81 | 95 |
| Coalition {C} | 28 | 31 | 35 | 34 | 39 |
| Coalition {AB} | 674 | 876 | 1130 | 1384 | 1876 |
| Coalition {AC} | 557 | 723 | 932 | 1141 | 1522 |
| Coalition {BC} | 95 | 105 | 118 | 115 | 134 |
| Coalition {ABC} | 759 | 986 | 1272 | 1557 | 2087 |

**Table 16.3** *Coalitional payoffs for energy swap and various investment arrangements*

| Investment arrangement | No investment | A invests | B invests | C invests | A B invest | A C invest | B C invest | A B C invest |
|---|---|---|---|---|---|---|---|---|
| Coalition {A} | 903 | 903 | 903 | 903 | 903 | 903 | 903 | 903 |
| Coalition {B} | 95 | 93 | 121 | 95 | 119 | 93 | 121 | 119 |
| Coalition {C} | 40 | 39 | 36 | 50 | 35 | 49 | 45 | 45 |
| Coalition {AB} | 1384 | 1411 | 1499 | 1384 | 1527 | 1411 | 1499 | 1527 |
| Coalition {AC} | 1141 | 1167 | 1126 | 1186 | 1151 | 1211 | 1166 | 1192 |
| Coalition {BC} | 135 | 132 | 157 | 145 | 155 | 142 | 167 | 164 |
| Coalition {ABC} | 1557 | 1586 | 1657 | 1602 | 1687 | 1631 | 1697 | 1727 |

Note: US$ billion with annual flow = 14,900.

It is clear that climate affects the distribution of payoffs among the basin riparians. One way to suggest an agreement for the allocation of a regional payoff is to apply one of the cooperative game theory allocation schemes – the Shapley Value – to distribute these.[8] The results are presented in Tables 16.4 and 16.5.

**Table 16.4** *Grand coalition payoffs (US$ billion) and Shapley allocation shares to the states without investments in water infrastructure*

| Climate | Grand coalition payoff | A | B | C |
|---|---|---|---|---|
| 6525 | 759 | 0.74 | 0.18 | 0.08 |
| 8900 | 986 | 0.75 | 0.17 | 0.07 |
| 11,900 | 1272 | 0.76 | 0.16 | 0.07 |
| 14,900 | 1557 | 0.73 | 0.18 | 0.08 |
| 20,725 | 2087 | 0.69 | 0.20 | 0.10 |

**Table 16.5** *Grand coalition payoffs (US$ billion) and Shapley allocation shares to the riparians with investments in water infrastructure*

| Investment option | Grand coalition payoff | A | B | C |
|---|---|---|---|---|
| No investment | 1557 | 0.75 | 0.17 | 0.08 |
| A invests | 1586 | 0.75 | 0.17 | 0.08 |
| B invests | 1657 | 0.73 | 0.20 | 0.07 |
| C invests | 1602 | 0.74 | 0.17 | 0.09 |
| A B invest | 1687 | 0.73 | 0.20 | 0.07 |
| A C invest | 1631 | 0.74 | 0.17 | 0.09 |
| B C invest | 1697 | 0.72 | 0.20 | 0.08 |
| A B C invest | 1727 | 0.72 | 0.20 | 0.08 |

As suggested in Table 16.4, the share of total payoff going to State A is highest in river flows with 8900 and 11,900MCM/year and lowest in the highest and lowest values. The opposite occurs in the case of States B and C. These states' shares in the regional payoff are highest under the lowest and highest flow scenarios. Further, for the 14,900MCM flow scenario expected in 2020,[9] and under energy swap agreements and investment scenarios (Table 16.5) it is clear that State A's incremental payoff, as a result of its investment, is nil or very small. In the case of States B and C, their highest payoffs are obtained through an investment scenario that does not include A.

From Figure 16.5 it can be seen that basin-wide cooperation levels in the Lara basin are hill-shaped with regard to the level of flow. In both very dry and very wet climates the incentive for the riparians to cooperate is relatively low. This finding has important ramifications for future climate impact on flow in the basin. Rather, I would argue that more sub-basin cooperation is apparent, based on the opportunities each partial coalition may find in the basin.

In summarizing the results from the Lara Basin model, Figure 16.6 presents the proposed structure of impact of flow variability on strategic alliance in the Lara Case. While the grand coalition on all basin riparians has the highest payoff, these payoff values are valid only under a range of flow scenarios around the long-term mean. As flow values increase or decrease, the payoff is reduced. In high and low flow levels, smaller coalitions (sub-coalitions) are more attractive even though they produce a lower payoff. However they are still higher than the grand coalition, and probably more acceptable to the coalition members.

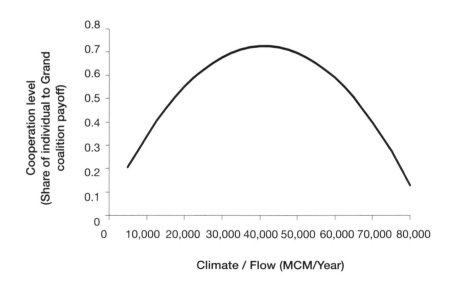

**Figure 16.5** *Cooperation level as a function of climate*

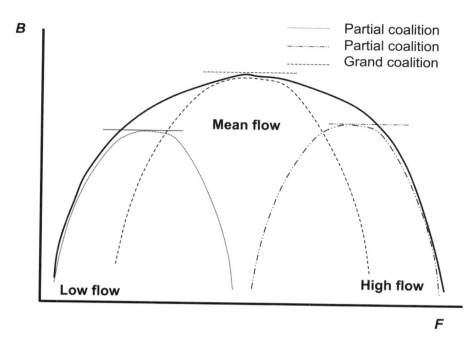

**Figure 16.6** *Switching coalitions as an adaptation to flow variability*

## Conclusion and further research needs

Many international treaties are signed under fixed water flow assumptions. With more evidence of increased future water variability, such treaties may be subject to instability, unless further cooperation is established. This chapter argues and demonstrates, through both anecdotal and illustrative modelling evidence, that under certain circumstances associated with the flow variability scenario and cooperation arrangements, as flow level departs from the long-term mean, attractiveness of cooperation moves away from the basin-wide to sub-basin arrangements. The practical idea behind this phenomenon is that with increased water supply variability, direct and indirect externality effects will impact the basin riparians. It will also be harder to cope with in a basin-wide arrangement compared with a sub-basin arrangement. With higher transaction costs for monitoring and managing situations of high water variability within a basin wide arrangement, riparians would find it much more attractive to pull resources from outside the basin to amend low flows, or to mitigate damages from high flows in specific locations in a much more efficient way.

Because of economic, infrastructure and geographical considerations, only part of the basin riparians can be engaged in such arrangements. Thus, even if the payoff might be lower than in the case where the mean flow allowed a basin-wide cooperation, sub-basin cooperation is still higher under extreme flow conditions, compared to that under mean long-term flow.

The analysis thus far is at an abstract level and does not take into account many needed prerequisite considerations. In addition to the engineering–economic considerations, one would have to incorporate institutional, economy-wide and equity considerations to allow for a balanced and comprehensive analysis of the cooperation options.

## Notes

1. The IPCC (2001) distinguishes between stream flow and runoff. In general terms, stream flow is the water within a river trunk ($m^3$/s) and runoff is the amount of precipitation that does not evaporate, expressed as an equivalent depth of water across the area of the catchment. Runoff can be regarded as stream flow divided by catchment area, although this is not the case in arid regions because of percolation.
2. The flow data used are taken from specific gauging stations that do not necessarily represent a full coverage of the basins.
3. By deterministic flow I mean flow distribution that is below a given variance.
4. Severe floods in 1987 and 1988 left 10 million people in Bangladesh homeless (Dinar et al, 2007).
5. The model code was developed by Daene McKinney and Rebecca Teasley. It can be run through software that is provided on a CD-ROM, with access to GAMS software that was generously provided by the GAMS Development Corporation.
6. Once the Aral Sea environmental degradation and its external impact on the basin riparians is included in the analysis, the propensity of the individual riparian states to form sub-basin coalitions, including individual ones will be even more enhanced.
7. State B and State C can also invest in storage on their land. I do not consider this option in the chapter.
8. The Shapley Value allocates to each state the average contribution of that state when joining a coalition of the other states, assuming that all state permutations are possible. For more on the Shapley Value see Shubik (1982).
9. For space-saving purposes, only one future climate scenario is presented. However, the results hold for the new low and high flows expected in 2020.

## References

Arora, V. K. and G. J. Boer (2001) 'Effects of simulated climate change on the hydrology of major river basins', *Journal of Geophysical Research*, vol 106, pp3335–3348

Bernhardt, E. S., M. A. Palmer, J. D. Allan and others (2005) 'Synthesizing US river restoration efforts', *Science*, vol 308, pp636–637

Brown C. and U. Lall (2006) 'Water and economic development: The role of variability and a framework for resilience', *Natural Resources Forum*, vol 30, pp306–317

Dinar, S. (2004) 'Water worries in Jordan and Israel: What may the future hold?' in A. Maraquina (ed.), *Environmental Challenges in the Mediteranean 2000–2050*, Kluwer Academic Press, Dordecht

Dinar, S. (2005) 'Treaty principles and patterns: Selected international water agreements as lessons for the resolution of the Syr Darya and Amu Darya Water Dispute', in H. Vogtmann and N. Dobretsov (eds), *Transboundary Water Resources: Strategies for Regional Security and Ecological Stability*, Springer, Berlin

Dinar. A., S. Dinar, S. McCaffrey and D. McKinney (2007) *Bridges Over Water: Understanding Transboundary Water Conflict Negotiation and Cooperation*, World Scientific Publishers, Singapore and New Jersey

Economic and Social Commission for Asia and the Pacific (ESCAP) (1997) *Regional Cooperation on Climate Change*, ESCAP, New York

Frederick, K. (ed.) (2002) *Water Resources and Climate Change*, Edward Elgar, Cheltenham, UK

Frederick, K., D. Major and E. Stakhiv (eds) (1997) *Climate Change and Water Resources Planning Criteria*, Kluwer Academic Publishers, Dordrecht

Gleditsch, N. P., R. Nordås and I. Salehyan (2007) *Climate Change and Conflict: The Migration Link*, Working Paper Series, International Peace Academy, New York, May

Gleick, P. H. and D. B. Adams (2000) *Water: The Potential Consequences if Climate Variability and Change for the Water Resources of the United States*, Report of the US Global Change Research Program, September

Haddadin, M. J. (2000) 'Negotiated resolution to the Jordan–Israel water conflict', *International Negotiation*, vol 5, pp263–288

Intergovernmental Panel on Climate Change (IPCC) (1996a) *Climate Change 1995: The Science of Climate Change*, Contribution of Working Group I to the Second Assessment Report of the Intergovernmental Panel on Climate Change, Cambridge University Press, Cambridge

Intergovernmental Panel on Climate Change (IPCC) (1996b) *Climate Change 1995: Impacts, Adaptation, and Mitigation of Climate Change, Scientific-Technological Analyses*, Contribution of Working Group II to the Second Assessment Report of the Intergovernmental Panel on Climate Change, Cambridge University Press, Cambridge

Intergovernmental Panel on Climate Change (IPCC) (2001) *Climate Change 2001: Impacts, Adaptation, and Vulnerability*, Contribution of Working Group II to the Third Assessment Report of the Intergovernmental Panel on Climate Change, Cambridge University Press, Cambridge

Intergovernmental Panel on Climate Change (IPCC) (2007) *Intergovernmental Panel on Climate Change Fourth Assessment Report, Climate Change 2007*, Synthesis Report, Summary for Policymakers, www.ipcc.ch/pdf/assessment-report/ar4/syr/ar4_syr_spm.pdf (accessed 11 November 2006)

Just, R. E. and S. Netanyahu (1998) 'International water resource conflict: Experience and potential', in R. E. Just and S. Netanyahu (eds), *Conflict and Cooperation on Transboundary Water Resources*, Kluwer Academic Publishers, Boston

Kahn, M. E. (2005) 'The death toll from natural disasters: The role of income, geography and institutions', *The Review of Economics and Statistics*, vol 87, no 2, pp271–284

Kilgour, M. D. and A. Dinar (2001) 'Flexible water sharing within an international river basin', *Environmental and Resource Economics*, vol 18, pp43–60

Miller, K. and D. Yates (with assistance from C. Roesch and D. Jan Stewart) (2005) *Climate Change and Water Resources: A Primer for Municipal Water Providers*, Awwa Research Foundation, Denver, CO

Miller, N. L., K. E. Bashford and E. Strem (2006) 'Changes in runoff', in J. B. Smith and R. Mendelsohn (eds), *The Impact of Climate Change on Regional Systems*, Edward Elgar, Cheltenham, UK

Milly, P. C. D., K. A. Dunne and A. V. Vecchia (2005) 'Global pattern of trends in stream flow and water availability in changing climate', *Nature*, vol 438, pp347–350

Mogaka H., S. Gichere, R. Davis and R. Hirji (2006) *Climate Variability and Water Resources Degradation in Kenya*. World Bank Working Paper No. 69. World Bank, Washington, DC

Nishat, A. and I. M. Faisal (2000) 'An assessment of the institutional mechanisms for water negotiations in the Ganges-Rahmaputra-Meghna system', *International Negotiation*, vol 5, pp289–310

Palmer, M. A., C. A. Reidy Liermann, C. Nilsson, M. Florke, J. Alcamo, P. S. Lake and N. Bond (2008) 'Climate change and the world's river basins: Anticipating management options', *Frontiers in Ecology and the Environment*, vol 6

Quiggin J. and J. Horowitz (2003) 'Costs of adjustment to climate change', *Australian Journal of Agricultural and Resource Economics*, vol 47, no 4, pp429–446

Riebsame, W. E., K. M. Strzepek, J. L. Wescoat Jr., R. Perritt, G. L. Gaile, J. Jacobs, R. Leichenko, C. Magadza, H. Phien, B. J. Urbiztondo, P. P. Restrepo, W. R. Rose, M. Saleh, L. H. Ti, C. Tucci and D. Yates (2002) 'Complex river basins', in K. Frederick (ed.), *Water Resources and Climate Change*, Edward Elgar, Cheltenham, UK

Shubik, M. (1982) *Game Theory in the Social Sciences: Concepts and Solutions*, The MIT Press, Cambridge, MA

Smith, J. B. and R. Mendelsohn (2006) *The Impact of Climate Change on Regional Systems*, Edward Elgar, Cheltenham, UK

Wolf, A. T., J. A. Natharius, J. J. Danielson, B. S. Ward and J. K. Fender (1999) 'International river basins of the world', *International Journal of Water Resources Development*, vol 15, no 4, pp387–427

# Index

aboriginal water rights  51
   *see also* Native American water rights
Accelerated and Shared Growth Initiative of South Africa (ASGISA)  210
Active Management Areas (AMAs) (Arizona)  69–72, 84
   groundwater transfers  49–50, 63
   safe-yield goals  70, 73, 78, 80, 85
adaptation to climate change  146–147, 152–154, 306
   European policy  156–158, 161–165
   international river basins  304
   national frameworks  158–160
   policy in Spain  166–168
   and sustainable development  154–155
   western US  87–88
adaptive management
   environmental flow programmes  119, 122
   irrigation forbearance  62
   water management institutions  77–78
Agenda 21  155
agriculture
   adaptation to climate change  158, 161–163
   benefits of water use in  176
   employment trends  130–131
   forbearance programmes  59–62
   resistance to change  76, 84, 105
   Spain  128, 129–131, 132
   *see also* irrigated agriculture
AGUA (Initiative for Water Management and Utilisation) (Spain)  127–128
Ak-Chin Settlement  58

All-American Canal  293
AMAs *see* Active Management Areas (AMAs) (Arizona)
Amazon River, hydropower potential  200
Amman  18–19, 20, 23, 182
   urbanization  33
   water quality  40, 181, 185
   water rationing  186
Amman Water and Sewage Authority  18, 19, 20
Amman–Zarqa basin  33, 38–39, 42
ANA (Brazilian water regulatory agency)  195, 196, 197
   River Basin Pollution Abatement Programme (PRODES)  198–199
Ananias, Patrus  190
Andalusia, devolution of competencies on Water affairs  127, 133
ANEEL (Brazilian electricity regulatory agency)  204–205
Aqaba  23, 42
Aquifer 23 (Spain)  138
aquifers
   artificial recharge  74
   Mesa San Luis  293
   urbanization and recharge areas  33, 37–38
   *see also* groundwater; transboundary groundwater aquifers
Arab Fund for Economic and Social Development  24
Aral Sea basin  305, 312–313, 315
Argentina, natural gas production  202
Arizona  3–4, 47–64, 67–88

addressing increasing management
    challenges 84–88
effects of climate change 59, 67–68,
    78, 87
environmental sustainability 76–77, 83
groundwater transfers 49–50
management institutions 68–73,
    77–78, 83–84
surface water management 71, 76
transboundary issues 83–84
transfers based on temporary land
    fallowing 59–62
tribal water rights settlements 50,
    51–58, 63, 81–82
value and price of water 80–81, 86–87
water bank 58, 63, 75
water quality issues 82
water supplies 68, 75–76, 85
see also Arizona Water Banking Authority
    (AWBA); assured water supply
    (AWS) (Arizona)
    Yuma Desalting Plant (YDP) 59–60,
        63, 283, 285
Arizona Department of Water Resources
    (ADWR) 50
Arizona Public Service Company v. John
    F. Long 51
Arizona Water Banking Authority
    (AWBA) 58, 63, 75
Arizona Water Settlements Act (AWSA)
    56–58, 81–82
Arora, V. and G. Boer 304–305
arsenic 82
artificial recharge 74
assured water supply (AWS) (Arizona)
    73–74, 79–80, 83, 86
Australia 4, 91–105
    future policy considerations 100–105
    historical, legal and institutional
        influences on policy 93–96, 105
    Native Title Act 51
    politics and policy inertia 91–92
    recent reform episodes 96–100
    water consumption 100–101
    see also Murray-Darling Basin
Australian Aborigines 93

autonomous adaptation 152–153
axiomatic approach to bargaining 259,
    279–280
Azraq basin 19, 29
Azraq Oasis 35, 37–38

Baltic Sea agreement 233
Bangladesh 310–311
bargaining power 237, 242, 243–246,
    247–248, 250, 280
bargaining problems 234–235
bargaining solutions 259–260, 270
bargaining theory 257, 258–260,
    273–275, 279–280
    model 264–268
    negotiation structure and strategy
        269–273
    see also formal negotiation models
BBBBEE (broad based black economic
    empowerment) 211
BECC (Border Environmental
    Cooperation Commission) 288,
    290, 291, 292
BEIF (Border Environmental
    Infrastructure Fund) 291, 292
Belluno Province 240
biodiversity
    effects of climate change 149
    river habitats 110, 113
bio-electricity 202
bird migration
    Arizona 71
    Jordan 37, 38, 39, 41, 42
Bishkek Treaty 313
Bolivia, natural gas production 202
Border 2012 Program 293
Border Environmental Cooperation
    Commission (BECC) 288, 290,
    291, 292
Border Environmental Infrastructure
    Fund (BEIF) 291, 292
boron concentration, Jordan River 40
Brazil 6, 189–206
    hydropower 199–205
    Sao Francisco River inter basin transfer
        project 190–196

Sao Paulo water supply 196–197
  sewage treatment 197–199
business initiatives, Australia 103
buy-back of surplus water, Australia 98, 99
buyouts of water rights, Spain 138–139

Caballo Reservoir 281, 282
CAGRD (Central Arizona Groundwater Replenishment District) 79–80
Calexico–Mexicali 287
CALFED 262, 272
California 7, 257–258, 260–262, 293
  analysed issues and stakeholders 264
  bargaining model 264–268
  current water policy debate 262–264
  negotiation structure and strategy 269–273
  regional water flows 261
CAP see Central Arizona Project (CAP)
Cappio, Luiz 189–190, 192, 195
CAP water 50, 53, 56–57, 58, 75
  vulnerability to shortage 59
Castille-La Mancha 133, 137
Catchment Management Agencies (CMAs) (South Africa) 219
Catchment Management Strategy, Inkomati Water Management Area 218, 219
Center for Strategic and International Studies (CSIS) 177
Central Arizona Groundwater Replenishment District (CAGRD) 79–80
Central Arizona Project (CAP) 50, 53, 63, 74, 80
  infrastructure 76, 80
  see also CAP water
Central Arizona Water Conservation District (CAWCD) 50
Central Water Authority (Jordan) 18
Chander-Tulkens cost sharing rule 287, 288
Cidacos Pilot project 139–141
Cienega de Santa Clara, Mexico 59–60
CIRCLE project 158

climate change 145–168, 301–302
  effects of in Arizona 59, 67–68, 78, 87
  effects in Australia 102, 104
  effects in Jordan 36–37
  effects in Spain 128, 150–152, 153, 166
  European context 148–150
  European policy 156–158, 161–165
  formal negotiation models 251
  implications for ecosystem rehabilitation 122
  and sustainable development 154–155
  see also adaptation to climate change; climate change and international water
climate change and international water 8, 301–322
  economic impacts of water supply variability 305–307
  strategic alliance analytical framework 307–310
  strategic alliance management options 310–321
  water runoff effects 303–305
Climate Change National Adaptation Plan (Spain) 160
Climate Impacts Programme (UKCIP) 160
co-decision making, Spain 125
collective negotiated decision making 233
  see also formal negotiation models
Colorado River 59–60
  Arizona water bank 58, 63, 75
  California allocation 264, 293
  effects of water diversions 85–86, 87
  Native American rights 51, 58
  predicted effects of climate change 67
  salinity 82, 283, 285
  transboundary issues 83, 281, 282, 284, 285
Common Agricultural Policy (CAP) 128, 129–130
  adaptation to climate change 161–163, 166–167
community involvement in deployment of environmental water 118, 120–121

compulsory licensing of water, South Africa  211–212, 215, 218, 221, 222–223, 223–224
computer simulations *see* Piave River Basin (PRB)
conflict *see* water conflicts
connectivity
    flow volumes  111
    physical barriers  112
consultation *see* participation in water management
Convention on Boundary Waters (US–Mexico)  281, 284
cost sharing, US–Mexico  286, 289–292
cost–benefit analysis, Ebro transfer  134–135
Council of Australian Governments (CoAG)  96–98
Criddle, Wayne  17
Crocodile River catchment  216, 218, 221, 225
cropping pattern changes, Spain  130
cultural attitudes to water  183–184

dams
    Brazil  199–205
    California  264
    Jordan  22
    Murray-Darling Basin  112
    Spain  127
Dead Sea  29, 31, 41
default outcomes  258
    California water policy debate  272–273, 275
default policies *see* disagreement policies
democratic government, and participation in water management  205–206, 262–264
desalinization  80
    Middle East  22, 312
    Spain  126, 127–128
    Yuma Desalting Plant (Arizona)  59–60, 63, 283, 285
disagreement policies  235, 236–237, 250
Domestic Water Corporation (Jordan)  18, 20, 21

downstream flow, river habitats  111
drinking water  19, 146, 182
    Arizona  82
    Jordan  32, 37, 181
    South Africa  209
    Spain  129
    US–Mexico negotiations on quality  285
drip irrigation technology  130
drought
    Brazil  191, 194
    economic impacts  305–306
    Jordan  17
    Murray-Darling Basin  118, 119
    Spain  150
dry-year options  48, 60, 63, 76
DWAF *see* Water Affairs and Forestry, Department of (DWAF) (South Africa)
DWR (Water Resources Department) (California)  266

Earthjustice  263, 273
East Canal Authority (Jordan)  18
Ebro water transfer  127, 128, 132–133, 134–135
ecological expertise, deployment of environmental water  118–119, 121
ecological role of water  110–111
ecological status
    EU WFD  126, 132
    Inkomati Water Management Area  216
economic and social development
    hydropower in Brazil  199–205
    Native Americans  56
    South Africa  210–211
    strategic importance of water  175–177
economics and water policy
    Australia  100
    cost-effectiveness  139–141
    impacts of water supply variability  305–307
    Jordan  22
    Spain  129, 130–131, 134–135
    US  48, 59, 61–62, 80–81

US–Mexico border 288–289, 291–292
  *see also* water tariffs
ecosystems *see* environmental degradation; environmental protection; Murray-Darling basin
EEA (European Environment Agency) 155
efficient use of water resources 210
  South Africa 212, 221, 224
effluent *see* treated municipal wastewater
Egypt 13
employment
  agriculture 176
  lost through environmental requirements 224, 225
endangered species 39, 41
  Arizona 71, 76–77
  San Francisco Bay-Delta 261, 273
Endangered Species Act (ESA) (US) 53, 77, 257–258, 273
energy policy
  Brazil 199–205
  European Union (EU) 161
energy swap agreements, Lara Basin model 315–318
energy-water nexus 80, 175
enforcement mechanisms
  Arizona 71, 72
  Jordan 23, 43
  South Africa 226
engineering solutions to water shortage
  Australia 92, 98, 99–100
  US 85
  *see also* Central Arizona Project (CAP)
  *see also* dams
environmental costs 136
environmental degradation
  Australia 96
  Bangladesh 311
  exacerbated by water transfers 48
  Jordan 29, 37–42
  South Africa 208
environmental impact assessment 43
environmental licensing process, hydropower in Brazil 203–204, 206
environmental protection
  and aboriginal water rights 52–53
  Jordan 31, 37–38, 40, 41, 42
  land fallowing 62
  options contracts 103
  San Francisco Bay-Delta 263, 273, 274, 275
  South Africa 216, 218, 221, 224, 225
  *see also* Murray-Darling Basin
environmental sustainability
  Arizona 76–77, 83
  Piave River Basin (PRB) 238
  South Africa 212
  *see also* sustainability of water resources
environmental use of water
  meeting ecological needs 114–118
  optimizing deployment 118–121
  recovery of water 113–114, 120
  San Francisco Bay-Delta 261
  Zuni Heaven (Arizona) 53
Environment Ministry (MOE) (Jordan) 31–32, 39, 43
equity in water allocation 182
  South Africa 211–212, 218, 221, 222, 226–227
ethanol 201
Euphrates River 19, 178, 179
European Commission (EC)
  Green Paper 'Adapting to climate change in Europe' 147, 156–158, 167
  initiative on drought and water scarcity 165
European Environment Agency (EEA) 155
European Union (EU)
  climate change policy 5, 156–158, 165, 166–167
  energy policy 161
  Floods Directive 164
  Gothenburg Sustainable Development Strategy (SDS) 154
  impact of climate change 148–149
  Lisbon Strategy 154
  national climate change adaptation frameworks 158–160

Seventh Framework Programme 165
  see also Common Agricultural Policy (CAP); European Commission (EC); Water Framework Directive (WFD)
evaluating climate change policy 165
evaporation losses and productive use of water 194
exchange of water rights
  Spain 126, 130, 136–138
  see also water trading
extreme weather events 145, 147, 150, 166

Farakka Barrage 310–311
Farm Advisory System (EU) 163, 167
Fernandez, L. 292
finite horizon strategic negotiation models 236–237
Finland, National Energy and Climate Strategy 158–160
flooding risk 150, 158
  economic impacts 305–306
  EU Floods Directive 164
  Spain 166
flow classes 112
flow events 112–113
FLOWS method 115, 118, 120
flow volumes
  connectivity 111
  cost-effectiveness in increasing 140, 141
  cues for ecological processes 111
  effects of modification 111–113
formal negotiation models 234–251
  assessing players' strategies and allocation rules 246–249
  model application 240–243
  Piave River Basin (PRB) 237–240
  policy implications 249–251
  results 243–246, 250
  underlying bargaining framework 235–237
  see also bargaining theory
fossil fuels 202–203
Freimuth, L. et al 36

Friant Dam 264
full-cost recovery prices 136

game theory xv–xviii, 1, 7–8, 279–280
  and consequences of climate change 302
  formal negotiations models 234, 236–237
  model application 240–246
  see also bargaining theory; US–Mexico border environmental management
Ganges-Brahmaputra-Meghna basin 305, 310–311
General Corporation for Environmental Protection (GCEP) (Jordan) 31
Gila River Indian Community (GRIC) 56–57, 81–82
golf course irrigation 51
Gómez, C. M. and A. Garrido 139–140
Good Agricultural and Environmental Condition (GAEC) standards 163
Good Neighbor Environmental Board (GNEB) 293
Gorai Restoration Project 311
greenhouse gases (GHGs) 145–146, 152
  and energy policy in Brazil 200–204
groundwater
  availability in South Africa 208
  impact of climate change 68, 303
  overdraft see over abstraction of groundwater
  price and value 86–87
  quality in Jordan 33, 38, 39
  resources in Jordan 20, 23, 34, 35–36, 37–38
  in Spain 125, 135, 136, 138–139
  transfers 49–50
  use in Australia 103
Groundwater Management Act (Arizona) 49–50, 63, 68–72, 83
  assured water supply (AWS) 73–74
  effectiveness of 77, 78, 81
  groundwater rights system 69, 72–73
Groundwater Transportation Act (Arizona) 49

Guadalquivir basin 127, 129, 133, 137
Guadiana basin 138, 139

Harmon Doctrine 281, 284
Hasbani tributary sub-basin 180
Hawaii, Water Code 51
health, Jordan 24, 32, 181
Health Ministry (Jordan) 32
household food gardens 220
Howard Humphry Consultants 19
Howard, John 91
human footprints, Murray-Darling Basin 111–113
human resources
  Jordan 17, 27
  South Africa 225–226
hydroelectricity
  Aral Sea basin 312
  Australia 94, 112
  Brazil 199–205
  Piave River Basin (PRB) 238, 240–241
hydrological cycle and climate change 301–302, 303–305

IBC (International Boundary Commission) 281, 284
IBWC (International Boundary and Water Commission) 284–288, 294
identified ecological needs approach 115–118
illegal wells
  Jordan 20, 23, 35, 37
  Spain 138
India 310–311
indigenous peoples see Australian Aborigines; Native American water rights; South Africa
industrial waste see pollution; water pollution
industrial water use
  economic and social benefits 177
  water transactions 51, 57
infrastructure
  California 257–258, 260–262, 263
  Central Arizona Project (CAP) 76, 80
  Jordan 22, 40
  see also US–Mexico border environmental management
Inkomati estuary 216
Inkomati river basin, Interim IncoMaputo Agreement 214–215
Inkomati Water Management Area 215–224
  Crocodile River catchment 216, 218, 221, 225
  demand for water 215–216
  Sand River catchment 221
  Water Allocation Reform 218–224, 227–229
innovative voluntary water transactions 47–49, 63–64
  Arizona water bank 58, 63, 75
  effluent transfers 51
  groundwater transfers 49–50
  involving indigenous peoples 50, 51–58
  transaction costs 61–62, 64
  transfers based on temporary land fallowing 59–62
  transfers of public project entitlements 50
instream flow rights, Arizona 77, 87
insurance 163
inter-basin groundwater transfers, Arizona 49
inter-basin transfers
  South Africa 208
  Spain 133, 137, 138
  see also Ebro water transfer
  see also Sao Francisco River inter basin transfer project; Sao Paulo (MRSP)
interconnected games 292–293, 294
Intergovernmental Panel on Climate Change (IPCC) 67, 87, 145, 152, 301–302, 303–304
Interim IncoMaputo Agreement 214–215, 218
International Boundary Commission (IBC) 281, 284
International Boundary and Water Commission (IBWC) 284–288, 294

international equity  212
International Monetary Fund (IMF)  21
international negotiations  233–234
    see also transboundary river systems; US–Mexico border environmental management
International Rivers Network (IRN)  200
international watercourses see transboundary river systems
Iraq  19, 178, 179
irrigated agriculture
    Arizona  48, 59–60
    Australia  92, 93–94, 98, 99, 100–102, 104
    benefits of  176
    California  260
    causing salinity downstream  179
    effects of climate change  148–149, 151
    Jordan  24–25, 26, 33, 43, 182–183, 184, 186
    Piave River Basin (PRB)  238, 239, 240
    problems with ag-to-urban transfers  76
    Spain  129–131, 136
    use of treated municipal wastewater  24–25, 43, 51, 183
    and water security in Brazil  194
    see also irrigation forbearance programmes
irrigation forbearance programmes  59–62
Israel
    disputed water resources  179, 180
    shared water resources with Jordan  16, 17, 23, 311–312
issue linkage  245, 269–270
    US–Mexico environmental negotiations  284, 292–293, 294
issue space  258, 274
    California water policy debate  269–270, 272
    see also multiple issue bargaining
Italy, Piave River Basin (PRB)  237–240
IUCN (International Union for Conservation of Nature)  39

Johnston, Eric  16, 17

Jordan, Hashemite Kingdom of  2–3, 6, 13–27, 29–43
    agricultural strategy (JAGS)  24–26
    ecosystems  37–42
    effects of climate change  36–37
    institutions and management  18, 19–21, 30–32
    issues in water allocation  181–183
    issues in water pricing  183–186
    rural employment  176
    shadow water  186
    shared water resources  15–17, 23, 179, 180–181, 311–312
    wastewater treatment and reuse  19, 20, 23, 24–25, 34–35, 42, 43, 183
    water policy  17–19, 21–24, 42–43
    water quality  33–34, 185
Jordan River Basin
    disputed territories  179
    environmental degradation  29, 39–40
    regional cooperation  15–16
    river water quality  180
    water flow  305, 311–312
Jordan Society for Sustainable Development (JSSD)  42
Jordan Valley Authority (JVA)  18, 19, 20, 21, 41, 185
Jordan Valley Commission  18
Jucar basin  138–139

King Abdullah Canal (KAC)  40
King Talal Dam (KTD)  40, 185
Kruger National Park  216, 218, 221
Kyoto Protocol  145, 146, 155
Kyrgyz Republic  312–313

land fallowing  60–62
landfill sites, Jordan  33
land use and water supply planning, Arizona  79–80, 86
Lara Basin model  313–321
Laredo–Nuevo Laredo  287–288
learning frameworks  119
leases
    Arizona Water Settlements Act (AWSA)  56–57

CAP allocations 50, 53, 56, 58
Lebanon, conflict over water quality 180
Long-Term Central Valley Project Operations Criteria and Plan Biological Opinion 263
Lower Colorado River Basin, agricultural forbearance programmes 59–60

Madeira River 203, 204
Maputo River 214, 215
market-based reforms, Australia 92, 97, 100, 103
Mediterranean region, climate change 150, 151, 160
Mesa San Luis aquifer 293
Metropolitan Water District (MWD) (California) 263–264
Mexico *see* US–Mexico border environmental management
migration from dry areas, Brazil 192
Miller, K. and D. Yates 303
mitigation of climate change 145–146, 152, 163, 306
Miyahuna (water company) 23
model outputs 267–268
Mozambique
    Interim IncoMaputo Agreement 214
    *see also* Inkomati Water Management Area
multilateral bargaining model 248–249, 250
multiple issue bargaining 245, 250
    definition of success 270–272, 274–275
    *see also* issue space
municipal wastewater *see* treated municipal wastewater
municipal water providers
    Arizona 73
    California 263–264
    Jordan 18–20, 21, 23, 186
Murray-Darling Basin 4, 93, 95, 109–122
    cap on water extractions 96, 114, 122
    deployment of environmental water 118–121
    ecological needs 110–111
    effects of climate change 102
    human footprint 111–113
    recovery of water for the ecosystem 113–118
Murray-Darling Basin Ministerial Council 96, 115–116
Murrumbidgee River valley 103, 113, 120

NADBank *see* North American Development Bank (NADBank)
Nash bargaining games 285, 286, 292
Nash, J. F. 259
national adaptation strategies 158–160
National Agenda (Jordan) 30–31
National Center for Agricultural Research and Technology Transfer (NCART) 24
National Hydrological Plans (NHP) (Spain) 127–128, 132, 133
National Planning Council (Jordan) 18, 19, 21
National Plan for Water Security (the Plan) (Australia) 91, 98, 99
national strategies for sustainable development (NSDS) 155
National Water Act (South Africa) 210, 211–212, 219, 220, 223
National Water Initiative (NWI) (Australia) 97, 99
National Water Resources Financial Assistance Act (Australia) 96
Native American water rights 47, 50, 51–58, 63, 81–82
Native Title Act (Australia) 51
natural gas production 202–203
Natural Resources Authority (Jordan) 18, 20
Navajo Reservation 52
Navarre 140
negotiation processes 234, 236–237
    computer simulations 237, 240–249
    policy implications 249–251
    *see also* bargaining theory
NetSyMoD framework 240

Nevada, Interstate Water Banking
    Agreement  58
New Mexico
    Caballo Reservoir  281, 282
    Zuni Reservation  53
New River  285, 287
New South Wales Murray Wetland
    Working Group (MWWG)
    120–121
New Zealand, Treaty of Waitangi  51
Nogales Wash river  285
non-cooperative multilateral bargaining
    model  7, 235, 238, 249–251, 258
North American Development Bank
    (NADBank)  288, 290, 291–292
Novo, P  130
nuclear power stations  51, 80

Offer of Public Purchase (OPA) (Spain)
    138–139
options contracts  103
Oregon  62
Orontes River  179
over abstraction of groundwater
    Arizona  53, 68, 74
    Jordan  33, 35–36, 37–38
    Spain  138–139
over-allocation of water
    Australia  98, 110, 121
    Colorado River  87
    South Africa  212, 216, 218, 219–220

Palestinians, displacement of populations
    13, 15, 185
Palestinian territories  179
participation in water management
    189–206
    and democratic government  205–206
    hydropower  199–205
    Inkomati Water Management Area
        218–219, 220–221, 222–223
    Murray-Darling Basin  118, 120–121
Sao Francisco River inter basin transfer
    project  189–196
    Sao Paulo water supply  196–197
    Spain  125

path dependencies and policy formulation
    99–100
Patricio, Steve  272
payoff functions  258
PEAG (Especial Plan of the Upper
    Guadiana)  138, 139
Perata, Pro Tem  271, 273
Phoenix, Arizona  53, 71
Piave River Basin (PRB)  7, 237–240
    model application  240–243, 246–249
    results of computer simulation
        243–246, 250
Piracicaba River basin  196–197
planning
    for integrated development and water
        supply  226
    for integrated land use and water
        supply  79–80, 86
players
    California water policy debate  258,
        265–266
    Piave River Basin (PRB)  240, 241–243
players' access probability  242, 243, 244,
    247–248
players' payoffs  234–235, 243, 258, 267,
    268
    and issue linkage  292
    Lara Basin model  318–320
    see also players' utility
players' utility  241, 242, 246–249, 258
    see also players' payoffs
policy-driven adaptation  152–154
political influence, formal negotiations
    models see players' access probability
polluter pays principle  198
pollution  147
    Jordan  39, 42
    US–Mexico border  292, 293
    see also water pollution
population increases  147
    Jordan  13, 21, 34, 185–186
    see also urbanization
poverty
    and lack of water security  194–195
    in South Africa  208–209
precautionary principle  205

precipitation and climate change 150–151, 152, 301–302, 303–304, 306
priorities in water allocation
　environmental water 126
　Jordan 22, 35, 37, 181–182
　proportional versus fixed share allocation 248
　South Africa 212, 221
　tribal water rights 81
private sector participation
　Australia 103
　Jordan 24, 25
professional expertise
　Spain 131–132
　see also ecological expertise
protest action
　dam-hating NGOs 200, 203–205
　Dom Luiz Cappio 189–190
public awareness
　Arizona 78–79
　Cidacos project 140
　climate change 147, 167
　EUWFD reports 136
　Inkomati Water Management Area 220
　Jordan 23, 43
　water pricing 184

Quantification Settlement Agreement 264
Quiggen, J. and J. Horowitz 306

race and access to water
　in South Africa 208–209, 211, 216
　see also Native American water rights
rain-fed agriculture 176, 181
　Jordan 26, 33
rainwater harvesting, Inkomati Water Management Area 220
Ramsar Convention 37
rationing of water 102, 186
Rausser, G. and L. Simon 235, 236
Reagan Administration, cost sharing 286
re-allocation of water 182, 183
　see also South Africa

regional cooperation, Middle East 15–16, 43
reservoirs, effect on downstream water temperatures 113
Riebsame, W et al 304
Rio de Janeiro Earth Summit 155
Rio Grande 281, 282, 284, 285, 287–288
risk attitudes in formal negotiations 246
risk management
　adaptation to climate change 152
　flooding 158
risks, assigning 97–98
River Basin Committees (Brazil) 198
River Basin Pollution Abatement Programme (PRODES) (Brazil) 198–199
river basins
　and climate change 304–305, 310
　see also transboundary river systems
river ecosystems 116, 121–122
　see also environmental protection; environmental use of water
River Reach programme 120
Royal Marine Conservation Society (Jordan) 42
Royal Society for the Conservation of Nature (Jordan) 38, 40, 41
Rubinstein, A. 259
runoff regimes, Spain 129
Rural Development Programmes (EU) 162, 163, 166–167
rural–urban water transfer
　Arizona 63, 76
　see also Groundwater Management Act (Arizona); innovative voluntary water transactions
　Australia 103, 104
　Jordan 19

Sacramento River basin 260
salinity
　in river water 40, 82, 83
　in water supplies 38, 39, 82
Salt River Project (Arizona) 68, 71
San Diego, agricultural forbearance programmes 62

San Diego–Tijuana, pollution 285, 286–287, 292
San Francisco Bay-Delta 260–262
  smelt 263, 273
San Joaquin River 260, 264
San Pedro River 71, 77
Santo Antonio hydro plant (Brazil) 204–205
Sao Francisco River Basin Revitalization 195
Sao Francisco River inter basin transfer project 190–196
  donor region 195
  lack of policy focus 195–196
  protest by Dom Luiz Cappio 189–190
  recipient region 192–194
Sao Paulo (MRSP), water supply 196–197
Schwarzenegger, Arnold 262–263, 269, 275
sea levels 150
seasonal flow inversion 112
sewage treatment
  Brazil 197–199
  US–Mexico negotiations 285–288
  *see also* treated municipal wastewater
shadow water 177, 186
shared water resources 233–234
  *see also* formal negotiation models; transboundary river systems
side payments 288–292, 294
Single Payment Scheme (EU) 162, 163
Snowy Mountain Scheme 94, 112
Social Network Analysis techniques 242
soil water 26, 176
solar power 200–201
South Africa 6–7, 207–229
  development and economic growth drivers 210–211
  international agreements 213–215, 225
  policing water allocations 226
  policy and legislative framework 211–212, 224–225
  staffing issues 225–226
  water availability 208–210
  *see also* Inkomati Water Management Area
South African Development Community (SADC), Revised Protocol on Shared Watercourses 208, 213, 214
Southern Arizona Water Rights Settlement Act 57–58
Southwest Consortium for Environmental Research and Policy (SCERP) 293
Spain 5, 125–142
  Autonomous Communities (ACs) 127, 132–133
  breakdown of consensus on water policy 131–133
  case studies 133–138
  changes in demand for agriculture 129–131
  Cidacos Pilot project 139–141
  drivers of change 128–131
  effects of climate change 128, 150–152, 153
  national climate change adaptation frameworks 160
  over-abstraction of groundwater 138–139
  vulnerability of water resources 166, 167
  water policy 125–128, 166–168
staffing issues
  Jordan 17, 27
  South Africa 225–226
stakeholders
  Active Management Areas in Arizona 72
  *see also* participation in water management
state bond issues, California 7, 257–258, 262–263, 269–270
storage of surplus water
  Arizona Water Banking Authority (AWBA) 58, 75
  artificial recharge 74
  and climate change 306
  Jordan 17
  Sacramento River 260
stranded irrigation assets 99

strategic behaviour *see* bargaining theory; game theory
strategic importance of water 175–177, 178
strategic models of bargaining 259, 279, 280
Stream-flow Management Plans (SFMPs) 115
successful negotiation, California water policy debate 270–272, 274–275
sustainability of water resources 146–147
  Arizona 75–76, 79–80, 83–84, 86
  Jordan 23, 34
  *see also* environmental sustainability; water stress
sustainable development
  and climate change 154–155
  EU agriculture 161–163
Swaziland, Interim IncoMaputo Agreement 214
Syria 13
  shared water resources with Jordan 16–17, 23, 179, 180–181
  shared water resources with Turkey 178, 179

Tagus–Segura aqueduct 137
Tajikistan 306, 312–313
tariffs *see* water tariffs
Tasmania 93
Tenerik, Lake 312
The Living Murray (TLM) programme 115–117, 119
third-party effects, water transfers 137
Three Way Agreement 262, 270
Tiberias, Lake 17, 181
Tigris, River 178
Tohono O'Odham Nation 57–58
tourism
  Jordan 22
  Piave River Basin (PRB) 238
  South Africa 216
trading partners
  game theory 280
  strategic importance of water 177, 186

transactions involving native peoples 51–56
transboundary environmental management *see* transboundary river systems; US–Mexico border environmental management
transboundary groundwater aquifers 177–178
transboundary river systems 177
  Australian states 94–95, 110
  coping with climate change 302, 304–305
  Jordan 15–17, 23
  South Africa 208, 212, 213–215, 222, 227
  Spain 127, 132–133
  strategic alliance analytical framework 307–310
  strategic alliance management options 310–321
  *see also* US–Mexico border environmental management; water conflicts
treated municipal wastewater
  Brazil 197–199
  Spain 135
  US–Mexico border 285–288
  *see also* wastewater treatment and reuse
Treviso province 240
tribal water rights *see* Native American water rights
Turkey, transboundary river systems 178, 179

unanimity requirement 236
uncertainty
  and climate change 87, 246
  formal negotiation models 237, 243, 246–249, 250
Underground Storage and Recovery programme (Arizona) 74
United Kingdom (UK), adaptation strategy 160
United Nations (UN)
  Convention on the Non-Navigational

Uses of International Water
    Courses  178, 213
  Framework Convention on Climate
    Change (UNFCCC)  36,
    145–146, 154–155
urbanization
  Jordan  33
  restricted by water supplies  70
  South Africa  208
  US–Mexican border  285, 288–289
  water leases  53
  see also municipal water providers
urban water policy, Australia  100,
  102–103
USAID (United States Agency for
  International Development)  21, 24,
  36
user pay principles  92
US–Mexico border environmental
    management  7–8, 59–60, 82–83,
    281–288, 293–294
  institutional arrangements  280–281,
    282–283
  interconnected games and issue linkage
    292–293
  policy design  288–292
Utah  60

value and price of water
  Arizona  80–81, 86–87
  see also water tariffs
Verde River  71, 77
Victoria  98, 104
  Stream-flow Management Plans
    (SFMPs)  115

Wadi Heedan  41
Wadi Mujib  41
Waitangi Treaty (New Zealand)  51
Wala Dam  41
Wanger, Oliver  263
wastewater treatment and reuse
  Arizona  51, 57–58, 68, 82
  Jordan  19, 20, 23, 24–25, 34–35, 42,
    43, 183
Water Act (Brazil)  197–198

Water Affairs and Forestry, Department of
    (DWAF) (South Africa)  211, 219,
    220, 223, 225
Water Allocation Framework (WAF),
  Inkomati Water Management Area
  219–223
water allocation issues  181–183
  Piave River basin (PRB)  238, 240,
    241–242
  see also equity in water allocation; over-
    allocation of water; priorities in
    water allocation; water allocation
    rules
Water Allocation Plans (WAPs) (South
  Africa)  218, 219
water allocation rules, fixed and
  proportional  237, 248–249, 251
Water Authority of Jordan (WAJ)  18, 20,
    21, 38
  reclaimed water  35
  water quality standards  32
water availability
  and climate change  150
  international comparison  30
water banks
  South Africa  223, 224
  Spain  137–138
water conflicts  175–187
  allocation issues  181–183
  issues in water pricing  183–186
  over compliance  23, 179
  over territory  179
  over water quality  179–180
  over water sharing  178
  prevention and resolution of  187
  repercussions  180–181
  strategic importance of water  175–
    177, 178
  see also California; Piave River Basin
    (PRB); US–Mexico border
    environmental management
water footprints, crop comparisons  130
Water Framework Directive (WFD)  126,
    132
  adaptation to climate change  163–164,
    166, 167

Article 5 reports 135–136
Article 9 136
Article 11 definitions 139–141
'Water for the Future' (Australia) 98, 99
Water and Irrigation Ministry (MWI) (Jordan) 18, 21, 22, 23, 27, 31
  groundwater abstraction 35–36
Water Law (WL) (Spain) 125–126, 136
water markets 48
water pollution
  Arizona 82
  Brazil 197
  Jordan 33, 39, 42
  Spain 129
  US–Mexico border 285, 288
  see also treated municipal wastewater
water property rights
  Australia 97, 99
  see also exchange of water rights
Water Protection Fund (Arizona) 77
water quality
  conflicts over 179–180, 181, 185
  effects of climate change 148
  river habitats 110
  Spanish water bodies 128, 129, 135
  urban water in Spain 135
water quality standards
  Jordan 32, 39, 40
  US and Mexico 287
Water Reform Framework (CoAG) 97
Water Resources Department (DWR) (California) 266
water restriction fatigue 102–103
water runoff effects 303–305
water stress, Jordan 13, 181–183
water tariffs
  Australia 99
  Brazil 197–198
  EU pricing policies 163–164
  Jordan 20, 23, 24, 26, 35, 183–186
  Spain 135, 136
water trading
  South Africa 223
  see also exchange of water rights
water transactions see innovative voluntary water transactions
Water Treaty (US–Mexico) 282, 284
Wazzani Springs 180
Wehda Dam 22, 23, 181
West Bank 15, 17, 179
wetlands
  Azraq Oasis 37–38
  Cienega de Santa Clara 59–60
  Jordan River Basin 39–40
  Zuni Heaven, Arizona 53
wildlife
  importance of downstream flow volumes 111
  see also bird migration; endangered species
wind power 201
World Bank, loans to Jordan 21, 22
World Summit for Sustainable Development (WSSD) 155
World Trade Organization (WTO) agreements 128, 129, 251

Yarmouk River 16–17, 22, 40
  conflicts over water 17, 23, 179, 180–181
Yuma Desalting Plant (YDP) 59–60, 63, 283, 285

Zaragoza conferences 1, 2
Zarqa River 29, 33, 39
Zukowski, S. and S. Meredith 119
Zuni Indian Tribe Water Rights Settlement Act 53, 56

HD1691 .D563 2009
Policy and strategic
behaviour in water resour
management